Springer Proceedings in Mathematics & Statistics

Volume 457

This book series features volumes composed of selected contributions from workshops and conferences in all areas of current research in mathematics and statistics, including data science, operations research and optimization. In addition to an overall evaluation of the interest, scientific quality, and timeliness of each proposal at the hands of the publisher, individual contributions are all refereed to the high quality standards of leading journals in the field. Thus, this series provides the research community with well-edited, authoritative reports on developments in the most exciting areas of mathematical and statistical research today.

François Willot · Justin Dirrenberger ·
Samuel Forest · Dominique Jeulin ·
Andrej V. Cherkaev
Editors

Continuum Models and Discrete Systems

CMDS-14, Paris, France, June 26–30, 2023

Springer

Editors
François Willot
Centre for Mathematical Morphology
Mines Paris-PSL
Paris, France

Centre for Materials Science
Mines Paris-PSL
Paris, France

Samuel Forest
Centre for Materials Science
Mines Paris-PSL-CNRS
Paris, France

Andrej V. Cherkaev
Department of Mathematics
University of Utah
Salt Lake City, UT, USA

Justin Dirrenberger
PIMM Lab, Arts and Métiers
CNAM-CNRS
Paris, France

Dominique Jeulin
Centre for Mathematical Morphology
Mines Paris-PSL
Paris, France

ISSN 2194-1009 ISSN 2194-1017 (electronic)
Springer Proceedings in Mathematics & Statistics
ISBN 978-3-031-58664-4 ISBN 978-3-031-58665-1 (eBook)
https://doi.org/10.1007/978-3-031-58665-1

Mathematics Subject Classification: 35B27, 74C05, 74D10, 74C10, 35Q60, 35R30, 35R05, 60-08, 60G55, 60G60

This Springer imprint is published by the registered company Springer Nature Switzerland AG
The registered company address is: Gewerbestrasse 11, 6330 Cham, Switzerland

Introduction

The following volume contains contributions submitted by participants in the 14th International Symposium on Continuum Models and Discrete Systems (CMDS14) that was jointly organized by the Conservatoire National des Arts et Métiers (CNAM) and Mines Paris PSL [3], and held in Paris, France on June 26–30, 2023. The workshop was sponsored by the Mécamat association [1], Mines Paris and CNAM. During the symposium, hosted at CNAM, next to Clément Ader's "aéroplane n$^{\underline{o}}$ 3", 69 talks and lectures were held and 10 posters were presented [2]. A broad range of topics were covered in the fields of mathematics, theoretical and applied mechanics, physics and engineering with sessions devoted to: homogenization theories, morphological modeling of microstructures, image analysis for material science, porous media, computational methods in mechanics and change of scales, multiphysics coupling, metamaterials, generalized continua, damage, plasticity and architectured materials.

The international scientific committee was composed by Profs. Swantje Bargmann (University of Wuppertal, Germany), David Bergman (Tel Aviv University, Israel), Laurence Brassart (University of Oxford, UK), Bikask Chakrabarti (Saha Institute of Nuclear Physics, India), Andrej Cherkaev (University of Utah, USA), Justin Dirrenberger (Conservatoire National des Arts et Métiers, France), Samuel Forest (Mines Paris, France), Lori Graham-Brady (Johns Hopkins University, USA), Dominique Jeulin, (Mines Paris, France), Djimédo Kondo (Sorbonne Université, France), Graeme Milton (University of Utah, USA), Leon Mishnaevsky Jr. (Technical University of Denmark, Denmark), Natasha Movchan (University of Liverpool, UK), Martin Ostoja-Starzewski (University of Illinois, USA) and François Willot (Mines Paris, France).

We thank all the participants for their active contributions and the lively discussions during the symposium (Figs. 1 and 2) and for the outstanding articles submitted to the present volume, as well as CNAM, Mines Paris, PSL, Armines, and the Mécamat association for their support in organizing the event. We are also grateful to Profs. Heiko Andrä, Luis Bonilla, Andrej Cherkaev, Orkun Furat, Samuel Forest, Junyan He, Martín Idiart, Iwona Jasiuk, Dominique Jeulin, Eugene Kogan, Martin Ostoja-Starzewski, Yves-Patrick Pellegrini, Ping Sheng, Julien Yvonnet, and especially Jean-François Ganghoffer, Lionel Gélébart, Toufik Kanit, Eugene Rabkin, Julia

Fig. 1 Some of the participants at the CMDS 14 symposium. Most photographs were taken by D. Jeulin

Fig. 2 Group photo of the CMDS 14 symposium

Yeomans and other anonymous referees for their invaluable help in reviewing the manuscripts.

Paris, France — François Willot
Paris, France — Justin Dirrenberger
Salt Lake City, USA — Andrej V. Cherkaev
Paris, France — Samuel Forest
Paris, France — Dominique Jeulin

References

1. Mécamat: http://www.mecamat.org and the group "Probabilistic Approaches for the Mechanics of Materials" https://mecamat.ensma.fr/les-groupes-de-travail/approches-probabilistes-en-mecanique-des-materiaux/. (2023). Accessed from 21 Dec 2023
2. CMDS14: Book of abstracts. https://people.cmm.minesparis.psl.eu/users/willot/cmds14-bookOfAbstracts.pdf. Accessed from 21 Dec 2023
3. Jeulin, D. and Forest, S., Proceedings of the international conference *Continuum Models and Discrete Systems 11* (CMDS11), Paris July 30th–August 2nd 2007, Presses de l'École des Mines de Paris, 2008

Contents

Probabilistic Learning Inference Constrained by an Uncertain Model and a Target: A General Method with Application to Elasticity Homogenization Without Scale Separation

Christian Soize

Abstract We present a probabilistic learning inference that assimilates data (target set) into a parameterized large stochastic computational model resulting from discretizing a stochastic boundary value problem (BVP). A target is imposed on a vector-valued random quantity of interest (QoI), observed as the stochastic solution of the BVP. The probabilistic inference estimates the posterior probability model, which is constrained both by the second-order moment of the random residue of the BVP stochastic equations and the target set composed of statistical moments of the QoI. We assume that evaluating a single realization of the BVP is computationally expensive, so the training dataset comprises only a few points differing from big data approaches. The presented application contributes to three-dimensional stochastic homogenization of heterogeneous linear elastic media, specifically when the mesoscale and macroscale are not separated.

Keywords Probabilistic learning on Manifolds PLoM · Stochastic homogenization · Random elastic medium · Random spectrum · No scale separation

1 Introduction

(i) Objective. This paper addresses an important question concerning probabilistic learning algorithms that allow data (target dataset) to be assimilated into predictive models. These models have a small training dataset and the target dataset consists of statistical moments of quantities of interest (QoI). It assumes that these statistical moments, such as mean values and second-order moments, were esti-

C. Soize (✉)
Université Gustave Eiffel, MSME UMR 8208 CNRS, 5 bd Descartes,
77454 Marne-la-Vallée, France
e-mail: christian.soize@univ-eiffel.fr

F. Willot et al. (eds.), *Continuum Models and Discrete Systems*,
Springer Proceedings in Mathematics & Statistics 457,
https://doi.org/10.1007/978-3-031-58665-1_1

mated using realizations (samples) that are no longer available. This situation often occurs when the data is lost, deteriorated, not commented on, no longer exists, or is inaccessible. However, some statistical moment values may have been published or provided in technical reports. Additional physics-based constraints are simultaneously considered, consisting of controlling the learning process with the model to minimize the mean-square norm of the random normalized residue of the stochastic partial differential equation of the boundary value problem (BVP). These two types of constraints are both defined by an implicit function denoted as $\mathbf{h}^c$, for which an algebraic representation is unavailable. They ensure that the algorithm aims to bring the statistical moments closer to their targets while maintaining a small residue of the BVP stochastic equations. The framework presented uses a computational model with a computationally expensive single evaluation. Therefore, the training set comprises a small number of samples (or points), differing from big data approaches. The presented general method for probabilistic learning inference is applied to three-dimensional stochastic homogenization, specifically in scenarios where the mesoscale and macroscale are not separated. It is worth noting that stochastic homogenization has been extensively studied in numerous works (e.g., [1, 5, 13, 35]), with particular attention given to analyzing the size of the representative volume element (RVE) (see [4, 19, 21, 24]), and the morphological models of random structures (see [11, 12]). The presented application focuses on the stochastic homogenization of a random linear elastic medium at the mesoscale, defined on the domain $\Omega \subset \mathbb{R}^3$ of the microstructure. It is assumed that this domain is not a RVE, which means that there is no scale separation between the mesoscale and the macroscale. Since there is no scale separation, the macroscale is in fact another mesoscale at a larger scale than the initial mesoscale, which is described by random effective/apparent quantities. For the construction of the posterior probability measure using the proposed probabilistic learning inference, the constraints are defined by both a target set consisting of given statistical moments of the random effective/apparent elasticity tensor and the second-order moment of the random normalized residue of the BVP stochastic equations.

(ii) A few words on the main methodologies allowing for addressing the problem. The main statistical methods for addressing this type of problems are the following. The *Bayesian method* estimates the posterior probability measure of the control parameter using prior probability and a target dataset of realizations for the quantity of interest (QoI). However, this article introduces a different hypothesis that does not necessitate knowledge of realizations, which are assumed to be unavailable. It is important to note that Bayesian inference can be challenging to use, especially in high dimensions [20]. The *maximum-likelihood method* is employed to estimate the hyperparameters of a probability measure using a target dataset of realizations. Similar to Bayesian inference, the maximum-likelihood method is an effective statistical approach for estimating hyperparameters, but it relies on realizations as the target. The *parameterized functional representations* of random quantities, whether in finite or infinite dimensions, such as spectral methods like polynomial chaos expansions ([7, 27, 34]), facilitate the integration of data into predictive models. Bayesian

inference, maximum likelihood, and least-square approaches are commonly used to identify the parameters of these representations. The *deep machine learning method*, which relies on Artificial Neural Networks (ANNs), has gained considerable recognition in diverse scientific and engineering disciplines. However, it is important to note that ANNs usually rely on large datasets for accurate parameter identification [22] and may not be straightforward to use within a comprehensive probabilistic framework, such as the one proposed in this paper.

(iii) Proposed methodology. In this work, we present a probabilistic learning method under implicit constraints. The Kullback-Leibler minimum principle [3, 14, 16] is employed to estimate the posterior probability measure that satisfies the constraints. This principle uses the prior probability measure and constraints related to statistical moments, given as targets. The Kullback-Leibler minimum principle has been used in previous works [26, 29, 31, 32]. The posterior probability measure is represented by an algebraic expression involving the prior probability measure and a Lagrange multiplier vector, $\boldsymbol{\lambda}$, associated with the function $\mathbf{h}^c$. The optimal value of the Lagrange multipliers, $\boldsymbol{\lambda}^{\text{sol}}$, is obtained as the limit of a sequence $\{\boldsymbol{\lambda}_i\}_i$, allowing for the construction of a sequence of probability measures, where the limit corresponds to the desired posterior probability measure. Due to the high-dimensional framework and a small training dataset, the Probabilistic Learning on Manifolds (PLoM) [28, 30, 33] is used to generate the constrained learned dataset for each value of $\boldsymbol{\lambda}_i$. The generation is based on a Markov Chain Monte Carlo (MCMC) method employing a nonlinear Itô stochastic differential equation (ISDE) associated with a nonlinear stochastic dissipative Hamiltonian dynamical system. The inclusion of a dissipative term allows for the rapid elimination of the transient part and facilitates the attainment of the stationary response associated with the invariant measure. Evaluating the drift vector of the ISDE involves computing the gradient of function $\mathbf{h}^c$ multiple times. As an algebraic expression of this gradient is not available, direct numerical calculations considering the high dimensionality are not feasible. In [26], a method is proposed to construct an explicit statistical surrogate model, $\mathbf{h}^N$, of implicit function $\mathbf{h}^c$. The surrogate model depends on the number N of points generated in the constrained learned dataset, and its gradient has an explicit algebraic representation.

(iv) Difficulties and novelties. In the broader context of the proposed methodology, this statistical inverse problem poses several challenges. Firstly, there is a significant computational cost associated with evaluating a large Stochastic Computational Model (SCM) resulting form the discretization of the stochastic BVP. Consequently, a limitation arises from the availability of a small training dataset. Secondly, the random parameters contribute to a high stochastic dimensionality. Additionally, the constraints are defined by an implicit, nonlinear mapping operating in a high-dimensional space. To overcome each of these challenges, the author provides a brief overview of the recently proposed probabilistic learning methodology [26]. This general method is applied to 3D elasticity homogenization without scale separation.

2 Formulation of the Probabilistic Learning Inference Using the Kullback-Leibler Divergence

(i) Stochastic boundary value problem and quantity of interest. All the random quantities are defined on a probability space $(\Theta, \mathcal{T}, \mathcal{P})$. We consider a stochastic elliptic BVP on the open bounded domain $\Omega \subset \mathbb{R}^d$ (for instance, $d = 3$), whose partial differential equation (PDE) is written as $\mathcal{N}(\mathbf{Y}, \mathbf{G}, \mathbf{W}) = \mathbf{0}$ *a.s.* The unknown is the non-Gaussian vector-valued field $\{\mathbf{Y}(\boldsymbol{\xi}), \boldsymbol{\xi} \in \Omega\}$ that satisfies the boundary conditions. The coefficients of the stochastic elliptic operator depend on a non-Gaussian second-order vector-valued random field $\mathbf{G}$ and on a random vector-valued control parameter $\mathbf{W}$. It is assumed that the weak formulation of this stochastic BVP admits a unique strong stochastic solution $\mathbf{Y} = \mathbf{f}(\mathbf{G}, \mathbf{W})$ that is a second-order random field. The mapping $\mathbf{f}$ is not explicitly known. An approximation is constructed, on the one hand by using the finite element method to discretize the weak formulation, and on the other hand by using the numerical method of Monte Carlo to approximate the strong stochastic solution. The quantity of interest (observation) is a second-order vector-valued random variable $\mathbf{Q} = \mathcal{O}(\mathbf{X})$ with $\mathbf{X} = (\mathbf{Y}, \mathbf{G}, \mathbf{W})$, and where the observation operator, $\mathcal{O}$, is a given measurable mapping. This parameterization is summarized in Fig. 1.

(ii) Construction of the small training dataset. A prior probability model of $\{\mathbf{G}, \mathbf{W}\}$, represented by the joint probability measure $P^{\text{prior}}_{\mathbf{G},\mathbf{W}}$, is constructed, along with its generator of independent realizations. The subscript "inlinealttextd$d$" designates the quantities related to the training dataset, which is constructed by using the Monte Carlo numerical simulation method. Let $\{\mathbf{g}_d^1, \ldots, \mathbf{g}_d^{N_d}\}$ and $\{\mathbf{w}_d^1, \ldots, \mathbf{w}_d^{N_d}\}$ independent realizations of $\{\mathbf{G}, \mathbf{W}\}$, generated using the prior probability model of $\{\mathbf{G}, \mathbf{W}\}$. Each realization $\mathbf{y}_d^j$ is computed by solving the weak formulation of the

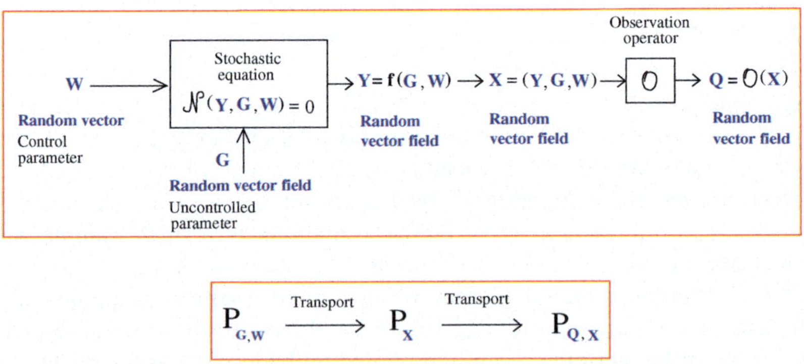

Fig. 1 Diagram illustrating the parameterization of the stochastic boundary value problem and the quantity of interest $\mathbf{Q}$. The prior probability model is defined by the joint probability measure $P_{\mathbf{G},\mathbf{W}}$, which is transported into the measure $P_{\mathbf{X}}$ and further transported into the joint probability measure $P_{\mathbf{Q},\mathbf{X}}$

PDE $\mathscr{N}(\mathbf{y}_d^j, \mathbf{g}_d^j, \mathbf{w}_d^j) = \mathbf{0}$ with the boundary conditions. Consequently, N_d independent realizations $\{\mathbf{y}_d^j, j = 1, \ldots, N_d\}$ of the random field $\mathbf{Y}$ are computed, where $\mathbf{y}_d^j = \mathbf{f}(\mathbf{g}_d^j, \mathbf{w}_d^j)$. The N_d independent realizations $\{\mathbf{q}_d^j, j = 1, \ldots, N_d\}$ of the random observation $\mathbf{Q}$ are deduced as $\mathbf{q}_d^j = \mathscr{O}(\mathbf{x}_d^j)$, where $\mathbf{x}_d^j = \{\mathbf{y}_d^j, \mathbf{g}_d^j, \mathbf{w}_d^j\}$ is the realization of $\mathbf{X} = \{\mathbf{Y}, \mathbf{G}, \mathbf{W}\}$. The training dataset is then composed of a small number N_d of points $\mathbf{x}_d^j$ for $j = 1, \ldots, N_d$, which represent N_d independent realizations of $\mathbf{X}$. It is assumed that the BVP can only be solved a limited number of times. This implies that the training dataset is a small dataset (in contrast to a big dataset). Therefore, the posterior model is constructed using a learning tool to generate the constrained learned realizations of $\mathbf{X}$ without solving the BVP, but solely using the training dataset.

(iii) Finite reduced-order representation and training dataset $\mathscr{D}_{N_d}(\boldsymbol{\eta})$. Let us assumed that $\mathbf{X} = \{\mathbf{Y}, \mathbf{G}, \mathbf{W}\}$ is a second-order random quantity (random fields and random variable) defined on $(\Theta, \mathscr{T}, \mathscr{P})$, and that its covariance operator is a Hilbert-Schmidt [6], symmetric, positive operator in a Hilbert space $\mathbb{X}$ equipped with the inner product $\langle \mathbb{x}, \mathbb{x}' \rangle_{\mathbb{X}}$. Therefore, $\mathbf{X}$ can be approximated by the truncated Karhunen-Loève expansion $\mathbf{X}^{(\nu)}$ [15, 17] of $\mathbf{X}$,

$$\mathbf{X}^{(\nu)} = \underline{\mathbf{x}} + \sum_{\alpha=1}^{\nu} \sqrt{\kappa_\alpha}\, \boldsymbol{\varphi}^\alpha\, H_\alpha \,, \tag{1}$$

where the eigenvalues of the covariance operator are $\kappa_1 \geq \cdots \geq \kappa_\nu \geq \cdots = 0$ with $\sum_{\alpha=1}^{+\infty} \kappa_\alpha^2 < +\infty$. The family of the eigenfunctions $\{\boldsymbol{\varphi}^\alpha\}_\alpha$ is a Hilbert basis of $\mathbb{X}$, where $\underline{\mathbf{x}} = E\{\mathbf{X}\}$, and where $\mathbf{H} = (H_1, \ldots, H_\nu)$ is a second-order, centered, $\mathbb{R}^\nu$-valued random variable whose covariance matrix is the identity matrix $[I_\nu]$ in $\mathbb{M}_\nu$. For $\alpha \in \{1, \ldots, \nu\}$, the component H_α is written as $H_\alpha = \kappa_\alpha^{-1/2} \langle \mathbf{X} - \underline{\mathbf{x}}, \boldsymbol{\varphi}^\alpha \rangle_{\mathbb{X}}$. The training set $\mathscr{D}_d$ related to $\mathbf{H}$ consists of the N_d independent realizations $\{\boldsymbol{\eta}_d^j, j = 1, \ldots, N_d\}$ such that $\boldsymbol{\eta}_d^j = \kappa_\alpha^{-1/2} \langle \mathbf{x}_d^j - \underline{\mathbf{x}}, \boldsymbol{\varphi}^\alpha \rangle_{\mathbb{X}}$. We assume that the kernel of the covariance operator is unknown. Therefore, we can only obtain an approximation of the covariance operator using an empirical estimator built with the N_d points $\{\mathbf{x}_d^j, j = 1, \ldots, N_d\}$. Under these conditions the largest value of ν will be $N_d - 1$ and the discretization of Eq. (1) will simply correspond to a normalization of the points that constitute the training set $\mathscr{D}_d$. If ν can be chosen as $N_d - 1$, then $\mathbf{X}^{(\nu)}$ is a reduced-order representation of $\mathbf{X}$.

(iv) Probabilistic learning inference under implicit constraints. We introduce the superscript "c" to designate the solution with the constraints, which corresponds to the posterior model. Let $\mathbf{X}^c = (\mathbf{Y}^c, \mathbf{G}^c, \mathbf{W}^c)$ and $\mathbf{Q}^c = \mathscr{O}(\mathbf{X}^c)$ be the posterior probability model of the prior probability model of $\mathbf{X} = (\mathbf{Y}, \mathbf{G}, \mathbf{W})$ with $\mathbf{Y} = \mathbf{f}(\mathbf{G}, \mathbf{W})$ and $\mathbf{Q} = \mathscr{O}(\mathbf{X})$, respectively. Let $R^c = r(\mathbf{X}^c)$ be the positive-valued random variable representing the norm of the random residue of the PDE. The probabilistic learning inference belongs to the class of the statistical inverse problems. The prior probability model of $\{\mathbf{G}, \mathbf{W}\}$ is given and we are interested in estimating the posterior probability model in order that the second-order moment $E\{(R^c)^2\}$ of R^c be as small

as possible and that some statistical moments of the posterior observations $\mathbf{Q}^c$ are close to some given targets. Using Eq. (1), the implicit function $\boldsymbol{\eta} \mapsto \mathbf{h}^c(\boldsymbol{\eta})$ from $\mathbb{R}^\nu$ into $\mathbb{R}^{n_c}$, which globally defines the constraints, yields the constraints equation,

$$E\{\mathbf{h}^c(\mathbf{H}^c)\} = \mathbf{b}^c \,, \tag{2}$$

in which $\mathbf{b}^c \in \mathbb{R}^{n_c}$ is the target and $\mathbf{h}^c$ is an implicit mapping.

(v) Formulation using the Kullback-Leibler divergence minimum principle. We use the Kullback-Leibler divergence minimum principle [3, 14, 16] for estimating the posterior probability measure $P_{\mathbf{H}^c}(d\boldsymbol{\eta}) = p_{\mathbf{H}^c}(\boldsymbol{\eta})\, d\boldsymbol{\eta}$ on $\mathbb{R}^\nu$ of the $\mathbb{R}^\nu$-valued random variable $\mathbf{H}^c = (H_1^c, \ldots, H_\nu^c)$. This estimation of $P_{\mathbf{H}^c}$ is performed using the prior probability measure $P_{\mathbf{H}}(d\boldsymbol{\eta}) = p_{\mathbf{H}}(\boldsymbol{\eta})\, d\boldsymbol{\eta}$ on $\mathbb{R}^\nu$ in which $p_{\mathbf{H}}$ is estimated with a modified (see (vi) below) Gaussian KDE method [2] using the points $\{\boldsymbol{\eta}_d^1, \ldots, \boldsymbol{\eta}_d^{N_d}\}$ of the training dataset $\mathscr{D}_d$, and using the constraint defined by Eq. (2). Therefore, the probability density function $p_{\mathbf{H}^c}$ on $\mathbb{R}^\nu$, which satisfies the constraint defined by Eq. (1) and which is closest to $p_{\mathbf{H}}$, is the solution of the optimization problem

$$p_{\mathbf{H}^c} = \arg \min_{p \in \mathscr{C}_{\mathrm{ad},p}} \int_{\mathbb{R}^\nu} p(\boldsymbol{\eta}) \log\left(\frac{p(\boldsymbol{\eta})}{p_{\mathbf{H}}(\boldsymbol{\eta})}\right) d\boldsymbol{\eta} \,, \tag{3}$$

in which the admissible set $\mathscr{C}_{\mathrm{ad},p}$ is defined by

$$\mathscr{C}_{\mathrm{ad},p} = \left\{\boldsymbol{\eta} \mapsto p(\boldsymbol{\eta}) : \mathbb{R}^\nu \rightarrow \mathbb{R}^+ \,, \int_{\mathbb{R}^\nu} p(\boldsymbol{\eta})\, d\boldsymbol{\eta} = 1 \,, \int_{\mathbb{R}^\nu} \mathbf{h}^c(\boldsymbol{\eta})\, p(\boldsymbol{\eta})\, d\boldsymbol{\eta} = \mathbf{b}^c\right\} . \tag{4}$$

(vi) Lagrange multipliers associated with the constraints and solution of the functional optimization problem. As explained in Sect. 1-(iii), the constraints within the admissible set $\mathscr{C}_{\mathrm{ad},p}$ are taken into account by introducing the Lagrange multipliers $\lambda_0 - 1$ with $\lambda_0 \in \mathbb{R}^+$ (associated with the normalization condition of the pdf) and $\boldsymbol{\lambda} \in \mathscr{C}_{\mathrm{ad},\boldsymbol{\lambda}} \subset \mathbb{R}^{n_c}$ (associated with the imposed constraints defined by the implicit function $\mathbf{h}^c$). It has been proven [26] that $\boldsymbol{\lambda}^{\mathrm{sol}}$ and the pdf $p_{\mathbf{H}^c}$ of $\mathbf{H}^c$ can be constructed as follows. For $1 \leq n_c \leq \nu$, let us assume that $\mathbf{h}^c \in C^1(\mathbb{R}^\nu, \mathbb{R}^{n_c})$ and there exist constants $\alpha > 0$, $\beta > 0$, $c_\alpha > 0$, and $c_\beta > 0$, independent of $\boldsymbol{\eta}$, such that for $\| \boldsymbol{\eta} \| \rightarrow +\infty$, $\| \mathbf{h}^c(\boldsymbol{\eta}) \| \ \leq c_\alpha \| \boldsymbol{\eta} \|^\alpha$ and $\| [\nabla_{\boldsymbol{\eta}} \mathbf{h}^c(\boldsymbol{\eta})] \|_F \ \leq c_\beta \| \boldsymbol{\eta} \|^\beta$ in which $[\nabla_{\boldsymbol{\eta}} \mathbf{h}^c(\boldsymbol{\eta})] \in \mathbb{M}_{\nu,n_c}$ with $[\nabla_{\boldsymbol{\eta}} \mathbf{h}^c(\boldsymbol{\eta})]_{\alpha k} = \partial h_k^c(\boldsymbol{\eta})/\partial \eta_\alpha$, and where $\| \,.\, \|_F$ is the Frobenius norm. Then, the admissible set $\mathscr{C}_{\mathrm{ad},\boldsymbol{\lambda}}$ of $\boldsymbol{\lambda}$ is a convex open subset of $\mathbb{R}^{n_c}$. For all $\boldsymbol{\lambda}$ in $\mathscr{C}_{\mathrm{ad},\boldsymbol{\lambda}}$, the pdf $p_{\mathbf{H}_{\boldsymbol{\lambda}}}$ of $\mathbf{H}_{\boldsymbol{\lambda}}$ can be written as

$$p_{\mathbf{H}_{\boldsymbol{\lambda}}}(\boldsymbol{\eta}\,; \boldsymbol{\lambda}) = c_0(\boldsymbol{\lambda})\, \zeta(\boldsymbol{\lambda}) \exp\{-\langle \boldsymbol{\lambda} \,, \mathbf{h}^c(\boldsymbol{\eta}) \rangle\} \,, \tag{5}$$

with $0 < c_0(\boldsymbol{\lambda}) < +\infty$ and where $\boldsymbol{\eta} \mapsto \zeta(\boldsymbol{\eta}) : \mathbb{R}^\nu \rightarrow \mathbb{R}^+$ is written as $\zeta(\boldsymbol{\eta}) = N_d^{-1} \sum_{j=1}^{N_d} \exp\{-\frac{1}{2\hat{s}^2} \| \frac{\hat{s}}{s} \boldsymbol{\eta}_d^j - \boldsymbol{\eta} \|^2\}$ where $\hat{s} = s \left(s^2 + (N_d - 1)/N_d\right)^{-1/2}$ with $s = (4/(N_d(2+\nu)))^{1/(\nu+4)}$. The $\mathbb{R}^{n_c}$-valued random variable $\mathbf{h}^c(\mathbf{H}_{\boldsymbol{\lambda}})$ is such that $E\{\| \mathbf{h}^c(\mathbf{H}_\lambda) \|^2\} < +\infty$. Let $\boldsymbol{\lambda} \mapsto \Gamma(\boldsymbol{\lambda}) : \mathscr{C}_{\text{ad},\lambda} \rightarrow \mathbb{R}$ be defined by $\Gamma(\boldsymbol{\lambda}) = \langle \boldsymbol{\lambda} \, , \mathbf{b}^c \rangle - \log c_0(\boldsymbol{\lambda})$ in which $\mathbf{b}^c \in \mathbb{R}^{n_c}$ is defined in Eq. (2). For all $\boldsymbol{\lambda}$ in $\mathscr{C}_{\text{ad},\lambda}$, we have

$$\nabla_{\!\lambda} \Gamma(\boldsymbol{\lambda}) = \mathbf{b}^c - E\{\mathbf{h}^c(\mathbf{H}_\lambda)\} \in \mathbb{R}^{n_c} \quad , \quad [\Gamma''(\boldsymbol{\lambda})] = [\text{cov}\{\mathbf{h}^c(\mathbf{H}_\lambda)\}] \in \mathbb{M}^+_{n_c} \, , \tag{6}$$

where $[\Gamma''(\boldsymbol{\lambda})]$ is the positive-definite covariance matrix of $\mathbf{h}^c(\mathbf{H}_\lambda)$, which is such that $[\Gamma''(\boldsymbol{\lambda})]_{kk'} = \partial^2\Gamma(\boldsymbol{\lambda})/\partial\lambda_k\partial\lambda_{k'}$. Function Γ is a strictly convex function on $\mathscr{C}_{\text{ad},\lambda}$. There is a unique solution $\boldsymbol{\lambda}^{\text{sol}}$ in $\mathscr{C}_{\text{ad},\lambda}$ of the convex optimization problem,

$$\boldsymbol{\lambda}^{\text{sol}} = \arg \min_{\boldsymbol{\lambda} \in \mathscr{C}_{\text{ad},\lambda}} \Gamma(\boldsymbol{\lambda}) \, . \tag{7}$$

If the following equation in $\boldsymbol{\lambda}$,

$$\nabla_{\!\lambda} \Gamma(\boldsymbol{\lambda}) = \mathbf{0}_{n_c} \, , \tag{8}$$

has a solution $\tilde{\boldsymbol{\lambda}}$ in $\mathscr{C}_{\text{ad},\lambda}$, then this solution is unique and we have $\boldsymbol{\lambda}^{\text{sol}} = \tilde{\boldsymbol{\lambda}}$. The pdf $p_{\mathbf{H}^c}$ of $\mathbf{H}^c$, which satisfies the constraint $E\{\mathbf{h}^c(\mathbf{H}^c)\} = \mathbf{b}^c$ is written as

$$p_{\mathbf{H}^c}(\boldsymbol{\eta}) = p_{\mathbf{H}_{\lambda^{\text{sol}}}}(\boldsymbol{\eta} \, ; \boldsymbol{\lambda}^{\text{sol}}) \quad , \quad \forall \boldsymbol{\eta} \in \mathbb{R}^\nu \, . \tag{9}$$

Constant $c_0(\boldsymbol{\lambda})$ cannot be computed when ν is large. As a result, the standard conjugate gradient algorithm cannot be used to solve the convex optimization problem defined by Eq. (7). Instead, $\boldsymbol{\lambda}^{\text{sol}}$ is obtained by solving Eq. (8) using a Newton iterative method, which necessitates estimating the mathematical expectations in Eq. (6). To accomplish this, a learned dataset $\mathscr{D}_{\text{learn}}(\boldsymbol{\eta}_\lambda) = \boldsymbol{\eta}_\lambda^1, \ldots, \boldsymbol{\eta}_\lambda^N$ is computed using the PLoM algorithm (refer to Sect. 1-(iii)) for any fixed value of $\boldsymbol{\lambda}$ in $\mathscr{C}_{\text{ad},\lambda}$ and for which the invariant measure admits the marginal probability measure $P_{\mathbf{H}_\lambda}(d\boldsymbol{\eta} \, ; \boldsymbol{\lambda}) = p_{\mathbf{H}_\lambda}(\boldsymbol{\eta} \, ; \boldsymbol{\lambda}) \, d\boldsymbol{\eta}$, defined by Eq. (5). This learned dataset comprises a large number $N \gg N_d$ of realizations of $\mathbf{H}_\lambda$. Since the implicit function $\mathbf{h}^c$ and its gradient must be evaluated a large number of times by the MCMC algorithm used by PLoM, an algebraic surrogate model of $\mathbf{h}^c$ must be constructed.

(vii) Statistical surrogate model of the implicit function $\mathbf{h}^c$ *and convergence analysis.* Let $\mathbf{A}_\lambda = \mathbf{h}^c(\mathbf{H}_\lambda)$ be the $\mathbb{R}^{n_c}$-valued random variable whose N independent realizations $\mathbf{a}_\lambda^1, \ldots, \mathbf{a}_\lambda^N$ are such that $\mathbf{a}_\lambda^\ell = \mathbf{h}^c(\boldsymbol{\eta}_\lambda^\ell) \in \mathbb{R}^{n_c}$ for $\ell = 1, \ldots, N$. The surrogate model $\boldsymbol{\eta} \mapsto \mathbf{h}^N(\boldsymbol{\eta} \, ; \boldsymbol{\lambda}) : \mathbb{R}^\nu \rightarrow \mathbb{R}^{n_c}$ of $\mathbf{h}^c$ is defined [26], for all $\boldsymbol{\eta}$ in $\mathbb{R}^\nu$, by

$$\mathbf{h}^N(\boldsymbol{\eta} \, ; \boldsymbol{\lambda}) = \sum_{\ell=1}^N \mathbf{a}_\lambda^\ell \, \frac{\beta_{\boldsymbol{\eta}}^N(\boldsymbol{\eta}_\lambda^\ell)}{\sum_{\ell'=1}^N \beta_{\boldsymbol{\eta}}^N(\boldsymbol{\eta}_\lambda^{\ell'})} \, , \tag{10}$$

in which for all $\boldsymbol{\eta}$ and $\tilde{\boldsymbol{\eta}}$ in $\mathbb{R}^\nu$,

$$\beta_{\boldsymbol{\eta}}^N(\tilde{\boldsymbol{\eta}}) = \exp\{-\frac{1}{2s_{\text{SB}}^2} \| \tilde{\boldsymbol{\eta}} - \boldsymbol{\eta} \|_H^2\} \quad , \quad \| \tilde{\boldsymbol{\eta}} - \boldsymbol{\eta} \|_H^2 = \langle [\sigma_{\mathbf{H}_\lambda}]^{-2}(\tilde{\boldsymbol{\eta}} - \boldsymbol{\eta}) \, , \tilde{\boldsymbol{\eta}} - \boldsymbol{\eta} \rangle \, , \tag{11}$$

in which $[\sigma_{\mathbf{H}_\lambda}]$ is the diagonal positive-definite matrix in $\mathbb{M}_\nu^+$ such that $[\sigma_{\mathbf{H}_\lambda}]_{\alpha\alpha}$ is the standard deviation of the real-valued random variable $H_{\boldsymbol{\lambda},\alpha}$, estimated using $\mathscr{D}_{\rm learn}(\boldsymbol{\eta}_{\boldsymbol{\lambda}})$, and where $s_{\rm SB}$ is the Silverman bandwidth $s_{\rm SB} = (4/\{N(2+n_c+\nu)\})^{1/(n_c+\nu+4)}$. Propositions 2–4 of [26] demonstrate the convergence of the sequence of MCMC generator of PLoM using this statistical surrogate model.

3 Application to 3D Stochastic Homogenization of Heterogeneous Material with Random Spectrum and Without Scale Separation

The details of this application and additional results can be found in [26]. The non-separation of the mesoscale with the macroscale means that the macroscale is another mesoscale at larger scale with random effective/apparent elastic properties. The posterior probability distribution is constructed using the general methodology presented in Sect. 2, where the constraints are defined as follows:

- a target set made up of given "experimental" statistical moments of the random effective/apparent elasticity tensor $\mathbb{C}^{\rm eff}$.
- the second-order moment of the random normalized residue of the random equations of the stochastic computational model.

These constraints ensure that the probabilistic learning algorithm aims to bring the statistical moments closer to their target values while maintaining a small residue of the random equations of the stochastic computational model.

In this application, the primary objective is to identify, in cases where the mesoscopic and macroscopic scales are not separable, the hyperparameters of a prior probabilistic model for the random apparent elasticity tensor field $\mathbb{C}$ at the mesoscopic scale. This identification is achieved through the solution of a statistical inverse problem using "experimental data", which are available at the macroscopic scale. These experimental data concern the statistical moments of the effective/apparent elasticity tensor $\mathbb{C}^{\rm eff}$ at the macroscopic scale. The only constraints imposed on the construction of the mesoscopic prior probabilistic model for $\mathbb{C}$ are twofold: first, it must consider all mathematical and physical properties of $\mathbb{C}$ at the mesoscopic scale, including symmetries and positivity; second, the parameterization of $\mathbb{C}$ must allow the generation of a sufficiently large (random) family for $\mathbb{C}^{\rm eff}$ through stochastic homogenization, encompassing the experimental effective/apparent elasticity tensor $\mathbb{C}^{\rm exp}$ at the macroscopic scale. These conditions enable the creation of a predictive statistical model based on experimental data. Such a probabilistic model, based on experimentation, is particularly valuable. Imagine conducting experimental tests on a specimen of a heterogeneous material subjected to deterministic imposed forces/displacements, where deformations are measured (e.g., a concrete specimen). If the specimen is smaller than a representative elementary volume (RVE), the measured macroscopic deformations are random, rendering the effective tensor $\mathbb{C}^{\rm eff}$ not deterministic but random (apparent tensor). Once the probabilistic model for $\mathbb{C}$ has been experimentally identified, the updated meso-macro probabilistic model can be used to simulate the deterministic effective tensor through numerical simulations on an RVE.

(i) Stochastic elliptic boundary value problem at mesoscale. We consider the stochastic homogenization of a heterogeneous linear elastic microstructure, which is modeled at the mesoscale rather than the microscale, and it occupies the 3D bounded open domain $\Omega =]0\,,1\,[\times]0\,,1\,[\times]0\,,0.1\,[\subset \mathbb{R}^3$ (square thick plate) with boundary $\partial\Omega$. For all m and r in $\{1, 2, 3\}$ the $\mathbb{R}^3$-valued displacement random field $\{\mathbf{Y}^{mr}(\boldsymbol{\xi}) \in \Omega\}$ at mesoscale is defined on $(\Theta, \mathcal{T}, \mathcal{P})$, is indexed by Ω, and satisfies the stochastic boundary value problem associated with the stochastic homogenization of a random elastic medium at mesoscale without scale separation, such that for $i = 1, 2, 3$, and almost surely,

$$-\frac{\partial}{\partial \xi_j}\big(\mathbb{C}_{ijpq}(\boldsymbol{\xi})\,\varepsilon_{pq}(\mathbf{Y}^{mr}(\boldsymbol{\xi}))\big) = \mathbf{0}_3 \quad , \quad \forall \boldsymbol{\xi} \in \Omega\,, \tag{12}$$

$$\mathbf{Y}^{mr}(\boldsymbol{\xi}) - \mathbf{y}_0^{mr}(\boldsymbol{\xi}), \quad \forall \boldsymbol{\xi} \in \partial\Omega\,, \tag{13}$$

in which the strain tensor is $\varepsilon_{pq}(\mathbf{y}) = (\partial y_p/\partial \xi_q + \partial y_q/\partial \xi_p)/2$ for all differentiable function $\boldsymbol{\xi} \mapsto \mathbf{y}(\boldsymbol{\xi})$ and where the components of $\mathbf{y}_0^{mr}(\boldsymbol{\xi})$ are defined by $y_{0,j}^{mr}(\boldsymbol{\xi}) = (\delta_{jm}\,\xi_r + \delta_{jr}\,\xi_m)/2$, in which δ_{jm} is the Kronecker symbol. At mesoscale, the linear elastic heterogeneous medium is described by the random apparent elasticity field $\{\mathbb{C}(\boldsymbol{\xi}), \boldsymbol{\xi} \in \mathbb{R}^3\}$, which is a non-Gaussian fourth-order tensor-valued random field $\mathbb{C} = \{\mathbb{C}_{ijpq}\}_{ijpq}$ with i, j, p, and q in $\{1, 2, 3\}$, defined on $(\Theta, \mathcal{T}, \mathcal{P})$. The stochastic homogenization consists, for i, j, m, and r in $\{1, 2, 3\}$, in analyzing at macroscale the component $\mathbb{C}^{\text{eff}}_{ijmr}$ of the random effective/apparent elasticity tensor $\{\mathbb{C}^{\text{eff}}_{ijmr}\}_{ijmr}$, which is defined by

$$\mathbb{C}^{\text{eff}}_{ijmr} = \frac{1}{|\Omega|}\int_\Omega \mathbb{C}_{ijpq}(\boldsymbol{\xi})\,\varepsilon_{pq}(\mathbf{Y}^{mr}(\boldsymbol{\xi}))\,d\boldsymbol{\xi}\,, \tag{14}$$

in which $\mathbf{Y}^{mr}$ is the $\mathbb{R}^3$-valued random field that satisfies Eqs. (12)–(13) and where $|\Omega| = \int_\Omega d\boldsymbol{\xi}$. The random effective/apparent elasticity tensor $\mathbb{C}^{\text{eff}}$ is symmetric and positive definite almost surely. If there was a scale separation, then the statistical fluctuations of this tensor would be negligible.

(ii) Prior probability model of the apparent elasticity field $\mathbb{C}$ at mesoscale. It is essential to note that the prior probability model of the apparent elastic field is not carried out at the microscopic scale of the microstructure but rather at the mesoscopic scale. This means that no properties specific to any symmetry class of the elastic field at the microscale are utilized. We refer the reader to [9, 10] for the implementation of symmetry classes in mesoscopic random models of the random apparent elasticity field and to [12, 18] for microscopic random models of the microstructure when it can be described in terms of constituents. Regarding the construction of the mesoscopic prior probabilistic model for the random apparent elasticity field $\mathbb{C} = \{\mathbb{C}(\boldsymbol{\xi}), \boldsymbol{\xi} \in \Omega\}$, it is essential to note that describing only the

correlation structure of $\mathbb{C}$ is insufficient. This is because $\mathbb{C}$ is non-Gaussian due to its positivity. To solve the nonlinear statistical inverse problem, it is necessary to build the probability measure of $\mathbb{C}$ and its associated random generator. Such a construction has been proposed in [23, 24] for cases involving anisotropic statistical fluctuations around the mean model at the mesoscale. Its extension to the case involving a random spectral measure, as applied here, is detailed in [25]. The mesoscopic prior probability model of $\mathbb{C}$ is defined as the restriction to Ω of a non-Gaussian, homogeneous, mean-square continuous, fourth-order tensor-valued random field indexed by $\mathbb{R}^3$. The mesoscopic mean model, also known as the mesoscopic nominal model, is isotropic and characterized by $\underline{C}_{\text{bulk}} = 1.09 \times 10^{11}\,\text{N/m}^2$ and $\underline{C}_{\text{shear}} = 6.85 \times 10^{10}\,\text{N/m}^2$. The mesoscopic statistical fluctuations around the mean mesoscopic model are anisotropic. The parameterization of $\mathbb{C}$ is expressed as $\mathbb{C}(\boldsymbol{\xi}) = \mathbb{c}(\mathbf{G}(\boldsymbol{\xi}), \underline{\mathbf{z}})$ for $\boldsymbol{\xi} \in \Omega$, where $\mathbb{c}$ is a tensor-valued function on $\mathbb{R}^{21} \times \mathbb{R}^3$, and $\mathbb{G} = \{\mathbf{G}(\boldsymbol{\xi}), \boldsymbol{\xi} \in \mathbb{R}^3\}$ is a non-Gaussian, second-order, homogeneous, mean-square continuous, $\mathbb{R}^{21}$-valued random field, indexed by $\mathbb{R}^3$. Here, $\underline{\mathbf{z}} = (\underline{C}_{\text{bulk}}, \underline{C}_{\text{shear}}, \underline{\delta}_{\mathbb{C}})$. The non-Gaussian probability measure of $\mathbb{G}$ is fully determined through statistical conditioning of a Gaussian $\mathbb{R}^{21}$-valued random field. The mapping $\mathbb{c}$ is constructed using the maximum entropy principle under constraints, ensuring that the properties of symmetry and positivity of the non-Gaussian random tensor $\mathbb{C}(\boldsymbol{\xi})$ are satisfied at any point $\boldsymbol{\xi}$. The correlation structure of the non-Gaussian field $\mathbb{C}$ is determined by the mapping $\mathbb{c}$ from the correlation structure of $\mathbb{G}$. The random matrix spectral measure of $\mathbb{G}$ is defined in [25] based on a prior algebraic model with several hyperparameters: (1) the level of uncertainties on the spectral measure that is controlled by a dispersion parameter $\delta_s = 0.1$, (2) the level of mesoscopic statistical fluctuations of the random medium that is governed by the dispersion coefficient $\underline{\delta}_{\mathbb{C}} = 0.3$, (3) and finally, three spatial correlations lengths denoted as $\underline{\mathbf{L}}_c = (\underline{L}_{c1}, \underline{L}_{c2}, \underline{L}_{c3})$. It is worth noting that the proposed model allows for the introduction of a total of 3×21 spatial correlation lengths, representing the most general parameterization.

(iii) Spatial correlation lengths and scale separation. Cases, SC1, SC2, and SC3 of the correlation lengths are considered for analyzing the level of scale separation:

$SC1$: $\underline{\mathbf{L}}_c = (0.1, 0.1, 0.1)$ (partial separation in ξ_1, ξ_2, not in ξ_3).

$SC2$: $\underline{\mathbf{L}}_c = (0.3, 0.3, 0.1)$ (not separated in ξ_1, ξ_2, and ξ_3).

$SC3$: $\underline{\mathbf{L}}_c = (0.5, 0.5, 0.2)$ (strongly not separated in ξ_1,ξ_2, and ξ_3).

(iv) Random control parameter for the probabilistic learning inference. The components of the $\mathbb{R}^3$-valued random control parameter $\mathbf{W}$ are $(\log C_{\text{bulk}}, \log C_{\text{shear}}, \log \delta_{\mathbb{C}})$ in which C_{bulk} and C_{shear} are Gamma independent random variables (see [8]) with mean values $\underline{C}_{\text{bulk}}, \underline{C}_{\text{shear}}$ and coefficients of variation $\delta_{\text{bulk}} = 0.5$, $\delta_{\text{shear}} = 0.25$, and where $\delta_{\mathbb{C}}$ is a uniform random variable on $[0.1\,, 0.5]$.

(v) Stochastic computational model. The finite element method is used for discretizing the weak formulation of the stochastic BVP and yields the stochastic equations on $\mathbb{R}^{n_y}$ where $n_y = 52\,215$. The discretization of the vector-valued random field $\mathbf{G}$ yields a $\mathbb{R}^{n_g}$-valued random variable where $n_g = 3\,626\,800$. The $\mathbb{R}^{n_w}$-valued random control parameter, with $n_w = 3$.

(vi) Training dataset and finite reduced-order representation. The mesoscopic prior probability model and the stochastic computational model are used to construct $N_d = 50$ realizations $\{\mathbf{x}_d^j, j = 1, \ldots, N_d\}$ of the random vector $\mathbf{X}$ with values in $\mathbb{R}^{n_x}$ where $n_x = 6\,n_y + n_g + n_w = 3\,942\,093$. The dimension ν of the reduced representation $\mathbf{X}^{(\nu)}$ of $\mathbf{X}$ constructed by PCA is $\nu = N_d - 1 = 49$.

(vii) Random normalized residue. For $\boldsymbol{\lambda}$ fixed in $\mathcal{C}_{\text{ad},\boldsymbol{\lambda}}$, the dimensionless positive-valued random variable $R_{\boldsymbol{\lambda}}^c$ represents the norm of the vector-valued random residue of the stochastic equations. A realization ℓ is denoted as $\rho_{\boldsymbol{\lambda}}^\ell = \hat{\rho}_{\boldsymbol{\lambda}}^\ell / \underline{\hat{\rho}}_0$, where $\underline{\hat{\rho}}_0$ is an adapted constant for normalization (refer to [26]). This ensures that the constraint equation for the random residue can be expressed as $E\{(R_{\boldsymbol{\lambda}}^c)^2\} = b_R^c$, with $b_R^c = 1$.

(viii) Statistical moments and their targets. Using the Voigt notation, let $[\mathbb{C}^{\text{eff}}]$ be the second-order $\mathbb{M}_6^+$-valued random variable associated with the random tensor $\mathbb{C}^{\text{eff}}$. The first statistical moment of interest is the mean value $[\,\underline{\mathbb{C}}^{\text{eff}}] = E\{\,[\mathbb{C}^{\text{eff}}]\,\} \in \mathbb{M}_6^+$ of random matrix $[\mathbb{C}^{\text{eff}}]$ while its target counterpart is the given matrix $[\,\underline{\mathbb{C}}^{\text{exp}}] \in \mathbb{M}_6^+$. The corresponding constraint equation, which allows the mean value to be fitted, will then be written as $E\{\,[\mathbb{C}^{\text{eff}}]\,\} = [\,\underline{\mathbb{C}}^{\text{exp}}]$. The second statistical moment of interest is the coefficient of dispersion $\delta^{\,\text{eff}}$ of random matrix $[\mathbb{C}^{\text{eff}}]$ and its target counterpart $\delta^{\,\text{exp}}$, which allows the level of statistical fluctuations to be fitted. We then introduce the positive-valued random variable $\Delta_2^{\text{eff}} = \|\,[\mathbb{C}^{\text{eff}}] - [\,\underline{\mathbb{C}}^{\text{eff}}]\,\|_F^2 / \|\,[\,\underline{\mathbb{C}}^{\text{eff}}]\,\|_F^2$. Let $\delta^{\,\text{eff}}$ be defined by $\delta^{\,\text{eff}} = (E\{\Delta_2^{\text{eff}}\})^{1/2}$. The constraint equation to control the statistical fluctuations are then written as $\delta^{\,\text{eff}} = \delta^{\,\text{exp}}$. It should be noted that, if $\delta^{\,\text{eff}}$ goes to zero, then the statistical fluctuations (represented by $(\Delta_2^{\text{eff}})^{1/2}$) of $[\mathbb{C}^{\text{eff}}]$ around $[\,\underline{\mathbb{C}}^{\text{eff}}]$ goes to zero.

(ix) Second-order statistics of the random residue and of the effective/apparent elasticity matrix estimated with the constrained learned dataset. For cases SC1, SC2, and SC3, Table 1 gives the posterior statistics computed with the constrained learned set for $N = 10\,000$ (subscript "c"), the prior statistics computed with the training set (subscript "d"), and the targets (superscript "exp"). It can be read,

(a) the second-order moment $E\{R_c^2\}$ of the random normalized residue R_c and the corresponding target b_R^c.
(b) the Frobenius norm $\|\,[\,\underline{\mathbb{C}}^{\text{eff}}]\,\|_F$ of the mean value of the random effective/apparent elasticity matrix.
(c) the coefficient of dispersion $\delta_{\text{ML}}^{\text{eff}} = \{\max_{\delta_2}\, p_{\Delta_2^{\text{eff}}}(\delta_2)\}^{1/2}$ in which $p_{\Delta_2^{\text{eff}}}$ is the pdf of random variable Δ_2^{eff}.

(x) Discussion about the presented results. Table 1 demonstrates that the constrained learned dataset significantly modifies the mesoscopic prior probability model. The posterior statistics align well with the targets. These results confirm controlled residue and small deviations from the reference (training) for the optimal solution. The mean value of the random effective/apparent elasticity matrix achieves the target. When there is no scale separation, there is a significant coefficient of dispersion, which increases as the spatial correlation lengths of the random apparent elasticity field at the mesoscale grow. There are statistical fluctuations in the effective/apparent elasticity tensor at the macroscale. Table 1 further demonstrates that the

Table 1 Posterior statistics with the constrained learned set for $N = 10\,000$ (subscript "c"), prior statistics with the training dataset (subscript "d"), and targets (supercript "exp")

	SC1	SC2	SC3
$E\{R_c^2\}$	1.2938	1.2687	1.2413
b_R^c	1	1	1
$\Vert\,[\underline{\mathbb{C}}_d^{\rm eff}]\,\Vert_F \times 10^{11}$	4.2106	4.1925	4.1943
$\Vert\,[\underline{\mathbb{C}}_c^{\rm eff}]\,\Vert_F \times 10^{11}$	4.6294	4.6923	4.6816
$\Vert\,[\mathbb{C}^{\rm exp}]\,\Vert_F \times 10^{11}$	4.6317	4.6549	4.6706
$\delta_{d,{\rm ML}}^{\rm eff}$	0.2257	0.2469	0.2701
$\delta_{c,{\rm ML}}^{\rm eff}$	0.1329	0.1476	0.1671
$\delta^{\rm exp}$	0.0946	0.1374	0.1825

constraints in place ensure that the probabilistic learning algorithm strives to align the statistical moments with their target values, while simultaneously maintaining a negligible residue of the random equations within the stochastic computational model.

References

1. Allaire, G.: Homogenization and two-scale convergence. SIAM J. Math. Anal. **23**(6), 1482–1518 (1992). https://doi.org/10.1137/0523084
2. Bowman, A., Azzalini, A.: Applied Smoothing Techniques for Data Analysis: The Kernel Approach With S-Plus Illustrations, vol. 18. Oxford University Press, Oxford: Clarendon Press, New York (1997). https://doi.org/10.1007/s001800000033
3. Cover, T.M., Thomas, J.A.: Elements of Information Theory, 2nd edn. Wiley, Hoboken (2006)
4. Dirrenberger, J., Forest, S., Jeulin, D.: Towards gigantic rve sizes for 3d stochastic fibrous networks. Int. J. Solids Struct. **51**(2), 359–376 (2014). https://doi.org/10.1016/j.ijsolstr.2013.10.011
5. Forest, S.: Homogenization methods and mechanics of generalized continua-part 2. Theoret. Appl. Mech. **28–29**, 113–143 (2002)
6. Gelfand, I.M., Vilenkin, N.I.: Les Distributions. Tome 4. Application de l'Analyse Harmonique. Dunod (1967). https://doi.org/10.1016/0375-9474(67)90547-7
7. Ghanem, R., Spanos, P.D.: Stochastic Finite Elements: A Spectral Approach. Springer, New York (1991)
8. Guilleminot, J., Soize, C.: On the statistical dependence for the components of random elasticity tensors exhibiting material symmetry properties. J. Elast. **111**(2), 109–130 (2013). https://doi.org/10.1007/s10659-012-9396-z
9. Guilleminot, J., Soize, C.: Stochastic model and generator for random fields with symmetry properties: application to the mesoscopic modeling of elastic random media. Multiscale Model. & Simul. (A SIAM Interdiscip. J.) **11**(3), 840–870 (2013). https://doi.org/10.1137/120898346
10. Guilleminot, J., Soize, C.: Non-Gaussian random fields in multiscale mechanics of heterogeneous materials. In: Altenbach H., Öchsner A. (eds.) Encyclopedia of Continuum Mechanics, pp. 1826–1834 (2020). https://doi.org/10.1007/978-3-662-55771-6_68
11. Jeulin, D.: Random texture models for material structures. Stat. Comput. **10**(2), 121–132 (2000). https://doi.org/10.1023/A:1008942325749

12. Jeulin, D.: Morphological Models of Random Structures. Springer Cham (2022). https://doi.org/10.1007/978-3-030-75452-5
13. Jeulin, D., Ostoja-Starzewski, M.: Mechanics of Random and Multiscale Microstructures. Springer (2001)
14. Kapur, J.N., Kesavan, H.K.: Entropy Optimization Principles with Applications. Academic, San Diego (1992)
15. Karhunen, K.: Ueber lineare methoden in der wahrscheinlichkeitsrechnung. Amer. Acad. Sci., Fennicade, Ser. A, I **37**, 3–79 (1947)
16. Kullback, S., Leibler, R.A.: On information and sufficiency. Ann. Math. Stat. **22**(1), 79–86 (1951). https://doi.org/10.1214/aoms/1177729694
17. Loève, M.: Functions aleatoires du second ordre. In: Lévy, P. (ed.) Processus Stochastiques et Mouvement Brownien, pp. 366–420. Gauthier-Villars, Paris (1948)
18. Malyarenko, A., Ostoja-Starzewski, M.: Tensor-Valued Random Fields for Continuum Physics. Cambridge University Press (2018)
19. Ostoja-Starzewski, M.: Material spatial randomness: From statistical to representative volume element. Probab. Eng. Mech. **21**(2), 112–132 (2006). https://doi.org/10.1016/j.probengmech.2005.07.007
20. Owhadi, H., Scovel, C., Sullivan, T.: On the brittleness of Bayesian inference. SIAM Rev. **57**(4), 566–582 (2015). https://doi.org/10.1137/130938633
21. Sab, K., Nedjar, B.: Periodization of random media and representative volume element size for linear composites. Comptes Rendus Mécanique **333**(2), 187–195 (2005). https://doi.org/10.1016/j.crme.2004.10.003
22. Schmidhuber, J.: Deep learning in neural networks: an overview. Neural Netw. **61**, 85–117 (2015). https://doi.org/10.1016/j.neunet.2014.09.003
23. Soize, C.: Non Gaussian positive-definite matrix-valued random fields for elliptic stochastic partial differential operators. Comput. Methods Appl. Mech. Eng. **195**(1–3), 26–64 (2006). https://doi.org/10.1016/j.cma.2004.12.014
24. Soize, C.: Tensor-valued random fields for meso-scale stochastic model of anisotropic elastic microstructure and probabilistic analysis of representative volume element size. Probab. Eng. Mech. **23**(2–3), 307–323 (2008). https://doi.org/10.1016/j.probengmech.2007.12.019
25. Soize, C.: Stochastic elliptic operators defined by non-Gaussian random fields with uncertain spectrum. Am. Math. Soc. J. Theory Prob. Math. Stat. **105**, 113–136 (2021). https://doi.org/10.1090/tpms/1159
26. Soize, C.: Probabilistic learning inference of boundary value problem with uncertainties based on Kullback-Leibler divergence under implicit constraints. Comput. Methods Appl. Mech. Eng. **395**, 115078 (2022). https://doi.org/10.1016/j.cma.2022.115078
27. Soize, C., Ghanem, R.: Physical systems with random uncertainties: chaos representations with arbitrary probability measure. SIAM J. Sci. Comput. **26**(2), 395–410 (2004). https://doi.org/10.1137/S1064827503424505
28. Soize, C., Ghanem, R.: Data-driven probability concentration and sampling on manifold. J. Comput. Phys. **321**, 242–258 (2016). https://doi.org/10.1016/j.jcp.2016.05.044
29. Soize, C., Ghanem, R.: Physics-constrained non-Gaussian probabilistic learning on manifolds. Int. J. Numer. Meth. Eng. **121**(1), 110–145 (2020). https://doi.org/10.1002/nme.6202
30. Soize, C., Ghanem, R.: Probabilistic learning on manifolds. Found. Data Sci. **2**(3), 279–307 (2020). https://doi.org/10.3934/fods.2020013
31. Soize, C., Ghanem, R.: Probabilistic learning on manifolds constrained by nonlinear partial differential equations for small datasets. arXiv:2010.14324 [stat.ML] pp. 1–38 (2020). URL http://arxiv.org/abs/2010.14324
32. Soize, C., Ghanem, R.: Probabilistic learning on manifolds constrained by nonlinear partial differential equations for small datasets. Comput. Methods Appl. Mech. Eng. **380**, 113777 (2021). https://doi.org/10.1016/j.cma.2021.113777
33. Soize, C., Ghanem, R.: Probabilistic learning on manifolds (PLoM) with partition. Int. J. Numer. Meth. Eng. **123**(1), 268–290 (2022). https://doi.org/10.1002/nme.6856

34. Tipireddy, R., Ghanem, R.: Basis adaptation in homogeneous chaos spaces. J. Comput. Phys. **259**, 304–317 (2014). https://doi.org/10.1016/j.jcp.2013.12.009
35. Torquato, S., Stell, G.: Microstructure of two-phase random media. v. the n-point matrix probability functions for impenetrable spheres. J. Chem. Phys. **82**(2), 980–987 (1985). https://doi.org/10.1063/1.448475

Tensor Random Fields

Martin Ostoja-Starzewski

Abstract Mechanics and physics of random media strongly suggest that stochastic PDEs and stochastic finite element methods require mesoscale tensor-valued random fields (TRFs) of constitutive laws with locally anisotropic fluctuations. Such models are also useful when there is interest in fields of dependent quantities (velocity, strain, stress...) that need to be constrained by the balance laws (of mass, momentum...); examples are irrotational and solenoidal TRFs. In this article, we review the canonical forms of general correlation structures of second-order, mean-square continuous, wide-sense homogeneous and isotropic TRFs of ranks 1, ... , 4 in 3d. Besides "conventional" covariances, this approach can be used to construct TRFs with fractal and Hurst (long-range memory) characteristics. The current research extends the earlier work on scalar-valued RFs (including random processes) in vibration problems, rods and beams with random properties under random loadings, as well as elastodynamics, wavefronts, fracture, homogenization of random media, and contact mechanics.

Keywords Random fields, fractals · Hurst exponent · Correlation structure · Stochastic mechanics

1 Motivation

Take a sheet of paper and hold it against the light. You will see grayscale fluctuations, i.e. a spatially non-uniform opacity of paper. This inhomogeneity reflects the so-called formation of paper which involves a random placement of fibers (Fig. 1a) in the presence of van der Waals forces, chemical additives, and turbulence on a paper machine. The paper web is continuously created in a high-speed process involving millions of fibers per second at speeds up to 15 m/s and widths up to 10 m. Roughly speaking, flocs comprising thousands of cellulose fibers result from a 'frozen-in' turbulence of water-fiber suspension streaks on a paper machine. In theory, one could

M. Ostoja-Starzewski (✉)
University of Illinois at Urbana-Champaign, Champaign, USA
e-mail: martinos@illinois.edu

F. Willot et al. (eds.), *Continuum Models and Discrete Systems*,
Springer Proceedings in Mathematics & Statistics 457,
https://doi.org/10.1007/978-3-031-58665-1_2

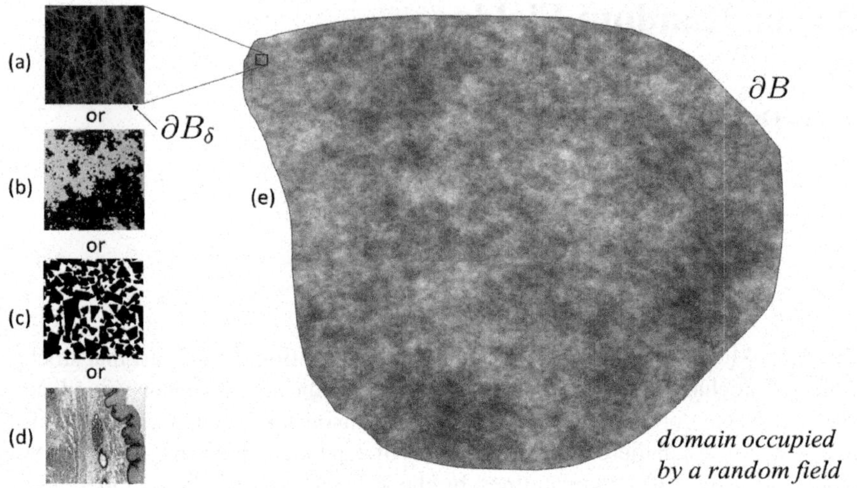

Fig. 1 Left column: A mesoscale domain placed in **a** a random cellulose network, or **b** a two-phase composite, or **c** a tungsten-carbide modeled by a Boolean model of Poisson polyhedra, or **d** grey matter in human brain. In each case **a–d**, the domain of random microstructure serves as a *statistical volume element* (SVE) of the *mesoscale random field*, a realization of which is shown in **e**. As the mesoscale [i.e., the ratio of domain size to a typical microstructural lengthscale (like grain size)] in **a**...**d** increases, the fluctuations in **e** vanish and a homogeneous continuum is attained

achieve a homogeneous medium if the fibers had very well-controlled lengths and were placed in a perfectly periodic manner analogous to a textile. Thus, cellulose fiber paper is a random quasi-2d medium, described in a continuum sense by a random field of mass density (Fig. 1e)

$$\left\{\rho\left(\mathbf{x},\omega\right);\mathbf{x}\in\mathbb{R}^2,\omega\in\Omega\right\}. \tag{1}$$

Here $\mathbf{x}$ is the location in the plane of paper and ω indicates one realization of the mass density field. Just like the opacity, paper's linear elastic (and other) properties are described by a random field of the in-plane stiffness (elasticity) tensor.

$$\left\{\mathsf{C}\left(\mathbf{x},\omega\right);\mathbf{x}\in\mathbb{R}^2,\omega\in\Omega\right\}. \tag{2}$$

The basic issue relative to Eqs. (1) and (2) is the randomness on a finite *mesoscale* $\delta := L/d$ on which homogenization is conducted; L is the mesoscale domain and d is the mean fiber length. Thus, strictly speaking, in (1) and (2) we should have $\rho\left(\mathbf{x},\omega,\delta\right)$ and $\mathsf{C}\left(\mathbf{x},\omega,\delta\right)$. If δ is finite, we have a statistical continuum (Fig. 1e). In effect, paper is a paradigm of mesoscale models of random media formed in manufacturing processes similar to, say, a slab of metal formed in continuous casting. Myriad other production as well as natural processes—except those resulting in perfect systems (such as pure crystals)—involve some degree of randomness, Fig. 1b–

d. This randomness decreases as we move the paper sheet away from our eyes, i.e. as we increase the observation length scale L relative to a typical microscale d. Hence, we have the limits

$$\lim_{\delta \to \infty} \rho(\mathbf{x}, \omega, \delta) = \rho^{eff}, \quad \lim_{\delta \to \infty} \mathsf{C}(\mathbf{x}, \omega, \delta) = \mathsf{C}^{eff}, \tag{3}$$

corresponding to a conventional, deterministic continuum.

Depending on the random microstructure and constitutive response, the upscaling to ρ^{eff} and C^{eff} may require the size of the representative volume element (RVE) to be too coarse for deterministic treatment, and this is why scalar- and tensor-valued models of random materials are needed. For example, the RVE is at least $5\,\text{cm} \times 5\,\text{cm}$ in paper [23]. Hence, we have *mesoscale tensor-valued random fields* (TRFs) based on a statistical volume element (SVE) [1, 23] providing inputs for *stochastic boundary-value problems* (SBVPs) in myriad problems in natural sciences and engineering.

The mesoscale TRFs should and can be informed by experimental measurements aided by computer simulations of random microstructures with the help of mathematical morphology models, e.g. [2, 3]. The most consistent strategy, when determining the mesoscale constitutive responses and scaling laws of transport/mechanical/electromagnetic properties is to require the satisfaction of the Hill-Mandel condition ensuring the agreement between energetic and mechanical interpretations; e.g. [4].

The study of resulting statistical continuum mechanics models may further be complicated by the presence of fractal and Hurst effects. However, these effects can be handled by our TRFs as they are mean-square continuous; fractal porous media fall outside the scope of this article and they require a different approach [5, 6]. Fractality is understood as a geometric pattern in time and/or space containing detailed structure at many (arbitrarily) small scales, usually with a fractal dimension strictly exceeding the topological dimension. Physical as well as numerically simulated fractals display such properties within lower and upper cut-offs and, therefore, strictly speaking, they should be called *pre-fractals* to distinguish them from mathematical fractals. The Hurst characteristic refers to a long spatial (resp., temporal) memory of a random field (resp., process).

As an example of strong effects that material spatial randomness may cause, consider the anti-plane response of a linear elastic half-plane to an impact force, Fig. 2. While plot (a) gives the response of an idealized homogeneous medium, Fig. 2b–d show that this Lamb problem responds differently when the half-plane has a random mass density and locally isotropic or anisotropic stiffness: the wavefront gets progressively distorted and attenuated through scattering on all the heterogeneities. The SBVPs (b), (c), and (d) get progressively more challenging as we go from scalar-valued to tensor-valued RFs. Also note: (i) it is impossible to obtain stochastic evolutions [whose snapshots are displayed in Fig. 2b–d by working with deterministic elastodynamics set up on some effective homogeneous medium, and (ii) there exists no effective medium theory for homogenization on TRFs, especially in dynamic

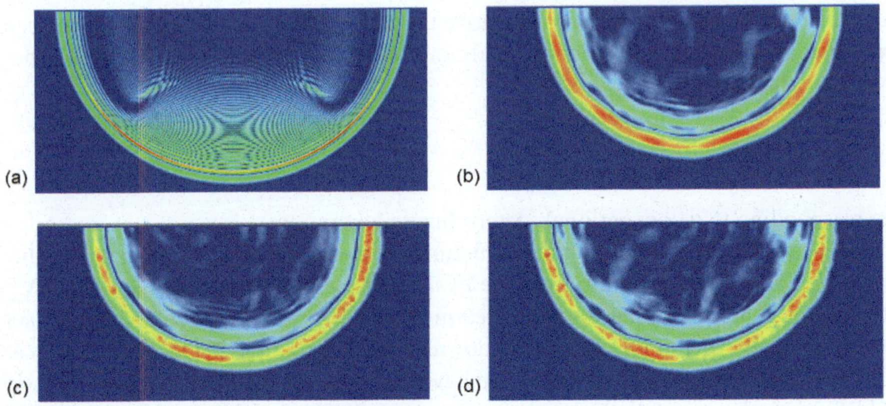

Fig. 2 Snapshots of von Mises stress patterns in anti-plane responses of a half-plane for several Lamb problems (with triangular-in-time impact in the middle of top edge) set up on **a** homogeneous model and three cases of spatial randomness: **b** mass density, **c** mass density and locally isotropic stiffness tensor, and **d** mass density and locally anisotropic stiffness tensor [7]

problems with fractal-and-Hurst effects. See [8, 9] for similar studies of other Lamb problems.

Some key questions arising with TRFs in the context of stochastic mechanics are: How can TRF models with local anisotropic fluctuations be constructed? What are the canonical correlation structures of statistically stationary and isotropic TRFs of ranks 1...4 in 3d? How can such models be subjected to restrictions dictated by the balance laws of mechanics? How can one introduce fractal-and-Hurst effects in such TRFs?

2 Restrictions on TRFs

In continuum mechanics, one is interested in two types of RFs: (i) dependent fields, i.e. those governed by field equations, and (ii) those representing constitutive properties of specific random media. In the first case, the correlation function (itself a tensor of twice higher rank) is subject to constraints such as the equilibrium equation or strain-displacement relation. In the second case, the stiffness tensor must be positive-definite.

2.1 Dependent Random Fields

Fields of this type must be set up so as to satisfy the governing field equations, rather than being assumed ad hoc without reference to mechanics and physics. There are many physical situations where the assumptions of statistical stationarity and isotropy apply, e.g. mass conservation restricting random velocity fields in incompressible media, heat flux in stationary heat conduction, compatibility of a random displacement field with a random strain field, linear momentum balance equations implying zero-divergence stress fields or the curl of distortion field equal to the density of dislocations, zero-curl condition restricting the electrostatic or gravitational fields in random media. Overall, while the literature on scalar random fields is vast (e.g. [10, 11]), the works on vector RFs have typically been specialized to statistical fluid turbulence [12, 13], whereas systematic research on TRFs of ranks 1...4 has been progressing only recently [14–18].

2.2 Random Fields of Constitutive Properties

A simple paradigm of SPDE [19] and stochastic finite elements (SFE) [20] is written in terms of a hyperbolic equation for anti-plane (scalar) displacement u:

$$\nabla \cdot (C(\mathbf{x}, \omega)\nabla u) = \rho \ddot{u}, \qquad \mathbf{x} \in \mathbb{R}^2, \quad \omega \in \Omega. \tag{4}$$

Here $C(\cdot, \omega)$ are spatial realizations of elastic shear modulus μ from the Ω space of a second-order scalar random field (RF) over a material domain in 2d or 3d. Equation (4) is well justified for a piecewise-constant description of realizations of a random medium such as a multiphase composite made of locally *isotropic* grains, Fig. 2a, b. However, in the case of an SBVP set up on a coarser level (mesoscale) having continuous realizations of properties, a rank 2 *tensor-valued random field* (TRF) of material properties, linked to actual material microstructures (Fig. 2c) is much more appropriate, e.g. [21–23]. Extensive scale-dependent, micromechanics-homogenization studies have shown that the anisotropy in myriad different (not only elastic) constitutive behaviors goes to zero as the mesoscale increases: mesoscale anisotropy accompanies the randomness. Hence, to truly allow for anisotropy at the SVE level, (4) should be replaced by

$$\nabla \cdot (\mathbf{C}(\mathbf{x}, \omega) \cdot \nabla u) = \rho \ddot{u}, \qquad \mathbf{x} \in \mathbb{R}^2, \quad \omega \in \Omega, \tag{5}$$

where $\mathbf{C}$ is the 2nd-rank TRF set up on a mesoscale, Fig. 2c. This type of model, instead of $\mathbf{C}(\cdot, \omega) = C(\cdot, \omega)\mathbf{I}$, is sorely needed in SPDE and SFE methods.

Moving to the 2d or 3d elasticity, if one assumes local material isotropy, a simple way to introduce material spatial randomness is to take the pair of Lamé constants (λ, μ) as a "vector random field", resulting in a generalization of the classical Navier equation for the displacement vector field $\mathbf{u}$:

$$\mu \nabla^2 \mathbf{u} + (\lambda + \mu)\nabla(\nabla \cdot \mathbf{u}) + \nabla \mu \cdot (\nabla \mathbf{u} + (\nabla \mathbf{u})^{\top}) + \nabla \lambda \nabla \cdot \mathbf{u} = \rho \ddot{\mathbf{u}}, \qquad \mathbf{x} \in \mathbb{R}, \quad \omega \in \Omega. \tag{6}$$

However, also here, the micromechanics-homogenization studies show that, with spatial gradients present, any realization of a RF in Fig. 2c involves some degree of anisotropy. Thus, in analogy to (4)–(5), (6) should be replaced by

$$\nabla\cdot(\mathsf{C}(\mathbf{x},\omega)\cdot\nabla\mathbf{u}) = \rho\ddot{\mathbf{u}}, \qquad \mathbf{x}\in\mathbb{R}^d, \quad \omega\in\Omega, \tag{7}$$

where the stiffness C $(= C_{ijkl}\mathbf{e}_i\otimes\mathbf{e}_j\otimes\mathbf{e}_k\otimes\mathbf{e}_l)$ is a 4th rank TRF; $\mathbf{e}_i$... are basis vectors. At any scale finitely larger than the microstructural scale, C is almost surely anisotropic, typically triclinic. $\mathbf{u}$, $\ddot{\mathbf{u}}$

The same arguments as above apply in case one prefers to work with field equations directly in the language of stresses. Then the *Ignaczak equation of elastodynamics* [24] is applicable:

$$\widehat{\nabla}\left[\rho^{-1}\sigma\right] = \mathbf{S}(\mathbf{x},\omega) : \ddot{\sigma}, \qquad \mathbf{x}\in\mathbb{R}^d, \quad \omega\in\Omega, \tag{8}$$

where, in case one is interested in random media, S stands for a rank 4 TRF of compliance. Interestingly, this formulation avoids the gradients of compliance but introduces gradients of mass density, which is the reverse of what the displacement language formulation requires.

Deterministic or random vector-valued body force fields may additionally be inserted in Eqs. (4)–(8).

3 Background on Tensor Random Fields (TRFs)

3.1 Stationary and Isotropic TRFs

We begin with a few basic notions of TRFs. A random field $\mathbf{T}$ is defined over a probability space $(\Omega,\mathcal{F},\mathsf{P})$ and a subset D of $\mathbb{R}^3$, where Ω is a sample space, $\mathcal{F}$ is a σ-field (as rich as needed in a particular problem), and P is a probability measure. Formally, $\mathbf{T}$ takes values in V, a finite-dimensional real Hilbert space. Thus, the TRF is $\mathbf{T}(\mathbf{x},\omega)\colon \mathbb{R}^3\times\Omega\to V$, such that, for any fixed $\mathbf{x}_0\in B$, the function $\mathbf{T}(\mathbf{x}_0,\omega)\colon \Omega\to V$ is measurable. As commonly done in probability theory, $\mathbf{T}(\omega)$ denotes a single realization of TRF, while $\mathbf{T}(\mathbf{x})$ stands for a random tensor at a given spatial location $\mathbf{x}$. Note:

(i) $\mathbf{T}$ takes values in $\mathsf{S}^2(\mathbb{R}^3)$, the space of symmetric rank 2 tensors over $\mathbb{R}^3$.
(ii) $\mathbf{T}$ is second-order: $\left\langle\|\mathbf{T}(\mathbf{x})\|^2\right\rangle<\infty$, where $\|\cdot\|$ is the norm on $\mathsf{S}^2(\mathbb{R}^3)$.
(iii) $\mathbf{T}$ is mean-square continuous: $\lim_{\mathbf{x}\to\mathbf{x}_0}\left\langle\|C(\mathbf{x})-C(\mathbf{x}_0)\|^2\right\rangle=0$, $\mathbf{x}_0\in\mathbb{R}^3$.

With $\langle\cdot\rangle$ denoting the statistical average, $\langle\mathbf{T}(\mathbf{x})\rangle$ is the mean of this TRF and

$$\mathbf{R}(\mathbf{x},\mathbf{y}) := \left\langle[\mathbf{T}(\mathbf{x})-\langle\mathbf{T}(\mathbf{x})\rangle]\otimes\left[\mathbf{T}(\mathbf{y})-\langle\mathbf{T}(\mathbf{y})\rangle\right]\right\rangle \tag{9}$$

is the two-point covariance function of $\mathbf{T}$. Assuming these two functions to be invariant with respect to arbitrary translations implies $\langle \mathbf{T}(\mathbf{x}) \rangle \in V$ is constant in space, while $R(\mathbf{x}, \mathbf{y}) \in V \otimes V$ depends only on the difference $\mathbf{r} = \mathbf{y} - \mathbf{x}$, i.e. $\mathbf{T}$ is said to be statistically stationary (or wide-sense homogeneous). For a zero-mean TRF, we simply work with the correlation in (9).

A further restriction follows if one assumes a *statistical isotropy*, which must be distinguished from the "conventional" isotropy of a tensor mentioned above, such as encountered in, say, all continuum mechanics courses. To this end, let $K = \mathrm{O}(3)$ be the group of rotations and reflections in $\mathbb{R}^3$, and let (V, γ) be an orthogonal representation of K. Assume that for all $k \in K$ and for all $\mathbf{x} \in \mathbb{R}^3$ we have

$$\langle \mathbf{T}(k\mathbf{r}) \rangle = \gamma(k) \langle \mathbf{T}(\mathbf{r}) \rangle, \quad \mathbf{R}(k\mathbf{r}) = \gamma(k)\mathbf{R}(\mathbf{r})\gamma^{-1}(k). \tag{10}$$

Such a field is called statistically isotropic. Note:

1. Condition (10) requires the invariance of the correlation function with respect to any rotation of the coordinate system, accompanied by a simultaneous rotation of tensors separated by the vector $\mathbf{r}$.
2. In the well-known case of a scalar-valued RF, this condition reduces to the covariance function being dependent only on $r = \|\mathbf{r}\|$, rather than on the vector $\mathbf{r}$.
3. It is possible to formulate statistically isotropic TRFs without requiring statistical stationarity, e.g. [25].
4. TRFs lacking the statistical isotropy - those displaying an overall (meso and macroscale) preferred orientation in one direction (such as the machine-made paper, as opposed to a laboratory-made paper) - can be generated from the former ones by a transformation, e.g. stretching in one coordinate.

We now focus on statistically stationary and isotropic TRFs of rank 0, 1, ..., 4.

3.2 *From Scalar to Tensor-Valued Random Fields*

Scalar (rank 0) TRF: A statistically stationary and isotropic RF $\mathbf{T}$ is simply a scalar field T, specified by $\langle \mathbf{T} \rangle = const$ and the tensor $\mathbf{R}$ of (9) becomes a scalar function $R(\mathbf{x}, \mathbf{y}) = R(\mathbf{r})$ ($\mathbf{r} = \mathbf{y} - \mathbf{x}$). The relation between the covariance and its Fourier transform (the spectral density function) is well-known. There are many explicit models of the covariance $R(r), r = \|\mathbf{r}\|$, for wide-sense isotropic RFs, beginning with the Gaussian white noise (WN), having a covariance $R_{\mathrm{WN}}(r) := \delta(r)$, $r \geq 0$. Many different models exist, e.g. the Ornstein–Uhlenbeck (OU) random field with $R_{\mathrm{OU}}(r; \nu) := \frac{\nu}{2} \mathrm{e}^{-\nu r}, r \geq 0$, with $\nu > 0$ a scaling parameter and a parametrization such that $\lim_{\nu \to \infty} R_{\mathrm{OU}}(\cdot; \nu) = R_{\mathrm{WN}}(\cdot)$. More sophisticated covariance functions allow modeling of fractal and Hurst effects; see the next subsection.

Rank 1 TRF (vector-valued RF): V has the dimension 3. Then $\gamma(k) = k$, $\langle T_i(\mathbf{x})\rangle = 0$ and $R_{ij}(\mathbf{x}, \mathbf{y}) = \langle [T_i(\mathbf{x}) - \langle T_i(\mathbf{x})\rangle] [T_j(\mathbf{y}) - \langle T_j(\mathbf{y})\rangle]\rangle$ in the statistically isotropic case is represented in terms of two continuous functions $K_1, K_2 \colon [0, \infty) \to \mathbb{R}$ with $K_2(0) = 0$,

$$R_{ij}(\mathbf{r}) = \delta_{ij} K_1(r) + r_i r_j K_2(r) = \sum_{\alpha=1}^{2} L_{ij}^{\alpha}(\mathbf{r}) K_{\alpha}(r) . \quad (11)$$

Note that with two scalar functions K_1, K_2 one can then simulate a vector RF. See [15, 26] for a spectral expansion of the two-point covariance tensor (11).

Rank 2 TRF: $V = \mathsf{S}^2(\mathbb{R}^3)$ is the space of all symmetric rank 2 tensors over $\mathbb{R}^3$, and the representation is $\gamma(k)\mathbf{T} = k\mathbf{T}k^{-1}$. Then $C_{ij}(\mathbf{x}) = C\delta_{ij}$ with $C \in \mathbb{R}$, and the covariance

$$R_{ijlm}(\mathbf{x}, \mathbf{y}) := \langle [T_{ij}(\mathbf{x}) - \langle T_{ij}(\mathbf{x})\rangle] [T_{lm}(\mathbf{y}) - \langle T_{lm}(\mathbf{y})\rangle]\rangle . \quad (12)$$

Using the theory of invariants, one can prove that $R_{ijlm}(\mathbf{r})$ in the statistically stationary and isotropic case is represented in terms of 5 continuous functions $K_1, \ldots, K_5 \colon [0, \infty) \to \mathbb{R}$ with $K_3(0) = K_4(0) = K_5(0) = 0$, along with 5 corresponding tensor functions L_{ijlm}^{n}, as

$$R_{ijlm}(\mathbf{r}) = \sum_{\alpha=1}^{5} L_{ijlm}^{\alpha}(\mathbf{r}) K_{\alpha}(r) \quad \text{with } L_{ijlm}^{1} = \delta_{ij}\delta_{lm}, \ \ldots, \ L_{ijlm}^{5} = \frac{r_i r_j r_l r_m}{r^4} . \quad (13)$$

The rank 2 TRFs are needed to model a host of dependent fields in continuum mechanics/physics (stress, strain,...) as well as material properties in stochastic problems involving anti-plane elasticity in $\mathbb{R}^2$, conductivity or permeability in $\mathbb{R}^2$ or $\mathbb{R}^3$, electrostatics... Clearly, with 5 scalar functions $K_1, \ldots, K_5$ one can then simulate such a TRF. See [26] for a spectral expansion corresponding to (13).

Rank 3 TRF: Here $\mathsf{V} = \mathsf{S}^2(\mathbb{R}^3) \otimes \mathbb{R}^3$ and the corresponding covariance defined as

$$R_{ijklmn}(\mathbf{x}, \mathbf{y}) := \langle [T_{ijk}(\mathbf{x}) - \langle T_{ijk}(\mathbf{x})\rangle] [T_{lmn}(\mathbf{y}) - \langle T_{lmn}(\mathbf{y})\rangle]\rangle . \quad (14)$$

is given, in the statistically stationary and isotropic case [13], in terms of 21 pairs of functions $\{K_n, L_{ijklprst}^{n}\}$:

$$R_{ijklmn}(\mathbf{r}) = \sum_{\alpha=1}^{21} L_{ijklmn}^{\alpha}(\mathbf{r}) K_{\alpha}(r) . \quad (15)$$

The explicit forms of all $L_{ijklmn}^{n}(\mathbf{r})$ and applications are discussed in [16]. The rank 3 TRF is needed for spatially random piezoelectric properties and, by analogies, for several other coupled field problems in continuum mechanics/physics.

Rank 4 TRF: $V = \mathsf{S}^2(\mathsf{S}^2(\mathbb{R}^3))$ is the space of all symmetric rank 4 tensors over $\mathbb{R}^3$, such as the stiffness tensor mapping a symmetric strain into a symmetric stress tensor. It follows that the covariance

$$R_{ijklprst}(\mathbf{x}, \mathbf{y}) := \langle [T_{ijkl}(\mathbf{x}) - \langle T_{ijkl}(\mathbf{x}) \rangle] [T_{prst}(\mathbf{y}) - \langle T_{prst}(\mathbf{y}) \rangle] \rangle \tag{16}$$

is given, in the statistically stationary and isotropic case [13], in terms of 29 pairs of functions $\{K_n, L^n_{ijklprst}\}$:

$$R_{ijklprst}(\mathbf{r}) = \sum_{\alpha=1}^{29} L^{\alpha}_{ijklprst}(\mathbf{r}) K_{\alpha}(r) . \tag{17}$$

See [13] for explicit forms of all $L^n_{ijklprst}(\mathbf{r})$ and a corresponding spectral expansion.

3.3 *TRFs with Fractal and Hurst Effects*

Many correlation functions for RF modeling have been developed in mathematical statistics, e.g. [11, 27, 28]. In case one is interested in modeling media with fractal and Hurst random patterns, one may consider two correlation functions

$$\text{Cauchy: } R_C(r) := (1 + r^{\alpha})^{-\beta/\alpha}, \quad r \geq 0; \quad 0 < \alpha \leq 2, \quad \beta > 0, \tag{18}$$

$$\text{Dagum: } R_D(r) := 1 - \left(1 + r^{-\alpha}\right)^{-\beta/\alpha}, \quad r \geq 0; \quad 0 < \alpha < 1, \quad 0 < \beta \leq 1. \tag{19}$$

Both of them possess a particularly attractive property: independent choice of the fractal dimension D and the Hurst exponent H, i.e. no restriction to fractals that are self-similar in the sense of Mandelbrot. Thus, in general, $D + H \neq 2$. To better explain the roles of D and H, consider a stationary Gaussian RF over $\mathbb{R}$, having a covariance function $R_X(r)$. If there exists $\alpha \in (0, 2]$, such that, when $h \to 0$, $1 - R_X(h) \sim h^{\alpha}$, then a realization of X has a fractal dimension

$$D = n + 1 - \alpha/2, \tag{20}$$

with probability 1. On the other hand, if there exists $\beta \in (0, 1)$, such that, when $h \to \infty$, $R_X(h) \sim h^{-\beta}$, then X is said to have a long memory with a Hurst exponent

$$H = 1 - \beta/2. \tag{21}$$

Effectively, the parameters (α, β) of Cauchy (17) or Dagum (18) models can be chosen, from the Eqs. (18)–(19), according to the desired (D, H). Heuristically speaking, D describes local properties: higher D values indicate higher roughness of realizations. On the other hand, for $0 < H < 0.5$, the RF is said to be *anti-persistent*

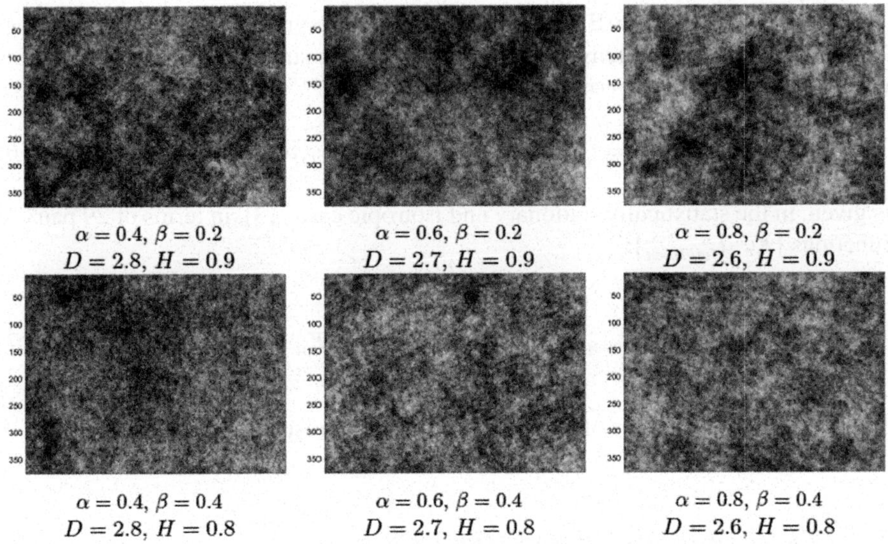

Fig. 3 Sample realizations of statistically stationary and isotropic RFs with Dagum covariance functions, over rectangular domains, for several (α, β) pairs with corresponding (D, H). The simulation of each case takes seconds on a laptop

(a decrease between two consecutive values will likely be followed by an increase) while for $0.5 < H < 1$, the RF is said to be *persistent* (an increase is likely followed by another increase); for $H = 0.5$, there is no preference for a decrease or increase following any particular value (like a basic random walk).

The effects of D and H are displayed in Fig. 3 which collects six sample realizations of RFs generated according to the Dagum covariance function.

So far, the Cauchy and Dagum scalar-valued RF models have been extensively employed in a range of applications in mechanics: vibration problems, rods and beams with random properties under random loadings, elastodynamics, wavefronts, fracture, buckling, homogenization of random media, statistical turbulence, and contact mechanics. It is important to note that mathematical statistics provides several alternatives to Cauchy and Dagum models which also possess fractal and Hurst effects—see e.g. [29–31]. Those models have not yet been employed in mechanics.

The scalar-valued RFs can be employed for the construction of TRFs with fractal and Hurst effects through a polyadic representation of tensors [32]. For example, a simple construction is offered by a dyadic product

$$\mathbf{T} = \mathbf{a} \otimes \mathbf{b}, \quad T_{ij} = a_i b_j, \quad i, j = 2, 3. \tag{22}$$

where all a_i and b_i components are scalar RFs. This TRF can be symmetrized to generate a rank 2 TRF taking values in the space $\mathsf{S}^2\left(\mathbb{R}^2\right)$ which, in turn, would provide input into an SPDE of anti-plane elasticity with local anisotropy: Eq. (2).

Indeed, the special case with the vectors $\mathbf{a} = a_i \mathbf{e}_i$ and $\mathbf{b} = a_i \mathbf{e}_i$ fully correlated was employed in [7], with the results indicating that taking a random medium with a stiffness TRF $\boldsymbol{C}$ with local anisotropy [Fig. 1d] leads to stronger fluctuations in responses to SBVPs than a random medium with a scalar-valued RF C, Fig. 1b, c. With an eye to polyadic constructions, it was recently shown that a product of two RFs with fractal and Hurst effects preserves these properties [33].

4 Some Applications

Field equations impose restrictions (or constraints) on the correlation structures of dependent RFs. As an example, consider a divergence-free vector RF in $\mathbb{R}^2$. In that case, the functions K_2 and K_1 of (11) must satisfy $rK_2'r + 2K_2 + K_1'/r = 0$, with prime denoting d/dr. Next, introduce two correlation functions

$$f(r) = \frac{\langle v_p(\mathbf{0})\, v_p(\boldsymbol{r})\rangle}{\langle v_p^2\rangle}, \quad g(r) = \frac{\langle v_n(\mathbf{0})\, v_n(\boldsymbol{r})\rangle}{\langle v_n^2\rangle}, \tag{23}$$

where p (or n) denotes parallel (resp., normal) vector components. Proceeding in the same manner as in turbulent incompressible flows in $\mathbb{R}^3$ [13], we arrive at the restriction

$$g = f + rf'. \tag{24}$$

On the other hand, consider a curl-free vector RF in $\mathbb{R}^2$. In that case, the functions K_2 and K_1 of (11) must satisfy $rK_1'r + 2K_1 + K_2'/r = 0$. In terms of the identically defined functions f and g, we now arrive at the restriction

$$f = g + rg'. \tag{25}$$

The mirror-type property between (24) and (25) reflects the fact that, in $\mathbb{R}^2$, the divergence and curl are operations inverse with respect to each other. Further, note that either f or g is typically accessible through experimental measurements or computer simulations [34], and this enables the determination of K_2 and K_1. This observation carries over to higher-rank TRFs where, given the presence of more K_n functions, a greater number of measurements or simulations is needed.

Likewise, assuming the displacement and strain fields to be statistically stationary and isotropic results in a direct relation between their covariances: the vector-valued RF of displacement cannot be chosen independently of the TRF of strain. Next, assuming a quasi-static situation results in a condition on the covariance of the stress field. There are special cases as well as generalizations of this situation, e.g. anti-plane mechanics (equivalent to in-plane conductivity), in-plane mechanics, 3d mechanics, and micropolar fields [14]. In all the cases, the applicable field equations (such as linear and angular momentum balances, and strain-displacement relations) lead to consequences for all the fields involved.

Acknowledgements We thank an anonymous reviewer for helpful comments on the first version of this article.

References

1. Ostoja-Starzewski, M.: Material spatial randomness: from statistical to representative volume element. Probab. Eng. Mech. **21**(2), 112–132 (2006)
2. Jeulin, D.: Advances in Theory and Application of Random Sets. World Scientific, Singapore (1997)
3. Jeulin, D.: Morphological Models of Random Structures. Springer (2021)
4. Sena, M.P., Ostoja-Starzewski, M., Costa, L.: Stiffness tensor random fields through upscal-ing of planar random materials. Probab. Eng. Mech. **34**, 131–156 (2013)
5. Li, J., Ostoja-Starzewski, M.: Thermo-poromechanics of fractal porous media. Phil. Trans. Roy. Soc. A **378**(2172), 20190288 (2020)
6. Li, J., Ostoja-Starzewski, M.: Micropolar mechanics of product fractal media. Phil. Proc. Roy. Soc. A **478**(2172), 20210770 (2022)
7. Zhang, X., Malyarenko, A., Porcu, E., Ostoja-Starzewski, M.: Elastodynamic problem on tensor random fields with fractal and Hurst effects. Meccanica **57**, 957–750 (2021)
8. Zhang, X., Nishawala, V., Ostoja-Starzewski, M.: Anti-plane shear Lamb's problem on random mass density fields with fractal and Hurst effects. Evolution Equations & Control Theory **8**(1), 231–246 (2019)
9. Zhang, X., Ostoja-Starzewski, M.: Impact force and moments problems on random mass density fields with fractal and Hurst characteristics. Phil. Trans. Roy. Soc. A **378**(2172), 20190591 (2020)
10. Adler, R.J.: The Geometry of Random Fields. Wiley, Chichester (1981)
11. Hristopulos, D.T.: Random Fields for Spatial Data Modeling: A Primer for Scientists and Engineers. Springer, Dordrecht (2020)
12. Robertson, H.P.: The invariant theory of isotropic turbulence. Proc. Camb. Phil. Soc. **36**, 209–223 (1940)
13. G.K. Batchelor. *The Theory of Homogeneous Turbulence*, Cambridge University Press (1953)
14. Ostoja-Starzewski, M., Shen, L., Malyarenko, A.: Tensor random fields in conductivity and classical or microcontinuum theories. Math. Mech. Solids **20**(4), 418–432 (2015)
15. Malyarenko, A., Ostoja-Starzewski, M.: Tensor-Valued Random Fields for Continuum Physics. Cambridge University Press (2019)
16. Malyarenko, A., Ostoja-Starzewski, M., Amiri-Hezaveh, A.: Random Fields of Piezoelectricity and Piezomagnetism. Springer (2020)
17. Malyarenko, A., Ostoja-Starzewski, M.: Tensor- and spinor-valued random fields with applications to continuum physics and cosmology. Probab. Surveys **20**, 1–86 (2023)
18. Tensor random fields in mechanics (2023) https://www.youtube.com/watch?v=zLBu14BY9EY
19. P.-L. Chow (2007) *Stochastic Partial Differential Equations*, Chapman & Hall CRC Press
20. Papadopoulos, V., Giovanis, D.G.: Stochastic Finite Element Methods. Springer (2018)
21. Ranganathan, S.I., Ostoja-Starzewski, M.: Scaling function, anisotropy and the size of RVE in elastic random polycrystals. J. Mech. Phys. Solids **56**, 2773–2791 (2008)
22. Ranganathan, S.I., Ostoja-Starzewski, M.: Towards scaling laws in random polycrystals. Int. J. Eng. Sci. **47**, 1322–1330 (2009)
23. Ostoja-Starzewski, M., Kale, S., Karimi, P., Malyarenko, A., Raghavan, B.V., Ranganathan, S.I., Zhang, J.: Scaling to RVE in random media. Adv. Appl. Mech. **49**, 111–211 (2016)
24. Ostoja-Starzewski, M.: Ignaczak equation of elastodynamics. Math. Mech. Solids. **24**(11), 3674–3713 (2018)

25. A. Malyarenko, M. Ostoja-Starzewski (2017) Fractal planetary rings: Energy inequalities and random field model, *Int. J. Modern Physics B* **31**, 1750236 (13 pages)
26. A. Malyarenko, M. Ostoja-Starzewski (2016) Spectral expansions of homogeneous and isotropic tensor-valued random fields, *ZAMP* **67**(3), paper 59
27. Christakos, G.: Random Field Models in Earth Sciences. Mineola, NY, Dover (2012)
28. Porcu, E., Montero, J.M., Schlather, M. (eds.): Advances and Challenges in Space-Time Modelling of Natural Events. Springer, Berlin (2012)
29. Barndorff-Nielsen, O.E., Leonenko, N.N.: Spectral Properties of Superpositions of Ornstein-Uhlenbeck Type Processes. Meth. Comput. Appl. Probab. **7**, 335–352 (2005)
30. Grahovac, D., Leonenko, N.N., Taqqu, M.S.: Limit theorems, scaling of moments and intermittency for integrated finite variance supOU processes. Stoch. Proc. Appl. **129**, 5113–5150 (2019)
31. Anh, V.V., Heyde, C.C., Leonenko, N.N.: Dynamic models of long-memory processes driven by Lévy noise. J. Appl. Probab. **39**, 730–747 (2002)
32. Malyarenko, A., Ostoja-Starzewski, M.: Polyadic random fields. ZAMP **73**, 204 (2022)
33. Y.S. Jetti, E. Porcu and M. Ostoja-Starzewski, New decouplers of fractal dimension and Hurst effects, *ZAMP* **74**123 (2023)
34. Laudani, R., Zhang, D., Faouzi, T., Porcu, E., Ostoja-Starzewski, M., Chamorro, L.P.: On streamwise velocity spectra model accounting for the fractal and long-memory effects. Phys. Fluids **33**, 035116 (2021)

Continuous Simulation of Heterogeneous Media: The Karhunen-Loève Approach Versus the Turning Bands Method

Christian Lantuéjoul, Xavier Freulon, Aleksandr Krivcenko, and Mike Pereira

Abstract This paper examines the feasibility of performing simulations of Gaussian random fields at all points of compact domains of $\mathbb{R}^d$. The simulations produced by the Karhunen-Loève method are compared with those obtained using the standard turning bands method. The interest of combining both methods is also tested.

Keywords Geostatistics · Gaussian random fields · Continuous simulations

1 Introduction

Upscaling/downscaling physical variables are commonly encountered problems in the Earth Sciences and the Environmental Sciences ([2, 12, 15]). For example, to model fluid flows through heterogeneous media, a fine scale description of its physical properties (e.g. porosity and permeability) is often the basis of three-dimensional coarse grids on which partial differential equations of the involved physics can be solved numerically. In order to build such descriptions, on regular or unstructured grids or on scattered points, Geostatistics interpret these variables as realizations of second order random fields, which allows to design algorithms for the synthesis of textures whose spatial characteristics are controlled by mean and covariance functions. The topic addressed in this paper is the continuous simulation of such random

C. Lantuéjoul (✉) · X. Freulon · M. Pereira
Centre for Geo-sciences and Geo-engineering, Mines Paris, PSL University, 35, rue Saint-Honoré, 77300 Fontainebleau Name, France
e-mail: christian.lantuejoul@minesparis.psl.eu

X. Freulon
e-mail: xavier.freulon@minesparis.psl.eu

M. Pereira
e-mail: mike.pereira@minesparis.psl.eu

A. Krivcenko
Télécom Paris, 19 place Marguerite Perey, CS 20031, 91123 Palaiseau Cedex, France
e-mail: krivcenko@telecom-paris.fr

F. Willot et al. (eds.), *Continuum Models and Discrete Systems*,
Springer Proceedings in Mathematics & Statistics 457,
https://doi.org/10.1007/978-3-031-58665-1_3

fields on a bounded domain of R^d with $d \geq 2$, i.e. to compute the value of a realization in a domain that contains an infinity, possibly non countable, of points. At first, this looks cranky, but an answer to that question consists of generating first a set of ingredients which capture the characteristics of the process on the simulation domain, and then using them to compute the value of the process at each point. This decomposition of the simulation algorithm into a stochastic initialization and a deterministic spreading allows to build numerical views of the same process at different scales, but still consistent, or to refine it locally to better capture physical singularities of the physics. The models considered in this paper are Gaussian random fields defined in the three-dimensional Euclidean space. In Sect. 2, three different methods to generate the basic ingredients are successively presented: the spectral method based on the Fourier analysis of their covariance function [13], a stereological device called the turning bands method which allows to simplify the three-dimensional problem into a set of unidimensional simulations [10], and finally the Karhunen-Loève decomposition of any covariance function defined on a compact set [7, 8]. As these methods are based on different principles, it is tempting to combine them to benefit from each other. This is illustrated with the continuous simulation of a Gaussian random field with an exponential covariance function using a combination of Karhunen-Loève and turning bands methods in Sect. 3. Finally, a general discussion and conclusions follow in Sect. 4. A sketch of the proof for the results presented in this paper is proposed in an Appendix.

2 Background

The problem addressed in this paper is to produce realizations of a Gaussian random field Z at each point of a compact domain D of $\mathbb{R}^d$. The statistical properties of Z are totally specified by its mean (supposed to be null), and its covariance function $Cov(Z(s), Z(t)) = C(s,t)$ (supposed to be continuous). In the case where Z is second order stationary, the covariance $C(s,t)$ does not explicitly depends on s and t, but only on the difference $s - t$.

2.1 *The Spectral Method*

Suppose here that C is second order stationary. Then Bochner theorem asserts that C is the Fourier transform of a positive measure $\mathcal{X}$ (spectral measure):

$$C(h) = \int_{\mathbb{R}^d} e^{i\,\langle h,u\rangle}\, d\mathcal{X}(u) \tag{1}$$

Assuming $C(0) = 1$, then $\mathcal{X}$ is a probability measure. It is not difficult to show that if Ω is a random spectral vector distributed like $\mathcal{X}$, and if Φ is an independent random phase that is uniform on $]0, 2\pi[$, then the random field Z defined by

$$Z(s) = \sqrt{2}\cos(\langle \Omega, s\rangle + \Phi) \qquad s \in \mathbb{R}^d$$

is centered with covariance function C. Note that Z is not Gaussian. However, by considering n independent copies of spectral vectors and random phases, the central limit theorem applies and shows that the random field

$$Z_n(s) = \frac{1}{\sqrt{n}} \sum_{i=1}^{n} \sqrt{2}\cos(\langle s, \Omega_i\rangle + \Phi_i) \qquad s \in D \tag{2}$$

is approximately Gaussian.

The spectral method is efficient provided that a limited number of spectral elements are required to ensure a good representativity of the spectral measure. Since the more regular the covariance function at the origin, the more concentrated the spectral measure, the spectral method is especially recommended for smooth covariance functions.

2.2 *The Turning Bands Method (TB)*

The turning bands method is a stereological device for reducing a pluridimensional simulation into unidimensional ones. Let us start with the polar decomposition of the spectral measure by putting

$$d\mathcal{X}(u) = d\mathcal{X}_\theta(\rho)\, d\varpi(\theta) \tag{3}$$

where (ρ, θ) are polar coordinates of u, $d\varpi(\theta)$ is the distribution of orientations of the spectral vectors, and $d\mathcal{X}_\theta$ is the conditional distribution of their moduli given their orientation. In what follows, it is convenient to take $\theta \in S_d^+$ (half a unit sphere in $\mathbb{R}^d$) and $\rho \in \mathbb{R}$, so that Bochner's formula (1) becomes

$$C(h) = \int_{S_d^+} \int_{\mathbb{R}} e^{i\rho\langle h,\theta\rangle} d\mathcal{X}_\theta(d\rho)\, d\varpi(\theta) \equiv \int_{S_d^+} C_\theta(\langle h, \theta\rangle)\, d\varpi(\theta) \tag{4}$$

where C_θ is a unidimensional covariance function.

This suggests the following approach. Consider at first n random directions $\Theta_1, \ldots, \Theta_n$ that are independently distributed like ϖ. Then, for each direction Θ_i a centered stochastic process X_i covariance C_{Θ_i} is devised. When n is large enough, the random field

$$Z_n(s) = \frac{1}{\sqrt{n}} \sum_{i=1}^{n} X_i(\langle s, \Theta_i \rangle) \qquad s \in D \tag{5}$$

is approximately a Gaussian random field.

This construction calls for a number questions. The calculation of the spectral measure and its polar decomposition is usually straightforward. For instance, if the covariance function C is a three-dimensional radial function, say $C(h) = C_3(|h|)$, then ϖ is uniform over S_3^+ and all the stochastic processes X_1, ..., X_n share the same covariance function given by[1]

$$C_1(r) = \frac{d}{dr}\,(rC_3(r)) \qquad r \geq 0 \tag{6}$$

To simulate the stochastic processes $X_1, \ldots, X_n$, one could think of using the previously seen spectral method. However, this would merely boil down to applying the spectral method in three dimensions. If C_1 is proportional to the autoconvolution $f \star \check{f}$ of some function f, then X can be simulated using a shot-noise model [11]. Note however that if f has an infinite support, care must be taken in the simulation procedure to avoid detrimental edge effects.

Similarly as for the spectral method, the random field Z_n produced by the TB method is only approximately Gaussian. Various criteria do exist to quantify how the spatial distribution of Z_n departs from that of a Gaussian random field. See [9] or [6] for more details.

2.3 The Karhunen-Loève Method (KL)

Let $C(s, t)$ be a covariance function that is positive definite on $D \times D$ but not necessarily stationary. Since C is continuous and D compact, then the following properties hold [3, 8]:

- The integral operator $T : L^2(D) \to L^2(D)$ defined by

$$T\varphi(t) = \int_D C(s, t)\,\varphi(s)\,ds \qquad t \in D, \quad \varphi \in L^2(D) \tag{7}$$

  admits a countable set of non-negative eigenvalues $\lambda_1 \geq \lambda_2 \geq \ldots \geq 0$.
- Any sequence $(\varphi_n)_{n\geq 1}$ of eigenfunctions associated with the eigenvalues constitutes an orthogonal basis of $L^2(D)$:

$$\int_D \varphi_n(t)\,\varphi_p(t)\,dt = \|\varphi_n\|^2\, 1_{n \neq p}. \tag{8}$$

---

[1] C_1 is called the band covariance attached to C_3.

– C admits the bivariate expansion

$$C(s,t) = \sum_{n=1}^{\infty} \lambda_n \frac{\varphi_n(s)\,\varphi_n(t)}{\|\varphi\|^2} \qquad s,t \in D. \tag{9}$$

As a immediate consequence of these properties, suppose that $(Y_n)_{n\geq 1}$ is a set of independent standard normal variables. Then

$$Z(t) = \sum_{n=1}^{\infty} \sqrt{\lambda_n}\, Y_n \frac{\varphi_n(t)}{\|\varphi\|} \qquad t \in D \tag{10}$$

is a Gaussian random field with covariance C.

In order to make it operating, formula (10) must be truncated. Note that the truncated random field remains Gaussian, so that no recourse to the Central Limit theorem is necessary. This a potent advantage of the KL method over the TB method regarding the choice of the truncation order n, as it rests only on the quality of approximation of the covariance function (or the variogram).

The main difficulty inherent to the KL method is the determination of the eigenvalues and eigenfunctions of the covariance function. Actually, only a few results are available in the literature [3].

3 Simulation of Three-Dimensional Gaussian Random Field with Exponential Covariance

There is no doubt that the determination problem encountered in the KL method would be alleviated if it was posed in $\mathbb{R}$ rather than in $\mathbb{R}^d$. Hence comes the idea of combining the KL and the TB methods. At first, stochastic processes are generated using the KL method along each band, then propagated to $\mathbb{R}^d$ using the TB method.

This idea is now illustrated by the simulation of a three-dimensional Gaussian random field with exponential covariance function. What remains to investigate is how to simulate stochastic processes with the band covariance

$$C_1(t) = e^{-a|t|}(1 - a|t|) \qquad t \in \mathbb{R}. \tag{11}$$

3.1 Simulation of C_1 Using a Shot-Noise Process

As a matter of fact, this is not exactly a shot-noise process but a variation of it that is proposed here. At first $\mathbb{R}$ is partitioned into intervals of length $\ell = l_0/a$ where l_0 is generated as a mixture of two Gamma distributions (parameters 2 and 3 with respective probabilities $1/3$ and $2/3$). A uniform offset on $[0, \ell[$ is introduced to make

Fig. 1 Simulation of a stochastic process with exponential band covariance using a shot-noise process

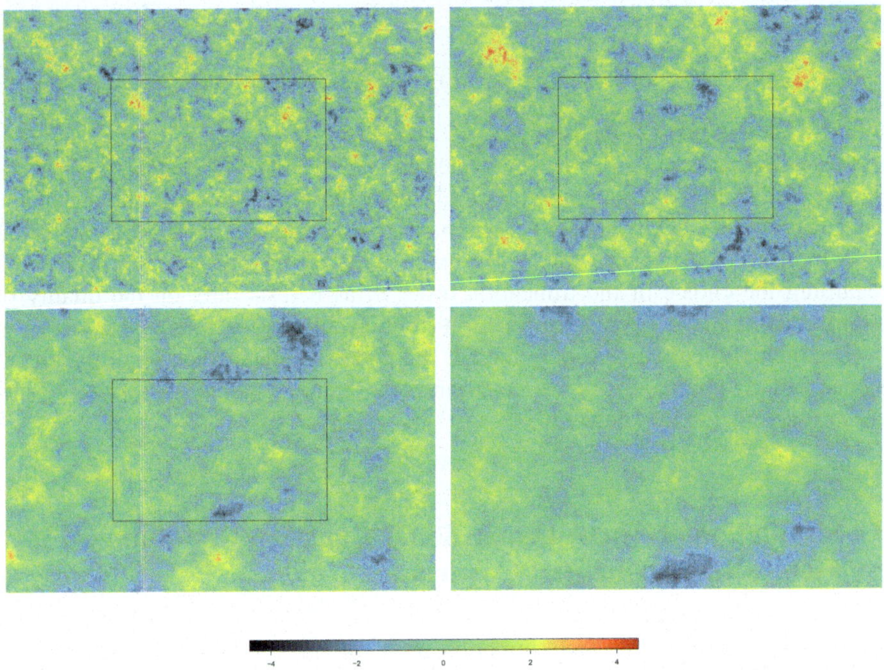

Fig. 2 Simulation of a Gaussian random field with exponential covariance (unit scale factor) using the TB method to combine shot-noise processes. The simulation is displayed at four different scales. From top to bottom and left to right, the four images are sorted in ascending order of their scale. The black rectangle in each image indicates the size of the next image

the partition stationary. Then the intervals are assigned independent segments of line that are either increasing or decreasing (same probability $1/2$). The lower and upper bounds of the segments are set to $\pm\sqrt{3}$ to ensure a unit variance. Figure 1 shows a realization of such a band process. This algorithm requires very few ingredients (a band direction, an interval width, an offset and a set of ± 1 values). It is very simple to implement and prevents any possible edge-effect.

This algorithm has been used to simulate a Gaussian random field with unit scale factor into a simulation domain of 60×40. $10,000$ bands have been used to get simulations at different scales (see Fig. 2).

3.2 Simulation of C_1 Using Karhunen-Loève Method

Since the three-dimensional simulation domain can be enclosed within a ball of radius b, the band processes need merely being simulated within a segment of length $2b$, say $[-b, +b]$. It can be easily shown that the KL expansion of the exponential band covariance over $[-b, +b]$ can be derived that over $[-1, +1]$ provided that the inverse scale factor a has been replaced by ab.

Accordingly, the implementation of the KL method required finding the eigenvalues $\lambda \geq 0$ and the eigenfunctions $\varphi \in L^2([-1, 1])$ that satisfy

$$\lambda\,\varphi(t) = \int_{-1}^{+1} e^{-a|s-t|}\,(1 - a|s-t|)\,\varphi(s)\,ds \qquad -1 \leq t \leq 1. \tag{12}$$

The solutions of this integral equation are as follows:

(i) Each eigenvalue λ can be written

$$\lambda = \frac{1}{a}\sin^2(2\theta) \qquad 0 < \theta < \frac{\pi}{4}. \tag{13}$$

(ii) Those θ values are of two types. Putting $\mu = \tan\theta$ and $\nu = \cot\theta$, the *even* values and the *odd* values are the respective solutions of the equations

$$\sin(a\mu + 2\theta)\,\cos(a\nu - 2\theta)\,\sin^2\theta = \sin(a\nu - 2\theta)\,\cos(a\mu + 2\theta)\,\cos^2\theta, \tag{14}$$

$$\sin(a\mu + 2\theta)\,\cos(a\nu - 2\theta)\,\cos^2\theta = \sin(a\nu - 2\theta)\,\cos(a\mu + 2\theta)\,\sin^2\theta. \tag{15}$$

(iii) The *even* eigenfunctions and the *odd* eigenfunctions can be respectively written

$$\varphi(t) = \sin(a\nu - 2\theta)\,\cos(a\mu t) + \sin(a\mu + 2\theta)\,\cos(a\nu t), \tag{16}$$

$$\varphi(t) = \cos(a\nu - 2\theta)\,\sin(a\mu t) + \cos(a\mu + 2\theta)\,\sin(a\nu t). \tag{17}$$

The reader will find a sketch of the proof in the Appendix.

The solutions of (14) and (15) are numerically tractable as the trigonometric functions are periodic (Fig. 3).

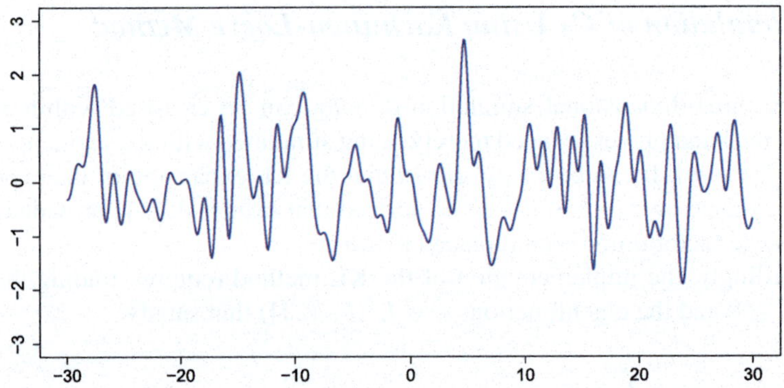

Fig. 3 Simulation of a Gaussian process using the KL method (100 eigen-elements)

3.3 Convergence

Starting from the representation (10), it is easy to derive that

$$\int_{-1}^{+1} C(t,t)\,dt = \sum_{n=1}^{\infty} \lambda_n. \tag{18}$$

This formula can be used for an experimental control all the more because $C(t,t) = 1$. The result $\sum_n \lambda_n = 2$ has been confirmed in the case $a = 1$ (cf. Fig. 4), even if the convergence seems to be rather slow. Indeed a correlation analysis based on the largest 500 eigenvalues (numerically computed) shows that the rate of convergence of λ_n to 0 is in n^{-2}.

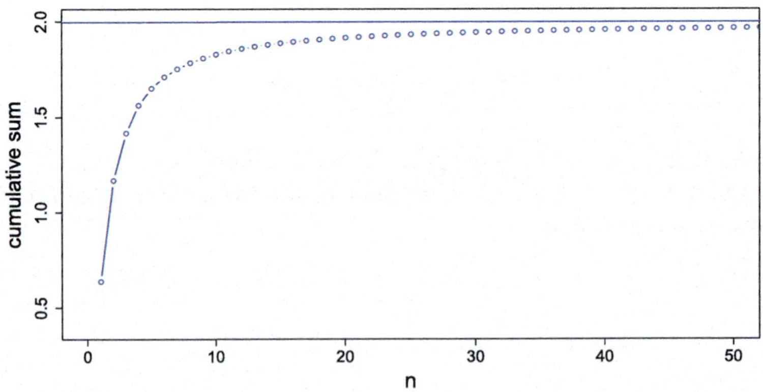

Fig. 4 Cumulative sum of eigenvalues

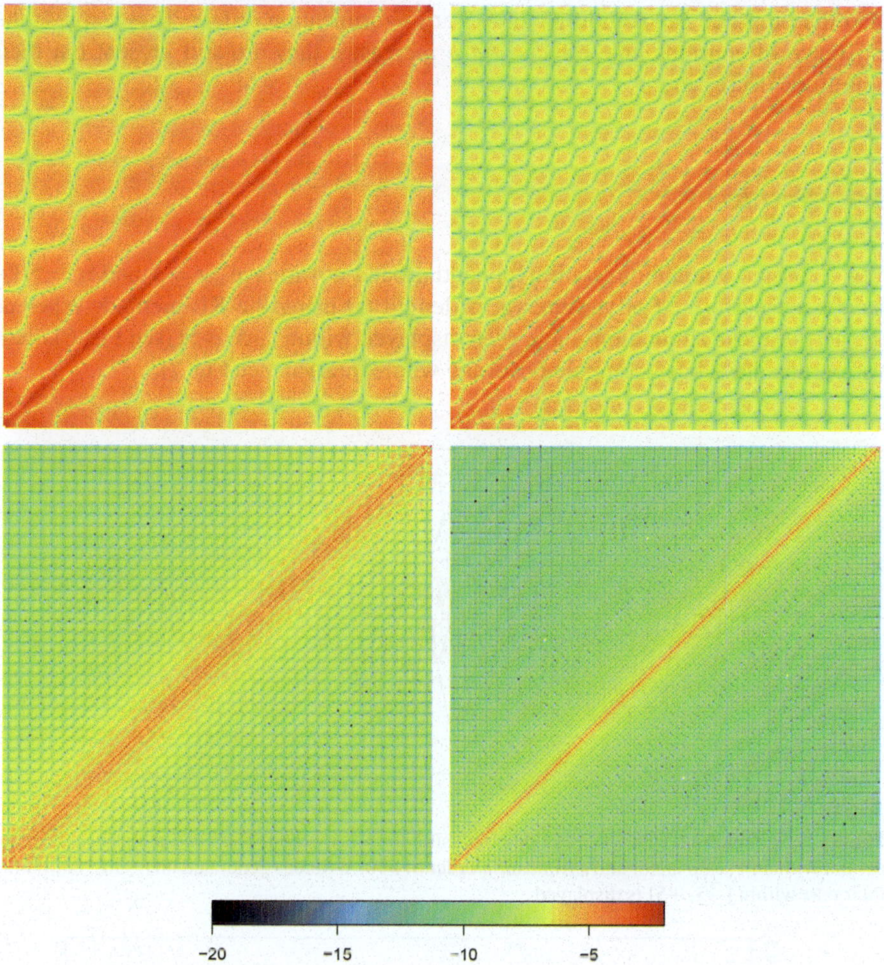

Fig. 5 Plotting the error images $\left(\varepsilon_n(s,t),\ -1 \leq s,t \leq 1\right)$ for different truncation orders. From top to bottom, and left to right, $n = 20$, 40, 60 and 80

Let $C_n(s,t)$ be the covariance deduced from $C(s,t)$ by truncation to the first n eigenvalues. Another possible test of convergence is to check how the covariance $C_n(s,t)$ converges to $C(s,t)$ as as n becomes very large. To do that, the following criterion has been introduced

$$\varepsilon_n(s,t) = \ln\left|C_n(s,t) - C(s,t)\right|. \tag{19}$$

Figure 5 shows the results obtained with $a = 1$ and 4 truncations of order 20, 40, 60 and 80. It appears that the criterion is maximal when $s = t$, which indicates that n must be much larger than 80 for the variance of Z to be properly retrieved. Note also the presence of "wiggling curves" at low truncation orders. They express the existence of spurious hole effects. As n increases, they are still present but their amplitude decreases.

3.4 Results

Ten Gaussian simulations with exponential band covariance ($a = 1$ and $n = 500$) have been performed on an interval of length 60 using the KL method. Whereas the part of those simulations have been displayed within $[-5, +5]$ (see Fig. 6), their variograms have been computed over the whole interval (see Fig. 7).

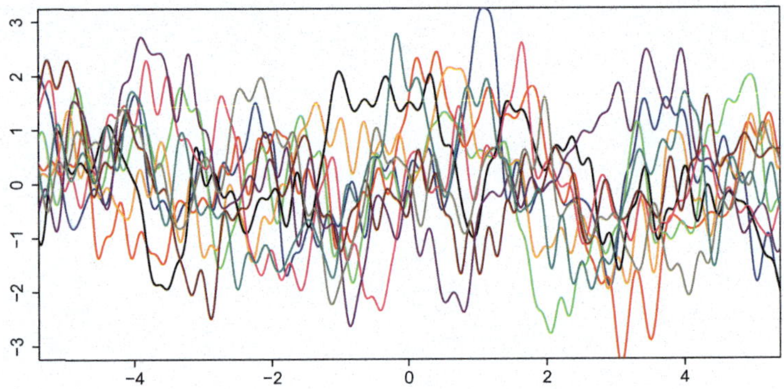

Fig. 6 Ten simulations of a Gaussian process using the KL method. The band covariance is exponential with unit scale factor and a truncation order of 500. Whereas generated on $[-30, +30]$, only their part within $[-5, +5]$ is displayed

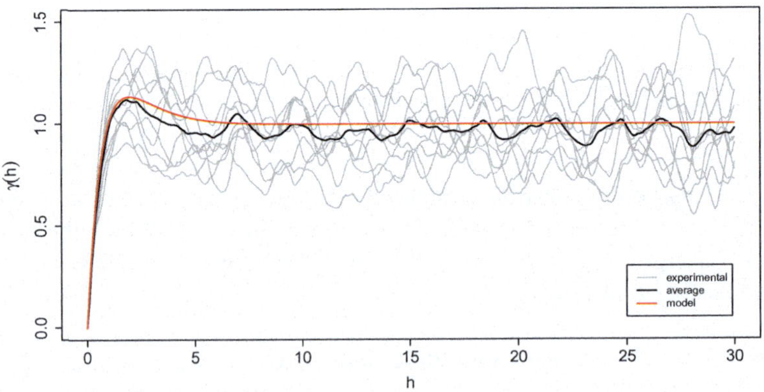

Fig. 7 Variograms of the simulations displayed on Fig. 6

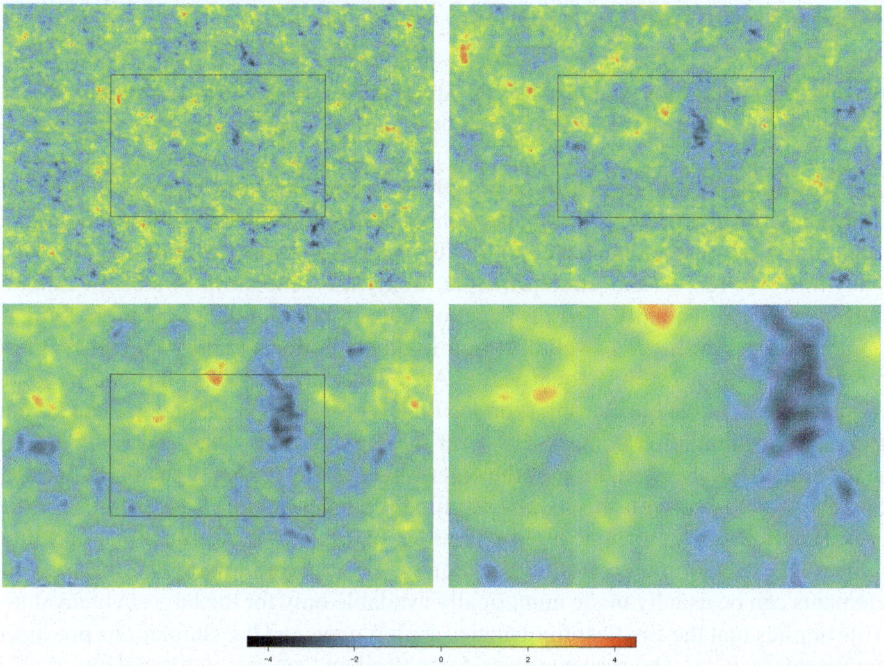

Fig. 8 Simulation of a Gaussian random field with exponential covariance (unit scale factor) using the TB method combined with KL processes. The simulation is displayed at four different scales. From top to bottom and left to right, the four images are sorted in ascending order of their scale. The black rectangle in each image indicates the size of the next image

The simulations produced are very smooth. Because of the truncation, they have been built as a finite linear combination of trigonometric functions that make them infinitely differentiable. However, the exponential band covariance has a linear behavior at the vicinity of the origin, from which it results that the realizations should be continuous but not differentiable.

The experimental variograms exhibit large statistical fluctuations, which is not amazing since its practical range (0.88) is far from negligible compared to the length of the simulation interval. On the average, the hole effect of the variogram model has been properly retrieved.

Regarding three-dimensional simulations, Fig. 8 displays the same exercise as in Fig. 2, up to the difference that the band simulations are performed using KL process rather than shot-noise processes. 400 turning bands have been used. Despite the relatively large number of eigen-elements considered per band (500), the resulting simulation looks still fuzzy. This results not only from the slow convergence of the eigenvalues toward 0, but also from the fact that the complete set of eigenvalues is not available, and the smallest ones that are responsible for the roughness of the simulations have been left out.

4 Discussion and Conclusions

Performing continuous simulations is a natural and pragmatic way of treating upscaling and downscaling problematics. Algorithms for such simulations have already been developed for various models of spatial extremes [4] or random sets [14]. The present paper shows that this is also the case for Gaussian random fields. Algorithms have been proposed starting from three different methods, namely the spectral method (S), the turning bands method (TB) and the Karhunen-Loève method (KL). They however rest on the same principle: firstly the generation of a set of random ingredients, and secondly their propagation at each point of the simulation domain.

Regarding their range of validity, their implementation and their efficiency, these three algorithms are not of equal value. As a generalization of S, TB can be made more efficient. Great hopes were founded on KL because it is adapted to the size and shape of the simulation domain. However these expectations have not been met. In many situations, the evaluation of the eigenvalues and eigenfunctions of the integral operator associated with the covariance function is an arduous task. Moreover, the eigenfunctions are often linear combinations of trigonometric functions and their numerical evaluation is long time consuming. Even more importantly, these eigen-elements can be usually made numerically available only for the largest eigenvalues. This implies that the smallest frequencies are left over, and the simulations produced are fuzzier than they should actually be. Such a salient feature is of limited importance for upscaling but of paramount importance for downscaling. An attempt has been made to simplify the evaluation of the eigen-elements. By combining KL and TB, this evaluation can be made unidimensional. This can certainly help in a number of situations, but no dramatic improvement is to be expected if all eigenvalues cannot be identified or they decay to 0 is rather slow (case of the exponential band covariance—see previous section).

Compared to KL, TB seems to have a limited range of validity as it was initially designed for the simulation of random fields models with stationary increments of any order [10]. More recently, importance sampling techniques have been introduced, not only to increase the TB efficiency [1], but also to perform TB simulations of general Gaussian random fields, stationary or not [5].

Appendix: Eigen-Elements of the Exponential Band Covariance

The problem addressed in this Appendix is the determination of the eigenvalues and the eigenfunctions of the integral operator

$$T\varphi(t) = \int_{-1}^{+1} e^{-a|s-t|}\left(1 - a|s-t|\right)\varphi(s)\,ds \qquad \varphi \in L^2\big([-1,+1]\big).$$

The main steps are presented here.

The starting equation can be put as follows. Assume that φ is an eigenfunction associated with the eigenvalue λ, then we have

$$\lambda\,\varphi(t) = \int_{-1}^{+1} e^{-a|s-t|}\,(1 - a|s-t|)\,\varphi(s)\,ds \qquad -1 \leq t \leq 1. \tag{20}$$

Differential Equation

By differentiating (20) twice, we obtain that

$$\lambda\,\frac{d^2\varphi}{dt^2}(t) = a(a\lambda - 4)\varphi(t) + 2a^2\,U\varphi(t)$$

where U is the integral operator defined by

$$U\varphi(t) = \int_{-1}^{+1} e^{-a|s-t|)}\,\varphi(s)\,ds.$$

Since

$$\frac{d^2U\varphi}{dt^2}(t) = -2a\varphi(t) + a^2\,U\varphi(t),$$

it follows that

$$\begin{aligned}\lambda\,\frac{d^4\varphi}{dt^4}(t) &= a(a\lambda - 4)\,\frac{d^2\varphi}{dt^2}(t) + 2a^2\,\frac{d^2U\varphi}{dt^2}(t)\\ &= a(a\lambda - 4)\,\frac{d^2\varphi}{dt^2}(t) + 2a^2\left(-2a\varphi(t) + a^2\,U\varphi(t)\right)\\ &= a(a\lambda - 4)\,\frac{d^2\varphi}{dt^2}(t) - 4a^3\,\varphi(t) + a^2\left(\lambda\frac{d^2\varphi}{dt^2}(t) - a(a\lambda - 4)\varphi(t)\right).\end{aligned}$$

It comes out that φ is a solution of a homogeneous differential of order 4

$$\lambda\,\frac{d^4\varphi}{dt^4}(t) - 2a(a\lambda - 2)\,\frac{d^2\varphi}{dt^2}(t) + a^4\lambda\varphi(t) = 0. \tag{21}$$

Solving the Differential Equation

The characteristic polynomial associated with this differential equation is

$$\lambda r^4 - 2a(a\lambda - 2)r^2 + a^4\lambda = 0. \tag{22}$$

This is a quadratic equation in r^2, and its discriminant is equal to $16a^4(1 - a\lambda)$. The sign of this discriminant is important because it specifies the shape of the eigenfunctions. Following the min-max principle, the largest eigenvalue satisfies

$$\lambda_1 = \max_{\|\varphi\|=1} \int_{-1}^{+1} \int_{-1}^{+1} e^{-a|s-t|} \left(1 - a|s-t|\right) \varphi(s)\, \varphi(t)\, ds\, dt.$$

Now, Cauchy-Schwarz inequality is applied and the resulting integral is explicitly calculated. We thus obtain that

$$\lambda_1 \leq \int_{-1}^{+1} \int_{-1}^{+1} e^{-2a|s-t|} \left(1 - a|s-t|\right)^2 ds\, dt = (e^{-4a} - 1 + 4a)\frac{1 + 8a^2}{4a^2} + 2 - 8a.$$

One readily shows that the right hand-side member is less than $1/a$ if and only if $e^{-4a}\,(1 + 8a^2) \leq 1$. This holds true because the left hand-side member is a monotonic decreasing function of a. Since $a > 0$, $\lambda_1 < 1/a$, which implies in turn that all eigenvalues are upper-bounded by $1/a$. Accordingly, the discriminant is positive and Eq. (14) has two real roots

$$r^2 = \frac{a}{\lambda}\left(\lambda a - 2 \pm 2\sqrt{1 - a\lambda}\right) = -\frac{a}{\lambda}\left(1 \pm \sqrt{1 - a\lambda}\right)^2.$$

The expression of these roots can be simplified by putting

$$a\lambda = \sin^2(2\theta) \qquad 0 < \theta \leq \frac{\pi}{4}, \tag{23}$$

which yields $r^2 = -a^2 \tan^2\theta$ or $r^2 = -a^2 \cot^2\theta$, then $r = \pm ia\tan\theta$ or $r = \pm ia\cot\theta$. Thus, the general solution of the differential equation (21) is

$$\varphi(t) = \alpha \sin(at\tan\theta) + \beta\cos(at\tan\theta) + \gamma\sin(at\cot\theta) + \delta\cos(at\cot\theta). \tag{24}$$

In the sequel, it will be occasionally useful to write $\mu = \tan\theta$ and $\nu = \cot\theta$.

Explicit Solutions

The next step is to identify which values of θ and which sets of coefficients α, β, γ and δ provide effective solutions to the integral equation (12).

Starting from the general solution of the differential equation (21), the explicit solutions are usually derived by adjoining initial conditions. In the present case, no such initial conditions do exist. This prompted us to introduce a set of compatibility relationships the explicit solutions are supposed to respect. There is a wide choice concerning these compatibility relationships. What has been considered here are the values taken by the eigenfunctions and their derivative at the origin, namely

$$\lambda(\beta + \delta) = \lambda\varphi(0) = \int_0^1 e^{-as}\,(1 - as)\,(\varphi + \check{\varphi})(s)\,ds, \tag{25}$$

$$\lambda a(\alpha\mu + \gamma\nu) = \lambda\frac{d\varphi}{dt}(0) = a\int_0^1 e^{-as}\,(2 - as)\,(\varphi - \check{\varphi})(s)\,ds, \tag{26}$$

$$-\lambda a^2(\beta\mu^2 + \delta\nu^2) = \lambda\frac{d^2\varphi}{dt^2}(0) = -4a\varphi(0) + a^2\int_0^1 e^{-as}\,(3 - as)\,(\varphi + \check{\varphi})(s)\,ds, \tag{27}$$

$$-\lambda a^3(\alpha\mu^3 + \gamma\nu^3) = \lambda\frac{d^3\varphi}{dt^3}(0) = -4a\varphi'(0) + a^3\int_0^1 e^{-as}\,(4 - as)\,(\varphi - \check{\varphi})(s)\,ds, \tag{28}$$

where $\check{\varphi}$ is the symmetric of φ around the origin. It can be noted that relations (25) and (27) provide even solutions, whereas relations (26) and (28) provide odd solutions.

Even Solutions

Replacing $\varphi + \check{\varphi}$ by $2\beta\cos(a\mu\,\cdot) + 2\delta\cos(a\nu\,\cdot)$ in Eqs. (25) and (27) lead to

$$\begin{aligned}
&\beta\,\cos^2\theta\,\cos(a\mu + 2\theta) - \delta\,\sin^2\theta\,\cos(a\nu - 2\theta) = 0,\\
&\beta\,\cos\theta\,\cos(a\mu + \theta) - \delta\,\sin\theta\,\sin(a\nu - \theta) = 0.
\end{aligned}$$

Let us modify the second equation by putting $a\mu + \theta = (a\mu + 2\theta) - \theta$ and $a\nu - \theta = (a\nu - 2\theta) + \theta$ and taking the difference withe first equation. We arrive to

$$\begin{aligned}
&\beta\,\cos^2\theta\,\cos(a\mu + 2\theta) - \delta\,\sin^2\theta\,\cos(a\nu - 2\theta) = 0,\\
&\beta\,\sin(a\mu + 2\theta) - \delta\,\sin(a\nu - 2\theta) = 0.
\end{aligned}$$

This homogeneous system of two equations β and δ admits a nontrivial solution only if its determinant vanishes. This yields

$$-\cos^2\theta\,\cos(a\mu + 2\theta)\,\sin(a\nu - 2\theta) + \sin^2\theta\,\sin(a\mu + 2\theta)\,\cos(a\nu - 2\theta) = 0.$$

Odd Solutions

Replacing $\varphi - \check{\varphi}$ by $2\alpha\sin(a\mu\,\cdot) + 2\gamma\sin(a\nu\,\cdot)$ in Eqs. (26) and (28) lead to

$$\begin{aligned}
&\alpha\,\cos^2\theta\,\sin(a\mu + 2\theta) - \gamma\,\sin^2\theta\,\sin(a\nu - 2\theta) = 0,\\
&\alpha\cos\theta\,\sin(a\mu + \theta) + \gamma\,\sin\theta\,\cos(a\nu - \theta) = 0.
\end{aligned}$$

Using the same approach as for the even case, we obtain the following homogeneous system of equations

$$\alpha \cos^2\theta \, \sin(a\mu + 2\theta) - \gamma \sin^2\theta \, \sin(a\nu - 2\theta) = 0,$$
$$\alpha \, \sin(a\mu + 2\theta) - \gamma \, \sin(a\nu - 2\theta) = 0.$$

and write that its determinant vanishes, which yields

$$-\cos^2\theta \, \sin(a\mu + 2\theta) \, \cos(a\nu - 2\theta) + \sin^2\theta \, \cos(a\mu + 2\theta) \, \sin(a\nu - 2\theta) = 0.$$

References

1. Arroyo, D., Emery, X.: Simulation of intrinsic random fields of order k with a continuous spectral algorithm. Stoch. Envirron. Res. Risk Assess **32**, 3245–3255 (2018)
2. Chilès, J., Delfiner, P.: Geostatistics: Modeling Spatial Uncertainty. Wiley, New York (1999)
3. Deheuvels, P., Martynov, G.: A karhunen-loève decomposition of a gaussian process generated by independent pairs of exponential random variables. J. Funct. Anal. **255**, 2363,2394 (2008)
4. Demongeot, M., Carnec, R., Chautru, E., Lantuéjoul, C.: Continuous simulation of storm processes. Extremes **25**, 363–387 (2002)
5. Emery, X., Arroyo, D.: On a continuous spectral algorithm for simulating non-stationary gaussian random fields. Stoc. Environ. Res. Risk Asses. Comput. Geosci. **32**, 905–919 (2018)
6. Emery, X., Lantuéjoul, C.: Tbsim: a computer program for conditional simulation of three-dimensional gaussian random fields via the turning bands method. Comput. Geosci. **32–10**, 1615–1628 (2006)
7. Ghanem, R., Spanos, P.: Stochastic Finite Elements: A Spectral Approach. Courier Corporation, North Chelmford (Ma) (2003)
8. Kac, M., Siegert, A.: An explicit representation of a stationary gaussian process. Ann. Math. Stat. **26**, 189–211 (1947)
9. Lantuéjoul, C.: Geostatistical Simulation: Models and Algorithms. Springer, Berlin (2002)
10. Matheron, G.: The intrinsic random functions and their applications. Adv. Appl. Prob. **5–3**, 439–468 (1973)
11. Matérn, B.: Spatial Variation. Springer, Berlin (1986)
12. Mourlanette, P., Biver, P., Renard, P., Noetinger, B., Caumon, G., Perrier, Y.A.: Direct simulation of non-additive properties on unstructured grids. Adv. Water Res. **143**, 103665 (2020)
13. Shinozuka, M., Jan, C.: Digital simulation of random processes and its applications. J. Sound Vib. **52–3**, 111–128 (1972)
14. Troncoso, A., Freulon, X., Lantuéjoul, C.: Sequential simulation of a conditional boolean model. Math. Geosciences (submitted) **54-2**, 389–411 (2022)
15. Zaytsev, V., Biver, P., Wackernagel, H., Allard, D.: Change-of-support models on irregular grids for geostatistical simulations. Math. Geosci. **48**, 353–369 (2016)

Morphological Models of Networks of Inter-granules Micropores: Identification and Validation Based on 2D and 3D Imaging

Léo Moutin, Christelle Duguay, Laurent Fayette, Andrew King, Renaud Masson, and Michel Bornert

Abstract In view of predicting the overall thermal conductivity of ceramics with a granular microstructure exhibiting complex networks of pores preferably located at the interfaces between granules, a simple microstructural model has been developed to capture the dominant features of such a specific morphology. Generated 3D virtual networks of pores are obtained as the intersection of a Voronoi tessellation, representing the granules, and a boolean model of monodisperse overlapping spheres [1]. The two non-dimensional parameters fully determining such networks have been first adjusted so as to reproduce at best some morphological properties extracted from 2D optical images of various real samples [2]. To further validate this model, several small samples of the same materials have in addition been imaged in 3D by synchrotron X-ray micro-computed tomography, with a specific setup adapted to the very strong attenuation of these ceramics. The consistency of the results measured in 2D on cross-sections of tomography images with those measured on optical

L. Moutin · C. Duguay · L. Fayette · R. Masson
CEA, DES, IRESNE, DEC, Cadarache, Saint-Paul-lez-Durance 13018, France
e-mail: christelle.duguay@cea.fr

L. Fayette
e-mail: laurent.fayette@cea.fr

R. Masson
e-mail: renaud.masson@cea.fr

A. King
Synchrotron Soleil, Psiché Beamline, L'Orme des Merisiers 91190, Saint-Aubin, France
e-mail: andrew.king@synchrotron-soleil.fr

M. Bornert (✉)
Laboratoire Navier, École des ponts, University Gustave Eiffel, CNRS UMR 8205, 6-8 avenue Blaise Pascal, Champs-sur-Marne 77455, Marne-la-vallée Cedex, France
e-mail: michel.bornert@enpc.fr

F. Willot et al. (eds.), *Continuum Models and Discrete Systems*,
Springer Proceedings in Mathematics & Statistics 457,
https://doi.org/10.1007/978-3-031-58665-1_4

microscopy images has been first checked. Next, qualitative comparisons between observed 3D pore networks and simulated ones with parameters adjusted from 2D observations have been complemented by more quantitative ones.

Keywords Ceramics · Porosity · Computed tomography · Image processing · Statistical morphology · Microstructure generation

1 Introduction

This work aims at predicting the thermal conductivity of porous uranium dioxide ceramics manufactured by sintering multi-crystalline granules. These materials exhibit complex networks of elongated pores preferably located at the interfaces between granules, with features depending on the manufacturing process. Because such microcrack-like pores act as strong barriers to heat propagation, accounting for the specific geometry of such pore networks is essential for accurate predictions.

Among the various methodologies for modelling microstructures (see e.g. the review in [3]), the adopted model, detailed in [1], is based on probabilistic approaches to generate random sets in 3D, in line with [4]. More specifically, following [5], it combines two random sets to constrain the geometric location of cracks, as in [6] for another case of cracked polycrystals. Following [7] who adopted a similar approach for cold spray coatings, the parameters of this 3D model could be adjusted in [2] so as to reproduce at best some morphological properties measured from 2D images from optical microscopy (OM) of real samples, appropriately filtered to extract the major thermal barriers at granules interfaces, without excessive details nor smoothing.

The 3D model finally obtained was thus tuned on 2D information only. The here-presented contribution extends its validation through comparisons with 3D images of the same real materials obtained by X-ray computed microtomography (XRCT). The experimental procedure to obtain and process these data and to perform these comparisons are presented in Sect. 4 after a brief review of the adopted morphological model (Sect. 2) and its tuning to various materials based on 2D data (Sect. 3).

2 A Simple Morphological Model of Inter-Granules Pore Networks

The adopted virtual microstructures describing networks of cracks located at inter-granules boundaries (IGB) are generated by combining two random sets [1]. The first one corresponds to the IGB and is obtained from a Voronoi tessellation of seeds resulting from the random sequential adsorption (RSA) of monodisperse spheres, with periodicity conditions.The used algorithm allows to almost reach its theoretical saturation with reasonably short simulation times, so that the obtained tessellation

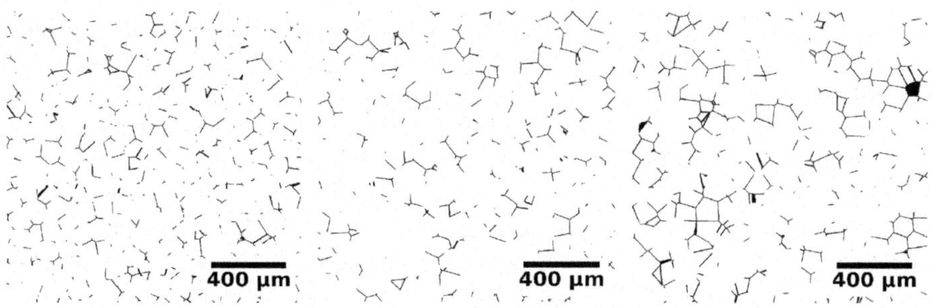

Fig. 1 Virtual microstructures obtained using the method introduced in [1] for various parameters choices. From left to right: $R_l/R_g = 0.7$, $R_l/R_g = 1.25$ and $R_l/R_g = 1.5$

exhibits rather homogeneous grain sizes. The second random set is obtained by randomly distributing monodisperse spheres, so-called delimiting spheres, over the simulation domain using a standard Poisson point process for their centers, providing a boolean model of spheres. The virtual microstructure of inter-granules pores is finally obtained by keeping the IGB not contained in the delimiting spheres.

This simple method has three input parameters: the radius R_g of the spheres of the RSA model, the radius R_l of the delimiting spheres and the density d_l of the Poisson process. Because of scale invariance, the effective thermal conductivity only depends on two non-dimensional parameters: the ratio R_l/R_g and the volume fraction of the boolean model defining the delimiting spheres. It can be computed through full-field simulations techniques, such as the FFT method (see implementation details in [1]). By varying these parameters, a wide variety of microstructures could be generated (illustrative examples are provided in Fig. 1) with significantly different effective thermal conductivities. The latter were also compared to those of uniformly weakened solids, with cracks randomly distributed over the bulk, at similar overall densities. It was observed that the concentration of pores on the IGB causes a significant reduction in effective thermal conductivity.

3 Tuning Parameters to Real Materials

A number of model materials with various microstructures have been fabricated to investigate the relationship between porosity and thermal properties. The manufacturing conditions were varied to obtain samples over a wide range of porosity. Among these many samples, focus will be given in the following to three of them, denoted C_1, C_2 and C_3. They exhibit increasing open porosity volume fractions: $c_1^o = 0.9\%$, $c_2^o = 1.1\%$ and $c_2^o = 3.45\%$, which cover the full porosity range of interest for the practical application. Due to their slender shape, interconnections of inter-granules pores in the volume are expected so that the open porosity is in practice well correlated with the inter-granules porosity. The aim is now to adjust the parameters of the simple model proposed in [1] to such real materials and then check its capability to reproduce sufficiently accurately the real porous networks, both in terms of their morphological characteristics and their thermal properties. We summarize in this section the methodology proposed in [2] to tune the parameters of the model so that it simulates a pore network representative of a given real ceramic.

First, based on 2D OM high definition images of large areas of samples properly prepared to reveal porosity, we characterize the porous networks of the studied ceramics in order to extract some descriptor parameters. To this end, an image processing protocol has been developed to clearly isolate and simplify the description of the objects of interest that are, in our case, the inter-granules pores. This image processing protocol consists of four main operations, to go from a raw OM image to a skeletonized (i.e. objects with one pixel thickness) binary image. They involve appropriate filtering to eliminate irrelevant details without over-smoothing (see [2] for details). The obtained topological skeleton encodes the main geometrical features of the inter-granule pore network and we focus on the following four quantities to characterize it. The density of the crack network of the ceramics is quantified using the length of the intercepts (l_{Inter}) [8]. To assess the size distribution of the inter-pore space, we use the principles of Matheron's granulometry [9]: we consider for each pixel in the solid phase the radius of the largest circle (r_{Mat}) that can be inserted between the skeleton and contains this pixel. The lengths of the cracks are characterized by estimating two quantities, namely the "lengths of the branches" (l_{bra}) and "the lengths of the primary branches" (l_{braPrim}) of the skeletons. The "branches" are defined as the linear objects obtained after eliminating the connection points where these objects meet in the skeleton and the "primary branches" the linear objects retained to quantify the geodesic diameter of the connected components of the skeleton. The cumulative statistical distributions of these quantities (H_{Inter}, H_{Mat}, H_{bra} and H_{braPrim}), weighted either by surface or number, are considered for a detailed description of the porous microstructures.[1]

This quantitative characterization of the porous network of a given ceramic provides the input data of an automated optimization procedure that aims at producing a tuned virtual microstructure, of which the above descriptor parameters, measured on 2D cross-sections of the 3D generated microstructures, are required to be as close as possible to their measured counterparts. The missmatch function to be minimized by this procedure is the average of the discrepancies for each of the four cumulative distributions (discretized over control points) with more weight given to the length of the branches and the primary branches, for practical reasons [2]. As a qualitative illustration of the results, 3D renderings of the obtained tuned virtual microstructures are given in Fig. 5. For the tested ceramics, the developed optimization algorithm gives satisfactory quantitative results as the cumulative distributions of the four descriptor parameters of the tuned microstructures and of the real ceramics are in good agreement (Fig. 2). One can however notice that the fitting of Matheron's 2D granulometry is less good than for the other quantities, especially for the larger scales. A possible

[1] Note that cumulative distributions are more convenient than distribution functions for subsequent optimization because they are more regular.

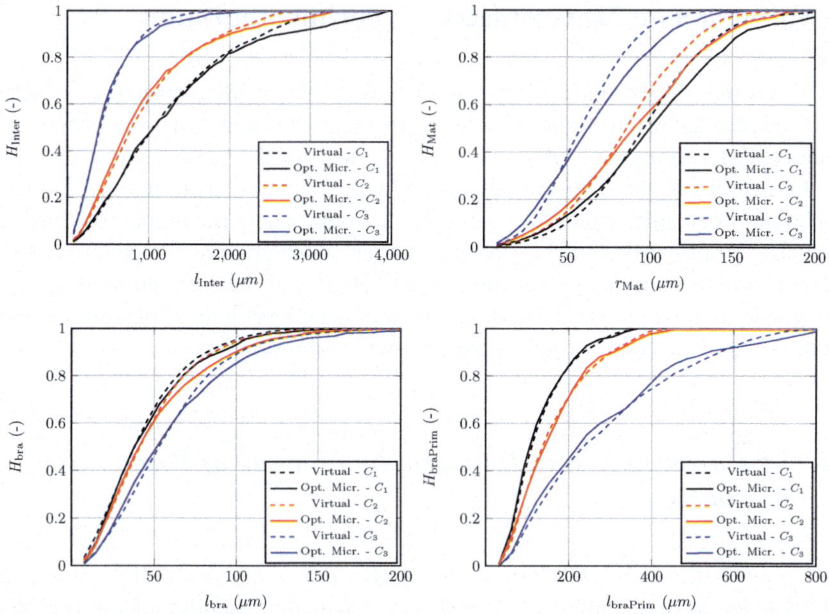

Fig. 2 Cumulative distributions associated with each descriptor parameter obtained on the reference ceramics and on the tuned microstructures (from [2])

explanation is that the selected scheme for the IGB generates a narrow grain size distribution and flat and rather regular boundaries. A modified point process for its seeds, or the use of Laguerre polyhedra or even locally anisotropic tessellations could provide the required degrees of freedom for a better fit, but at the price of a more complex model, involving more parameters.

To further validate this 2D identification process, full-field 2D numerical simulations (based on the FFT method) were performed on both processed OM images of the studied ceramics and cross-sections extracted from the tuned 3D microstructures. These 2D simulations are used as an additional validation of the fitted model, including not only the partial set of 2D morphological features used to fit the model, but also the full set of features determining the thermal properties of a 2D heterogeneous medium. Of course, this is not sufficient to validate the performance of the model in terms of full 3D thermal properties, but it does provide an additional and original means to confirm the similarity between the virtual and real 2D cross-sections. This validation was carried out on the effective 2D properties, but also and above all on the local field statistics in terms of distribution functions and spatial correlations. These simulations have allowed us to highlight a good correspondence in the 2D thermal behavior of the virtual and real microstructures [2].

4 Validation on 3D Geometric Experimental Data

The methodology for generating representative virtual microstructures described in [2] and recalled in the previous section is based on 2D images of the porous network of the ceramics of interest. Obviously, the relevance of the model in reproducing the 2D characteristics of the porous network and its efficiency in predicting 2D thermal properties is not sufficient to fully validate the relevance of the model for simulating the real (3D) properties of the ceramics under consideration. For this reason, and in order to obtain 3D images of the porous network of the real ceramics, an X-ray tomography campaign was carried out on the Psiché beamline at Synchrotron Soleil (Saint-Aubin, France), on some representative model samples.

4.1 Synchrotron X-Ray Tomography Imaging of Porous UO_2 Ceramics

Absorption X-ray computed tomography (XRCT) is a non destructive 3D imaging technique based on the attenuation of X-rays by matter. Radiographs of the sample along many directions, recorded by a 2D detector, serve as input data of a mathematical inversion algorithm which outputs a 3D image in which grey levels code the local attenuation coefficient of the constitutive material. This technique is well suited to image non absorbing pores in a matrix, as they induce a strong contrast, under the conditions that the spatial resolution of the radiographs is sufficiently high with respect to pore size. In addition, the energy of the probing X-rays needs to be adapted to the size of the sample and the attenuation properties of its constituents, which strongly increase with density and atomic number. The highly absorbing UO_2 ceramics under consideration (density of 10.96 g/cm^3) with inter-granules pores of micrometric thickness require thus rather non-conventional conditions for their XRCT imaging. They could be provided by the Psiché beamline, in particular thanks to its wiggler which generates a wide spectrum of radiations, with intense flux still available above 100 keV. Another difficulty to be faced was the radioactivity of the samples, which required to limit their size and to keep them in sealed aluminum sample holders all along the experiment, in order to comply with safety regulations.

To comply with all these constraints, small cylindrical samples were cored with a diameter of 1.2 mm and a height of 1.3 mm. Photons with energy below 65keV were removed with appropriate filters but the full high energy spectrum of the wiggler beam was kept. Under such conditions, the photon flux transmitted through the full diameter of the sample was maximal at about 90keV, but represented only about 5% of the incoming flux. With standard scanning procedures, this would have induced very strong contrast in the radiographs, and very poor signal to noise ratios to avoid saturation on their edges. To avoid this effect the beam profiling technique proposed in [10] was adopted. A specific filter made of copper which is roughly five times less attenuating than UO_2 in these conditions and in which five cylindrical holes with the same diameter as the sample and aligned with it were drilled, was placed in the beam before the sample so that the total attenuation was almost uniform over the radiographs. These could therefore be recorded with sufficiently low noise levels in reasonable times. Reference images of the filtered beam without the sample were recorded using a correspondingly shorter exposure time to avoid saturation in the centre of the image. The LuAg scintillator with a thickness of 50μm was imaged with a microscope with a $\times 10$ optical magnification by an ORCA digital camera with a 6.5μm pixel size. The 3D images, reconstructed with the PyHST software, had thus a voxel size of 0.65 μm. This is significantly smaller than the pixel size of 1.52μm of the OM images used for the 2D characterization of the pore network (see previous section), but the actual spatial resolution of the 3D images is indeed somewhat poorer than this nominal voxel size. Total scan time for a sample was about 90 min and 24 samples could be imaged during a four days campaign, among which samples of the three materials under consideration here. Several filtering operations, not detailed here, had to be applied to the obtained 2048^3 voxels images to remove ring artefacts and non-standard beam-hardening effects induced by the beam profiling technique.

4.2 Comparison Between OM Images and XRCT Cross-Sections

Before characterizing the porous network using the full 3D XRCT data, a preliminary study was carried out to check the consistency of the results measured in 2D on cross-sections of XRCT images with those measured on OM images. The 2D image processing protocol introduced in Sect. 3 to characterize OM images was therefore also applied to such XRCT cross-sections of same orientation. In order to threshold the XRCT images at a greylevel similar to the one used to binarize the OM images, the intrinsic grey levels of the matrix and pores phases were determined in both the XRCT and OM images. The threshold for XRCT images was then set at the same relative level between matrix and pores grey levels as the one used to segment the OM images. Figure 3 shows a cross-section of an XRCT and an OM image with the same observation window size. These images were taken on the C_3 ceramic and show the segmented pores, the inter-granules pore skeletons and their primary branches. The visual comparison between these images emphasizes the good qualitative consistency between the pore networks identified with both imaging techniques.

For a more quantitative comparison, two of the descriptor parameters introduced in Sect. 3 are considered again, namely the cumulative distribution functions associated with branches lengths and primary branches lengths. As XRCT windows are small, for the reasons explained in previous sections, some bias might be expected because of edge effects. This occurs when the size of the window is not large enough compared to the size of the objects being characterized, particularly when no special treatment is applied to objects that are cut at the edges. Such treatments are not necessary for the 2D analyses on the full OM images as they are sufficiently large. For simplicity

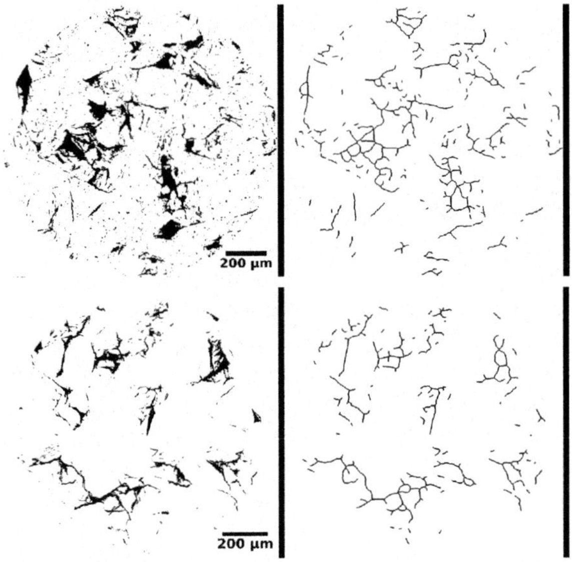

Fig. 3 Comparison of an OM image (top) and a radial section of a XRCT image (bottom) of ceramic C_3 using observation windows of same size and identical image processing protocols: segmented pores (left), inter-granules pore skeletons (centre) and their primary branches (right)

we did not implement them either for the smaller windows, but performed the same (biased) analysis on XRCT cross-sections and sub-windows of same size extracted randomly from the complete large OM images, as in Fig. 3, which ensures a valid quantitative comparison. For a better statistical representativity, the parameters are in addition measured over several windows: different sampling areas for the OM images and different slices for the XRCT ones.

Results are plotted in Fig. 4. For C_1 and C_3, the cumulative distribution functions H_{bra} and H_{braPrim} measured on the OM and XRCT images are close. This underlines the good quantitative coherence of the porous network observed by OM and XRCT. However, the comparison is less satisfactory for H_{braPrim} in the case of the C_2 ceramic. A possible explanation is that the small XRCT volumes may not adequately capture possible long range spatial variations in the morphology of the pore network. To quantify such spatial variations, we also measured the average differences between H_{bra} and H_{braPrim} obtained separately from all observation windows of the OM images. They are represented as uncertainty bars in Fig. 4. It is clear, especially for the primary branches of C_1 and C_2, that these variations are rather large, which lends weight to the above hypothesis: the XRCT result for H_{braPrim} for C_2 lies within the range of possible values measured by OM over similar sized windows. Note that the present discussion has been focused on the length of branches and primary branches, as these parameters are difficult to translate into 3D, but that similar results were obtained for the length of intercepts and Matheron's granulometry. These comparisons globaly confirm that 2D data extracted from OM surface measurements are consistent with

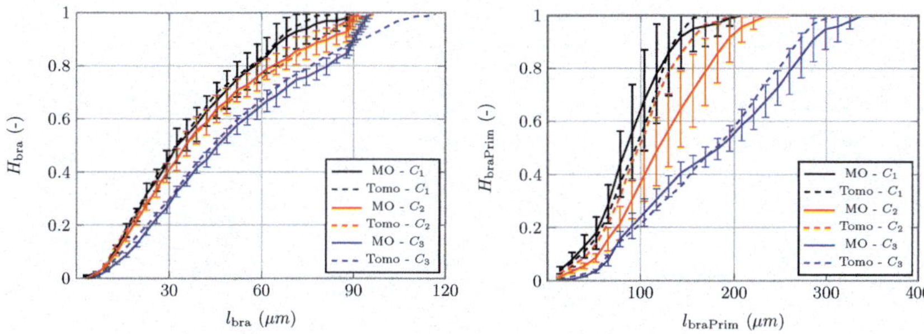

Fig. 4 Cumulative distribution functions associated with the length of branches (left) and the length of primary branches (right) measured on sub-windows of OM images reduced to the size of XRCT images and on sections of XRCT images. Bars quantify the spatial fluctuations of these functions when the observation window is randomly chosen in the large OM images

2D data extracted from the bulk of the material. There is in particular no significant bias due to surface preparation. On the other hand, they also show the limitations of XRCT images which are, at least for the considered ceramics, restricted to small window sizes and may not capture longer range effects.

4.3 3D Image Processing and Characterization

We now turn to the characterization of the pore network in the ceramics of interest using the 3D data of XRCT images. Thresholding in 3D is performed as in previous section, but the removal of the smaller connected objects not belonging to the inter-granules pore network is now performed based on their 3D properties. A criterion relative to their sphericity[2] is selected, based on the fact that the inter-granules pores are long and slender as opposed to the smaller isolated objects, which are more spherical. Note that a circularity criterion served to filter pores in OM images. By studying the evolution of the sphericity of the objects as a function of their volume, it appears that eliminating objects with a sphericity above 0.2 is a good criterion for isolating the inter-granules pores. The extent of the obtained pore network, depicted as 3D surface renderings in the bottom of Fig. 5, visually appears to be consistent with their open porosity ($\approx$ 1% for C_1 and C_2, and $\approx$ 3.5% for C_3).

In order to characterize the porous network in 3D, two of the descriptor parameters introduced in Sect. 3 are used again, namely the length of the intercepts (defined as in 2D but extended over 3D images) and Matheron's granulometry, which is however now based on inserted spheres and non longer disks. In this study, the samples are reduced to cubes of volume $V = 650 \times 650 \times 650\ \mu m^3$ for practical

[2] Sphericity is defined here in the sense given in the imageJ 3D Suite plugin documentation: it is the cubic root of the compactness C, which is computed from the volume V and the surface area S of an object, weighted so that it is 1 for a sphere ($C = 36\pi V^2/S^3$).

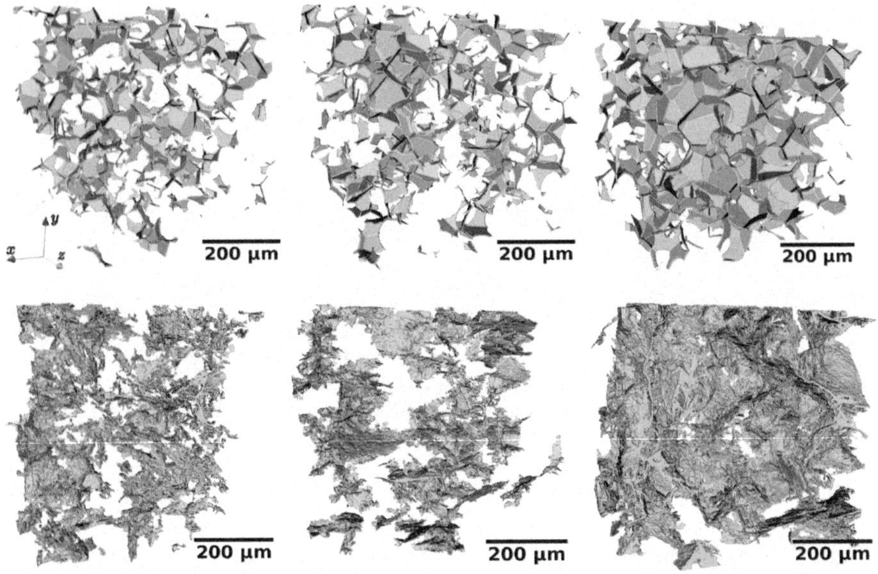

Fig. 5 3D illustrative views of local volumes ($\approx 650 \times 650 \times 650\ \mu m^3$) in the tuned virtual microstructures (top) and in the porous network extracted from XRCT images (bottom) associated respectively with the three ceramics C_1, C_2 and C_3 (from left to right)

reasons: simpler boundary conditions, reduction of computational time and memory requirements. Note also that the porous networks studied here using XRCT images are not skeletonized, mainly because it is not an easy task to perform in 3D. However, it turns out that the length of the intercepts and the radius of the Matheron spheres are hardly affected by skeletonization, since they aim at characterizing inter-pore spaces that are very wide compared to the thickness of the inter-granules pores.

The length of the intercepts is inherently a 1D data. Measuring this parameter in a 3D image therefore simply means measuring it along all possible directions, and in practice along the three axes of the images. Concerning Matheron's granulometry, the insertion of maximal spheres in the interpore spaces requires the full 3D distribution of pores. Since the investigated XRCT volumes are relatively small, edge effects are expected and need to be quantified. By studying the evolution of the measurement of the intercepts while progressively increasing the size of the observation window, we could show, by assuming periodicity and by attributing the size of the window to intercepts too large to be contained in the image, that the distribution functions obtained on reduced volumes remain consistent with those obtained on larger volumes, up to the size of the reduced volume. On the other hand, the calculation of the radius of Matheron spheres was only weakly dependent on window size, probably because the inserted spheres are strongly constrained by the presence of pores, and remain small with respect to window size for the tested range.

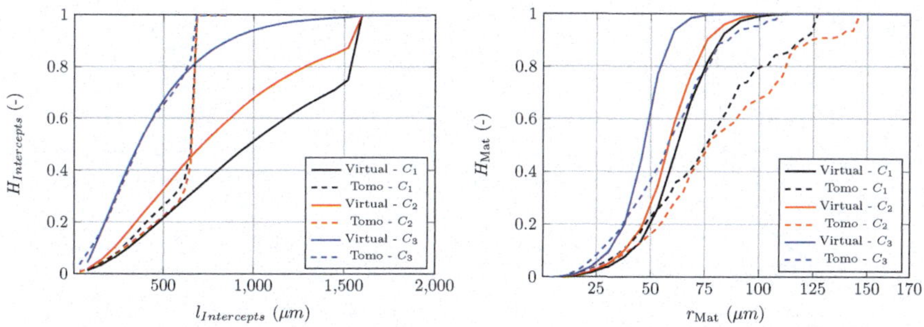

Fig. 6 Cumulative distribution functions H_{Inter} and H_{Mat} calculated on the XRCT images of the ceramics C_1, C_2 and C_3 as well as on the virtual microstructures associated with these ceramics, optimized on 2D data and simulated over $1556 \times 1556 \times 1556\ \mu\text{m}^3$ windows

Figure 6 shows the cumulative distribution functions H_{Inter} and H_{Mat} measured on the XRCT images and on the tuned virtual microstructures (depicted in the top line of Fig. 5). The presence of vertical asymptotes in H_{Inter} measured from the XRCT images results from the above described treatment of intercepts larger than the size of the study window. Up to this limit, the cumulative distribution functions associated with the length of the intercepts measured on the real and virtual microstructures are in very good agreement for the C_1 and C_3 ceramics but, againg, not for C_2. This confirms the results obtained in the previous section when comparing the branch and primary branch lengths measured in 2D on OM and XRCT images (Fig. 3) and might be explained by long range fluctuations in the C_2 ceramic. Regarding the distribution of the radius of the Matheron spheres, the discrepancy which was already observed in 2D (Sect. 3), is even more pronounced in 3D: virtual microstructures exhibit a too narrow distribution of Matheron spheres, with especially an under-representation of large spheres. This observation holds for all three ceramics and echoes the discussion in Sect. 3 about the limitations of the simple geometric model used to generate the virtual microstructures, in particular the IGB model, and the conclusions drawn from it, conclusions that are all the more valid in 3D.

Virtual microstructures are fitted on data obtained from large 2D fields of observations of the ceramics under study. They therefore incorporate some large scale variations in the descriptor parameters: in particular the distribution of the lengths of the intercepts is well fitted by the virtual microstructures up to 2500 μm for ceramic C_1 and almost 2000 μm for C_2 and C_3 (see Fig. 2). In order to put into perspective the characterization carried out by XRCT on relatively small volumes, we measured the length of the intercepts and the radius of Matheron's spheres in sub-volumes of size $650 \times 650 \times 650\ \mu\text{m}^3$, randomly extracted from the full volumes used for the optimization process of size $3113 \times 3113 \times 3113\ \mu\text{m}^3$, and quantified the statistical fluctuations of their distribution functions, represented as uncertainty bars on the plots in Fig. 7). As expected, when the length of the intercepts reaches the size of the sub-volumes, the vertical asymptotes are observed as for the XRCT images. Before this limit, the values obtained for the reduced volumes and the full volumes are con-

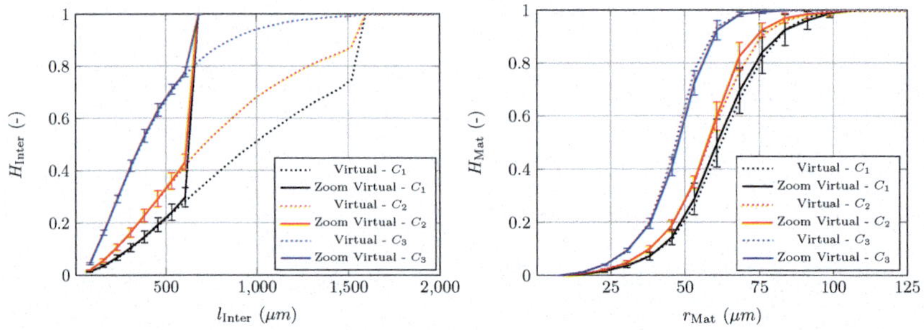

Fig. 7 Cumulative distribution functions H_{Inter} and H_{Mat} calculated on randomly chosen sub-volumes ($650 \times 650 \times 650\ \mu m^3$) and on large volumes ($1556 \times 1556 \times 1556\ \mu m^3$) of the optimized virtual microstructures associated with the ceramics of interest. Error bars quantify the fluctuations

sistent. This consistency is also found for the radius of the Matheron spheres, now over the whole range of values. The distributions obtained over reduced volumes are therefore in very good agreement with those measured over full volumes, up to the limit imposed by their size. Statistical fluctuations observed between the different small sub-volumes of the virtual microstructures remain small, for both H_{Inter} and H_{Mat}. In particular, for the ceramic C_2, these variations are not large enough to contain the cumulative distribution function associated with the length of the intercepts measured on the XRCT image.

5 Conclusions

Using the optimization process developed in [2], based on the method for generating virtual microstructures with inter-granules pores introduced in [1], 3D morphological models representative of the microstructures of UO_2 ceramics have been obtained. These models rely on a 2D characterization of the ceramics based on OM images. In order to confirm their relevance, a validation based on XRCT images, providing 3D informations on the microstructures, has been proposed here. This 3D characterization validates a posteriori the 2D characterization carried out to date, as well as the proposed 3D morphological model, at least for some of the selected morphological parameters and for the smaller scales. Larger scales are not accessible with XRCT images, because of the limited size of the samples that can be imaged.

Acknowledgements This study was funded by the CEA and École des ponts (Grant 5765). The microstructures and the FFT computations were produced with version 1.0 of the MEROPE software and version 1.8 of the TMFFT software (components of the PLEIADES platform), respectively. The XRCT investigations were run in April 2022 at Synchrotron Soleil in the context of proposal 20211112. The authors are grateful to M. Soulon from CEA, who prepared the core samples as well as to V. Bouineau from CEA and P. Aimedieu, J. Archez, C. Doré-Ossipyan and N. Du from Laboratoire Navier for their contributions to the X-ray tomography campaign.

References

1. Meynard, J., Ibrahim, M., Monnier, A., Bornert, M., Castelier, E., et al.: Effective properties of an isotropic solid weakened by micro-cracks located at inter-granular boundaries. Journal of the Mechanics and Physics of Solids (2022). https://doi.org/10.1016/j.jmps.2021.104647
2. Moutin, L., Meynard, J., Josien, M., Bornert, M., et al.: Realistic morphological models of weakly to strongly branched pore networks for the computation of effective properties. International Journal of Solids and Structures (2023). https://doi.org/10.1016/j.ijsolstr.2023.112249
3. Bargmann, S., Klusemann, B., Markmann, J., Schnabel, J.E., Schneider, K., Soyarslan, C., Wilmers, J.: Generation of 3D representative volume elements for heterogeneous materials: A review. Progress in Materials Science **96**, 322–384 (2018)
4. Jeulin, D.: Random texture models for material structures. Statistics and Computing (2000). https://doi.org/10.1023/A:1008942325749
5. Bilger, N., Auslender, F., Bornert, M., Michel, J.-C., Moulinec, H., Suquet, P., Zaoui, A.: Effect of a nonuniform distribution of voids on the plastic response of voided materials: a computational and statistical analysis. International Journal of Solids and Structures **42**(2), 517–538 (2005)
6. Gasnier, J.B., Willot, F., Trumel, H., Jeulin, D., Besson, J.: Thermoelastic properties of microcracked polycrystals. Part I: Adequacy of Fourier-based methods for cracked elastic bodies. International Journal of Solids and Structures **155**, 248–256 (2018)
7. Bortolussi, V., Figliuzzi, B., Willot, F., Faessel, M., Jeandin, M.: Morphological modeling of cold spray coatings. Image Analysis and Stereology **37**(2), 145–158 (2018)
8. Abrams, H.: Grain size measurement by the intercept method. Metallography (1971). https://doi.org/10.1016/0026-0800(71)90005-X
9. Matheron, G.: Éléments pour une théorie des milieux poreux. Masson, Paris (1967)
10. Sanchez, S., Fernandez, V., Pierce, S., Tafforeau, P.: Homogenization of sample absorption for the imaging of large and dense fossils with synchrotron microtomography. Nature Protocols (2013). https://doi.org/10.1038/nprot.2013.098

Collective Motion of Epithelial Cells in Tissues and the Active Vertex Model

Luis L. Bonilla

Abstract Epithelial tissues consist of cells attached to each other by junctions that may move collectively. We review results obtained by numerical simulations of vertex models in which cells are polygons that tile a region of the plane. In particular, we present results for wound healing assays, in which cells migrate to an empty space, and for antagonistic migration assays, in which cells from one tissue invade an adjacent tissue comprising cells of different type. We discuss coarse-graining the active vertex model to obtain continuum equations at macroscopic scales and review a selection of results on the anisotropy transition of homogeneous phases, their stability and the effects of shear.

Keywords Soft matter · Epithelial tissue · Active vertex model · Continuum model equations · Coarse-graining · Imperfect bifurcation

1 Introduction

The collective motion of cells is very important for physiological processes, such as organ growth and regeneration, tissue repair, wound healing [1–6] and morphogenesis [1, 7–9], as well as pathological processes like cancer [10–19] and retinopathies [20]. In tissues, the collective motion of epithelial cells, which are attached to each other by cell junctions, has been extensively investigated. Considered as materials, live tissues are very complex because they are made out of active cells that consume energy and exert forces between themselves and onto their surroundings [1–5]. In a tissue, cell junctions between individual cells are formed through specific transmembrane protein complexes (cadherins). The cell junctions physically link the actomyosin cortices of neighboring cells, enabling force transmission between them [1]. Changes of cell shapes, cellular divisions, rearrangements, and extrusions

L. L. Bonilla (✉)
Gregorio Millán Institute for Fluid Dynamics, Nanoscience and Industrial Mathematics and Department of Mathematics, Universidad Carlos III de Madrid, 28911 Leganés, Spain
e-mail: bonilla@ing.uc3m.es

F. Willot et al. (eds.), *Continuum Models and Discrete Systems*,
Springer Proceedings in Mathematics & Statistics 457,
https://doi.org/10.1007/978-3-031-58665-1_5

cause large-scale deformations of tissues. Within mechanobiology, the links between cellular processes, collective cellular motion and tissue deformations have been the subject of extensive theoretical and experimental investigations [5, 21–25]. Collective cell migration may exhibit fluid, solid or glass behavior with flocking and jamming/unjamming transitions that play an important role in metastasis [17, 18, 26]. In Sects. 2 and 3 of this chapter, we describe simple vertex models of confluent motion in monolayers of epithelial cells related to wound healing and invasion by cancerous cells [15], and the rigidity transition [27, 28] and excitability [29]. In Sect. 4, we review macroscopic continuum descriptions derived by coarse graining vertex models and consider the effects of shear on homogeneous phases [5, 30].

2 Active Vertex Model

The image of a *Drosophila* wing in Fig. 1a shows cell junctions visualized by fluorescent labeling. This motivates treating cells as polygons in a tesselation of the plane, as in Fig. 1b obtained by simulating the active vertex model (AVM). The vertex model considers the two-dimensional (2D) tissue as a mesh comprising polygons, edges and vertices and defines an energy which can be written in terms of the vertices. The configuration of the tissue is the one that minimizes the tissue energy. Active models introduce dynamics different from energy minimization to move the centers of the polygons, i.e., the cells. One possibility, the self-propelled Voronoi model, considers the Voronoi polygonal tesselation of the plane with the centers of the cells as seeds. The negative gradients of the tissue energy are nonlocal forces that, supplemented by active forces, define the dynamics of the cells. At each time step, the cell centers move according to the dynamics and then the Voronoi tesselation is recalculated [26]. The active vertex model considers the cell centers as the vertices of triangles of the Delaunay triangulation, which is the dual of the Voronoi tesselation. The cell centers

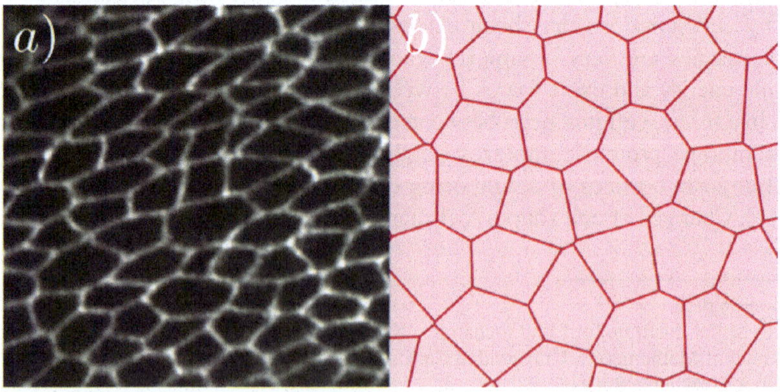

Fig. 1 **a** Apical view of the wing imaginal disc epithelium in the Drosophila embryo. **b** Snapshot of a two dimensional cell monolayer modelled by the active vertex model. Taken from [30]

move according to the appropriate dynamics, which may include negative gradients of the energy, active forces, and so on [31]. The Voronoi tesselation automatically follows its dual Delaunay triangulaltion and does not need to be calculated at each time step.

2.1 *Free Energy Function*

The properties and interactions of the cells in the tissue are implemented in the AVM by defining a free energy function that typically reads [26, 31]

$$\mathcal{F} = \sum_{\alpha=1}^{N} \left[\frac{\kappa_\alpha}{2}(A_\alpha - A_0^\alpha)^2 + \frac{\Gamma_\alpha}{2} P_\alpha^2 \right] + \sum_{<\mu\nu>} \Lambda_{\mu\nu} l_{\mu\nu}, \tag{1a}$$

$$\mathcal{F} = \frac{1}{2} \sum_{\alpha=1}^{N} \left[\kappa (A_\alpha - A_0)^2 + \Gamma (P_\alpha - P_0)^2 \right]. \tag{1b}$$

Here each cell is labeled by $\alpha = 1, \ldots, N$, and each vertex pair that shares a junction is designated by μ, ν. The first term implements an area elasticity, where $\kappa_\alpha > 0$ is the area modulus and $A_\alpha$is the cell area about its target value A_0^α. The second term is the perimeter contribution, where $\Gamma_\alpha > 0$ is the resistance of the cell to changing its perimeter P_α. This second term is absent in the earlier vertex models for the physics of foams [32]. In the last term, $\Lambda_{\mu\nu} = \gamma_c - \frac{\omega}{2} < 0$ is the line tension of the cell junctions of length $l_{\mu\nu}$ that results from the cortical tension γ_c along the contacts between cells and the cell-cell adhesion energy ω. The model allows for different cells to have different values of moduli κ_α, $\Lambda_{\mu\nu}$, and target areas if different cell types coexist. If there is only one type of cell, $\Lambda_{\mu\nu} = \Lambda$ is a constant and we can form a target perimeter $P_0^\alpha = -\Lambda/\Gamma_\alpha$. Then the two last terms of Eq. (1a) yield the second term in Eq. (1b). The shape index $p_0^\alpha = P_0^\alpha/\sqrt{A_0^\alpha}$ characterizes the ratio of the cell perimeter to the square root of its area. A critical shape index $p_0^* = 3.812$ separates solid-like and fluid-like behavior of the tissue [33]. With overdamped dynamics, the positions of all vertices change in the vertex model as the cells try to minimize the energy function for a given set of parameter values.

2.2 *Dynamics and Wound Healing and Antagonistic Migration Assays*

In the standard implementation of the AVM, the cells in the monolayer satisfy the following overdamped equations of motion [31]

$$\zeta \dot{\boldsymbol{r}}_\alpha = f_a \boldsymbol{n}_\alpha + \boldsymbol{F}_\alpha + \nu_\alpha, \tag{2a}$$

$$\zeta^r \dot{\theta}_\alpha = \boldsymbol{\tau}_\alpha \cdot \boldsymbol{N}_\alpha + \nu^r_\alpha. \tag{2b}$$

The unknowns $\boldsymbol{r}_\alpha$ and θ_α are, respectively, the positions of the centers of mass and the orientations of the cell polarity vector $\boldsymbol{n}_\alpha = (\cos\theta_\alpha, \sin\theta_\alpha)$ that marks the direction of motion. In Eq. (2a), $f_a \boldsymbol{n}_\alpha$ are active self-propulsion forces along $\boldsymbol{n}_\alpha$. $\boldsymbol{F}_\alpha$ are gradient forces arising from the free energy function, Eq. (1a), and ν_α and ν^r_α are stochastic forces (usually Gaussian white noises) representing random noises originating from intercellular processes and the environment. ζ and ζ^r are friction coefficients. In Eq. (2b), $\boldsymbol{\tau}_\alpha$ and $\boldsymbol{N}_\alpha$ are the torque (that acts on the vector $\boldsymbol{n}_\alpha$ and stems from cell-cell alignment models) and the normal unit vector to the cell monolayer (along the z-axis), respectively.

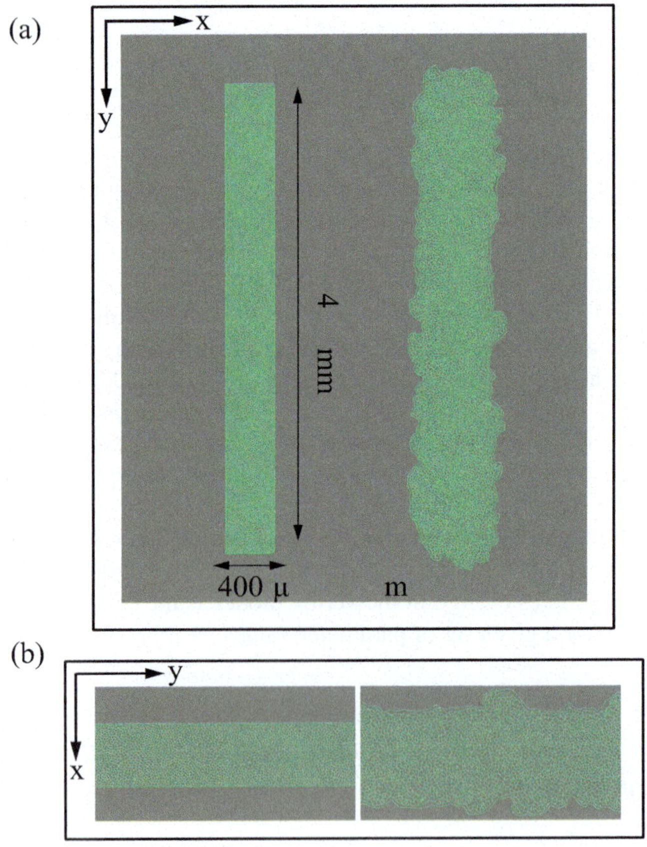

Fig. 2 Initial configuration and configuration after 20 h of stencil removal showing the formation of fingers according to the numerical simulation of the AVM. **a** Full view, **b** zoom. Initial box size is 1.6 mm^2, $P^0 = 10$, $A^0 = \pi$, and shape index $p^0 = 5.65$. Taken from [15]

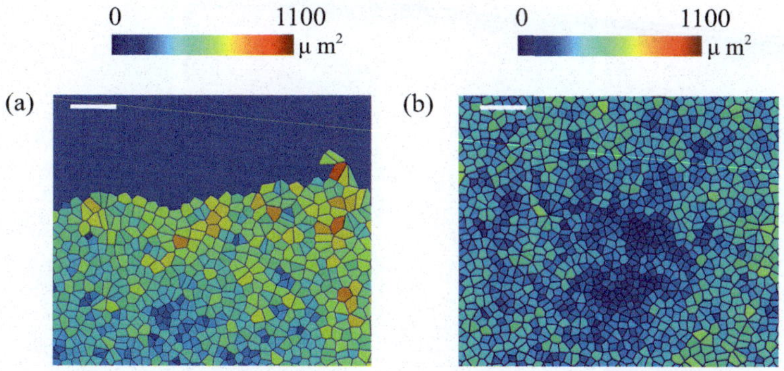

Fig. 3 Areas of cells during a simulation of a wound healing assay using the AVM with inertia: **a** Area of cells near the tissue boundary, **b** area of cells far from the boundary. Our simulations exhibit the same trend as measurements reported in Ref. [34]. Taken from [15]

In a wound healing assay, cells cultured in appropriate media are seeded in a culture insert placed on glass bottom plates. Cells are left to incubate overnight until they form a monolayer of cells sharing edges that is similar to Fig. 1a. Then the culture insert is removed and the cells migrate to the empty spaces forming a confluent tissue [14] and, typically, fingers form at its boundaries; see Fig. 2. It turns out that cells with larger area are far from the tissue boundaries in simulations of the AVM with dynamics given by Eqs. (2) whereas the opposite occurs in experiments [34]. However, numerical simulations of the AVM with dynamics that incorporates inertia as in the equations

$$\dot{\mathbf{r}}_\alpha = \mathbf{v}_\alpha, \quad \dot{\mathbf{v}}_\alpha = -\zeta\mathbf{v}_\alpha + \sum_{\langle\mu,\alpha\rangle}\left[\frac{\beta}{n_\alpha}(\mathbf{v}_\mu - \mathbf{v}_\alpha) + \mathbf{f}_{\mu\alpha}\right] + \boldsymbol{\varphi}_\alpha + \sigma_0\boldsymbol{\nu}_\alpha(t), \tag{3}$$

field configurations where the bigger cells are at the interface of the tissue with the extracellular domain as shown in Fig. 3. In Eq. (3), the sum is over the nearest neighbors of the vertex α of the Delaunay triangulation, n_α is the number of these neighbors, and ζ is a friction coefficient. The term containing the coefficient β tries to synchronize the velocity of the nearest neighbor cells that of the αth cell. $\mathbf{f}_{\alpha\mu} = -\nabla_{\mathbf{r}_\alpha}\mathcal{F}$ is the force per unit mass exerted by cell μ on cell α, $\boldsymbol{\varphi}_\alpha$ are active forces and $\sigma_0\boldsymbol{\nu}_\alpha(t)$ are white noises [15]. Inertia in Eq. (3) is supposed to be a collective effect because the mass of the cells is very small. Whether other mechanisms have the same effect of producing bigger cells in the interface remains to be explored.

In an antagonistic migration assay two populations of different cells (e.g., normal and transformed precancerous cells of the same type in the experiments of [14]) are cultured in adjacent wells separated by a stencil until they attach to each other and form a monolayer. Then the separation stencil is removed at the initial time. The cells move to occupy the empty space between them and collide with each other [14]. Figure 4 shows four snapshots of a simulation of the AVM with dynamics given

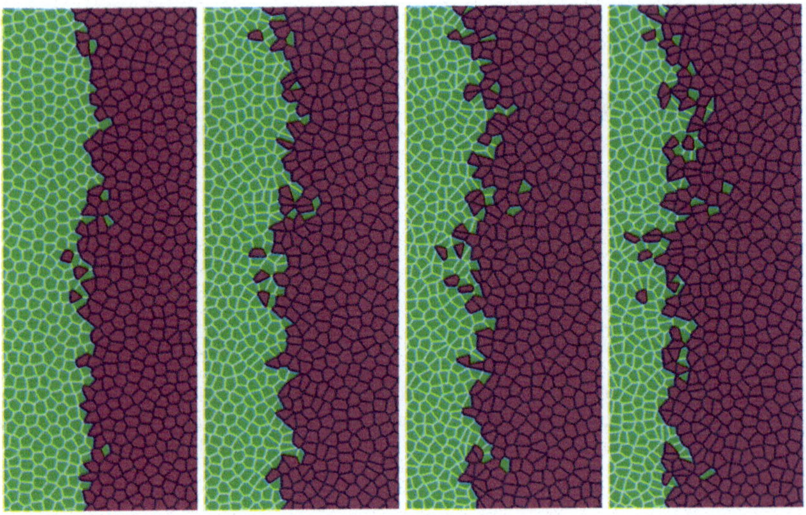

Fig. 4 Four snapshots of the simulation of an antagonistic migration assay using the AVM with inertia. The population of modified cells pushes back the population of normal cells while the interfaces mix, creating scattered islands of cells of different type. Taken from [15]

by Eq. (3) and appropriate values of the parameters that make transformed cells to be fluid-like whereas normal cells are solid-like [15]. The population with shape index above its critical value invades the space of the other (with subcritical shape index) and pushes it back creating a rough interface with island of one population inside the other. Topological data analysis allows to automatically group, track and classify the advancing interfaces of cellular aggregates obtained from experiments and from numerical simulations of the AVM [15].

3 Shear, Nonlinear Elasticity and Excitability

The behavior of epithelial tissue as described by vertex models exhibits interesting features if the tissue is under shear strain [27, 28] or the cell junctions have an intrinsic relaxation time [29].

Recently the response of a cellular monolayer comprising cells with homogeneous single cell parameters to shear has been studied by simulating the Voronoi vertex model with Lees-Edwards shear boundary conditions (flow direction along x, shear gradient along y) [27]. In this work, the shear strain γ is increased in small increments $\Delta\gamma$, while cell center positions are subject to an affine displacement $\Delta\mathbf{r}_\alpha = \Delta\gamma(y_\alpha, 0)$. Following each strain step, the free energy of Eq. (1b) is relaxed until the forces $\mathbf{f}_{\alpha\mu} = -\nabla_{\mathbf{r}_\alpha}\mathcal{F}$ are vanishingly small. For a simulation box size L, the shear stress is $\sigma = L^{-2}\sum_{\mu<\nu} T^x_{\mu\nu} l^y_{\mu\nu}$, where $\mathcal{T}_{\mu\nu} = \nabla_{\mathbf{l}_{\alpha\beta}}\mathcal{F} =$

$\Gamma[(P_\mu - P_0) + (P_\nu - P_0)]\mathbf{l}_{\mu\nu}/|\mathbf{l}_{\mu\nu}|$ is the line tension vector and $\mathbf{l}_{\mu\nu}$ is the vector of the junction $l_{\mu\nu}$ in Eq. (1a). Numerical simulations show that there is a solid-like to liquid-like transition at shape index $p_0^* = 3.812$ for zero shear γ. The shear modulus $G = \partial\sigma/\partial\gamma$ is zero for $p_0 > p_0^*$. For nonzero shear stress and supercritical shape index, $G = 0$ (liquid-like response) for $0 < \gamma < \gamma_C(p_0)$, and $G > 0$ for $\gamma > \gamma_C(p_0)$ (solid-like behavior). Thus, $\gamma = \gamma_C(p_0)$ marks a shear-induced rigidity transition. It turns out that the shear modulus G depends on true strain (defined through simulation results), stress and shape index but only two of these parameters are independent, so that, in the end, G depends on stress and shape index [27]. An *ad hoc* set of continuum partial differential equations (PDEs) for the cell perimeter, the cell orientation tensor and a plasticity variable produces the same phenomenology [28].

As explained in Sect. 2, inertia is key to explain why the area of boundary cells is larger than that of inner cells in a wound healing experiment. Given the small mass of cells, the origin of the collective inertia in Eq. (3) is unclear. Recently, the effect of including an intrinsic relaxation time for the cell junction in the vertex model has been explored [35]. This could be analogous to having a collective inertia although this needs to be more thoroughly explored. Experiments using optogenetic control of the cell junction length suggest that the rest length of the junction responds elastically to small deformations but it is permanently changed for strains beyond a certain threshold [35]. This means that the rest cell junction relaxes proportionally to a piecewise linear function of the strain $(l_{\mu\nu}^0 - l_{\mu\nu})/l_{\mu\nu}^0$ [35]. Combining this idea with the definition of inactive, active and refractory cell junctions, it is possible to attain pulses, traveling waves and spiral waves as solutions of an AVM [29]. These results show that having an intrinsic time scale for the relaxation of rest junction length of cells converts the epithelial tissue described by a vertex model into an excitable system.

4 Continuum Equations

Several successful continuum models of collective cell migration in tissues describe supracellular length and time scales and are more amenable to analysis than their subcellular and cellular counterparts. Their dynamics are governed by PDEs for fields such as average density, velocity, polarity, and nematic order parameter [5]. Systematically deriving these continuum models from subcellular and cellular models (e.g., vertex models) is an important but insufficiently explored task. In this section, we review coarse graining procedures that utilize Poisson brackets [30, 36].

Let us consider a fluid formed by deformable polygonal-shaped particles that is governed by the free energy of the AVM, Eq. (1a). Here we do not include active forces in our equations, which therefore will only reflect the effects of the fluid flow on cell shape and vice versa. On macroscopic scales, this fluid is described by coarse-grained hydrodynamic fields ϕ^a, which are the mass density ρ, momentum density $\mathbf{g}$ and the cell shape density tensor G. The latter accounts for the shape and elongation of the cells. The microscopic versions of the hydrodynamic fields are [36]

$$\hat{\phi}^\alpha(\mathbf{r},t)=\sum_{\alpha\mu}\phi^\alpha\delta(\mathbf{r}-\mathbf{r}^{\alpha\mu}(t)),\quad \phi^\alpha=m^\alpha,\mathbf{g}^\alpha,\quad G^\alpha_{ij}=\frac{1}{n}\sum_{\mu=1}^{n}\Delta x_i^{\alpha\mu}\Delta x_j^{\alpha\mu}. \tag{4}$$

Here $m^\alpha=1$ is the mass of cell α and $\mathbf{r}^{\alpha\mu}$ the position of vertex μ. The momentum of cell α is given by $\mathbf{g}^\alpha=m^\alpha\dot{\mathbf{r}}^\alpha$ and the shape tensor as specified via the vertices, with $\Delta x_i^{\alpha\mu}=\mathbf{r}^{\alpha\mu}-\mathbf{r}^\alpha$, where $\mathbf{r}^\alpha=\sum_\mu \mathbf{r}^{\alpha\mu}/n$ is the center of mass of the cell and Latin indices denote components. Note that the mass and momentum density fields are defined through the positions of the vertices in the delta function, while the shape tensor uses delta functions centered at the cell centers. Given microscopic Hamilton dynamics with Hamiltonian $\hat{\mathcal{H}}(\mathbf{r}^\alpha,\mathbf{g}^\alpha)$, the evolution of a microscopic function $\hat{\phi}(\mathbf{r},t)$ satisfies

$$\frac{\partial\hat{\phi}}{\partial t}=\{\hat{\mathcal{H}},\hat{\phi}\}=\sum_\alpha \frac{\partial\hat{\mathcal{H}}}{\partial g_j^\alpha}\frac{\partial\hat{\phi}}{\partial r_j^\alpha}-\frac{\partial\hat{\phi}}{\partial g_j^\alpha}\frac{\partial\hat{\mathcal{H}}}{\partial r_j^\alpha}. \tag{5}$$

The coarse-grained equation have a reactive part that comes directly from averaging the free energy and replacing the Poisson brackets in Eq. (5) by Poisson brackets of macroscopic fields, and a dissipative part that comes from interactions with the environment and consideration of other processes not included in that equation. Given a coarse-grained free energy, $\mathcal{F}$, which depends on the macroscopic fields $\phi^a=\rho,\mathbf{g},G$, the coarse-grained equations are [36]

$$\frac{\partial\phi^a}{\partial t}=\{\mathcal{F},\phi^a\}-\Gamma^{ab}\frac{\delta\mathcal{F}}{\delta\phi^b}, \tag{6a}$$

$$\{\mathcal{F},\phi^a\}=-\int d^2x'\mathcal{P}_{ab}(\mathbf{r},\mathbf{r}')\frac{\delta\mathcal{F}}{\delta\phi^b(\mathbf{r}')}, \tag{6b}$$

$$\mathcal{P}_{ab}(\mathbf{r},\mathbf{r}')=\{\phi^a(\mathbf{r}),\phi^b(\mathbf{r}')\}=-\mathcal{P}_{ba}(\mathbf{r}',\mathbf{r}), \tag{6c}$$

where summing over repeated indices is intended and we have dropped the time variable for now. The term $\{\mathcal{F},\phi^a\}$ in Eq. (6b) yields the reactive part of the evolution equations and the other term yields the dissipative part. The latter is proportional to the generalized forces, as defined by variations of the free energy, times kinetic coefficients Γ^{ab} that obey Onsager reciprocity relations near equilibrium [37]. The coarse-grained Poisson brackets $\{\phi^a,\phi^b\}$ have been calculated in Ref. [36].

For n-sided polygonal cells, the coarse-grained free energy can be split in a kinetic and a potential part , f_K and f_V, respectively:

$$\mathcal{F}(\phi^a)=\int d^2x(f_K+f_V), \tag{7a}$$

$$f_K=\sum_{i=1}^{2}\frac{g_i^2}{2\rho}=\sum_{i=1}^{2}\frac{1}{2}\rho v_i^2,\quad f_V=f_{\rm sc}+f_{\rm int}, \tag{7b}$$

$$f_{\text{int}} = \frac{K_G}{2}\left(\frac{\partial G_{kl}}{\partial x_i}\right)^2, \tag{7c}$$

$$f_{\text{sc}} = \frac{\kappa}{2}(\mu\sqrt{\det(G)} - A_0)^2 + \frac{\Gamma\nu^2}{2}\text{Tr}(G) + \Lambda\nu\sqrt{\text{Tr}(G)}, \tag{7d}$$

$$A_\alpha = \mu\sqrt{\det(G)}, \quad P_\alpha = \nu\sqrt{\text{Tr}(G)}, \tag{7e}$$

$$\mu = \frac{n}{2}\sin\left(\frac{2\pi}{n}\right), \quad \nu = n\sqrt{2}\sin\left(\frac{\pi}{n}\right). \tag{7f}$$

Here f_{sc} is the vertex free energy density obtained from Eq. (1a) with cell area and perimeter given by Eq. (7e) in terms of the coarse-grained shape tensor of components G_{ij}. We write the shape tensor as

$$G = M_0 e^{c\Theta} = M_0(\cosh c\, I + \sinh c\, \Theta), \tag{8a}$$

where M_0 and c are scalar fields, I is the identity and Θ is the traceless symmetric tensor field

$$\Theta = \begin{pmatrix} \cos 2\theta & \sin 2\theta \\ \sin 2\theta & -\cos 2\theta \end{pmatrix}. \tag{8b}$$

Note that $M_0^2 = \det(G)$ and $\Theta^2 = I$ hold. The field c measures the anisotropy of the cells, as $c = 0$ yields an isotropic shape tensor with $\text{Tr}(G) = 2\sqrt{\det(G)}$. The angle θ in Eq. (8b) determines the direction of the eigenvector corresponding to the largest eigenvalue of G [30]. In terms of $R = \text{Tr}(G)$, we can rewrite Eq. (8a) as

$$G = \frac{R}{2}(I + \tanh c\, \Theta), \quad \tilde{G} = \frac{R}{2}\tanh c\, \Theta, \tag{8c}$$

where the traceless tensor $\tilde{G}$ is the deviatoric part of the shape tensor. When $c \neq 0$, the coarse-grained cells are elongated ellipses that can be visualized by their long axis directed along the angle θ. The coarse-grained shape tensor has a form similar to the order parameter of nematic liquid crystals [38]. Thus, the fields R, $R\tanh c$ and θ describe the cell perimeter, the shape anisotropy and the director angle, respectively. We can rewrite the vertex free energy density Eq. (7d) in the following simple form:

$$f_{\text{sc}} = \frac{\kappa}{2}\left(\frac{\mu R}{2\cosh c} - A_0\right)^2 + \frac{\Gamma\nu^2 R}{2} + \Lambda\nu\sqrt{R}. \tag{9}$$

4.1 *Thermodynamic Stability*

The minima of the single-cell free energy, Eq. (9), correspond to stable homogeneous phases that solve

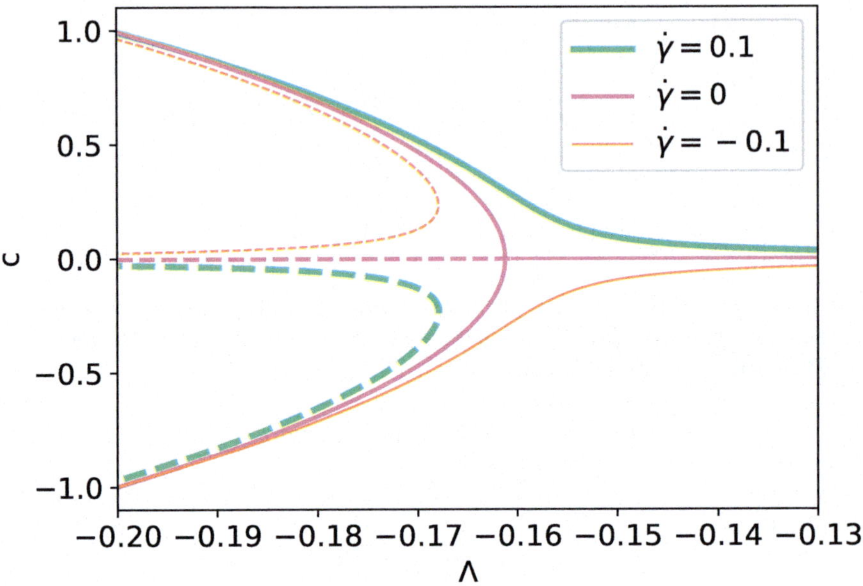

Fig. 5 Bifurcation diagrams for homogeneous phases in the presence of a shear flow $\dot{\gamma}$. The imperfect pitchfork bifurcation due to a shear flow is depicted on top of the pitchfork bifurcation represented by pink curves. Thick (green)/thin (orange) curves correspond to positive/negative shear flow, respectively. Continuous (stable) and dashed (unstable) curves follow from the linear stability analysis of [30]. Taken from [30]

$$\frac{\partial f_{sc}}{\partial R} = \frac{\kappa\mu}{2\cosh c}\left(\frac{\mu R}{2\cosh c} - A_0\right) + \frac{\nu}{2}\left(\Gamma\nu + \frac{\Lambda}{\sqrt{R}}\right) = 0, \tag{10a}$$

$$\frac{\partial f_{sc}}{\partial c} = -\frac{\kappa\mu R\sinh c}{2\cosh^2 c}\left(\frac{\mu R}{2\cosh c} - A_0\right) = 0. \tag{10b}$$

Solutions to Eq. (10a) exist only if the line tension Λ is non-positive (negative for anisotropic phases). In turn, Eq. (10b) has isotropic solutions $c = 0$ and $\pm c$-symmetric anisotropic solutions only for $R > 2A_0/\mu$. Isotropic solutions correspond to solid-like states whereas the elongated cells in anisotropic phases have a lower energy cost to change neighbors, thereby moving more easily, and consequently correspond to liquid-like states. Anisotropic solutions have $A = A_0$ and hence from Eq. (10a) it can be deduced that $\sqrt{R} = |\Lambda|/(\Gamma\nu)$, i.e., $P = |\Lambda|/\Gamma$. At $R = 2A_0/\mu$ and $A = A_0$, $c = 0$, that corresponds to a bifurcation value of the line tension Λ from which anisotropic phases issue forth. The corresponding bifurcation diagram is depicted in Fig. 5 for $\dot{\gamma} = 0$. The isotropic phase $c = 0$ is stable for $\Lambda > \Lambda^*$ whereas the anisotropic phases that bifurcate at Λ^* are stable for $\Lambda < \Lambda^*$. According to Eqs. (10), $|\Lambda^*|/(\Gamma\sqrt{A_0}) = P_0/\sqrt{A_0} = \sqrt{2\nu^2/\mu} = p_0^*$, and Λ^* corresponds to the critical shape index. Anisotropic homogeneous phases exist and are stable for $p_0 > p_0^*$ (liquid phase) whereas for $p_0 < p_0^*$ only the isotropic phase (solid phase) exists and is stable. Homogeneous phases are stable if and only if the Hessian matrix of f_{sc} is positive definite, which is equivalent to

$$\frac{\partial^2 f_{sc}}{\partial R^2} = \frac{\kappa\mu^2}{4\cosh^2 c} - \frac{\nu\Lambda}{4R^{\frac{3}{2}}} > 0\,, \quad \frac{\partial^2 f_{sc}}{\partial R^2}\frac{\partial^2 f_{sc}}{\partial c^2} - \left(\frac{\partial^2 f_{sc}}{\partial R \partial c}\right)^2 > 0\,, \tag{11a}$$

$$\frac{\partial^2 f_{sc}}{\partial c^2} = \frac{\kappa\mu R}{4\cosh^2 c}[2A_0 \cosh c(1 - 2\tanh^2 c) - \mu R(1 - 3\tanh^2 c)] > 0. \tag{11b}$$

4.2 *Coarse-Grained Hydrodynamic Equations and Shear Flow*

With the coarse-grained free energy (7), the coarse-grained Eqs. (6) become [30]

$$\frac{\partial \rho}{\partial t} + \nabla \cdot (\rho \mathbf{v}) = 0, \tag{12a}$$

$$\rho \frac{d}{dt} v_i = -\partial_i p + \partial_j (\sigma_{ij}^D + \sigma_{ij}^E + \sigma_{ij}^G), \tag{12b}$$

$$\frac{D}{Dt} G_{ij} = G_{ik} D_{kj} + D_{ik} G_{kj} - \Gamma_{ijkl} \frac{\delta \mathcal{F}}{\delta G_{kl}}, \tag{12c}$$

$$\Gamma_{ijkl} = \gamma_1 \frac{G_{ij} G_{kl}}{R^2} + \gamma_2 \frac{R^2 \Theta_{ij} \Theta_{kl}}{4\cosh^4 c} + \frac{\gamma_3}{8} \frac{\partial \Theta_{ij}}{\partial \theta} \frac{\partial \Theta_{kl}}{\partial \theta}. \tag{12d}$$

Here $\partial_i = \partial / \partial x_i$, and

$$\partial_i v_j = D_{ij} + \omega_{ij} + \frac{1}{2} \delta_{ij} \nabla \cdot \mathbf{v}, \tag{13a}$$

$$D_{ij} = \frac{1}{2}(\partial_i v_j + \partial_j v_i - \delta_{ij} \nabla \cdot \mathbf{v}), \tag{13b}$$

$$\omega_{ij} = \frac{1}{2}(\partial_i v_j - \partial_j v_i), \tag{13c}$$

$$\frac{d}{dt} = \frac{\partial}{\partial t} + \mathbf{v} \cdot \nabla, \quad \frac{D}{Dt} = \frac{d}{dt} - [\omega, \cdot], \tag{13d}$$

are the gradient of the average velocity, the deviatoric part of its symmetrization (the rate of strain tensor), the vorticity, the material derivative and the co-rotational derivative, respectively. In the latter, $[A, B]_{ij} = A_{ik} B_{kj} - B_{ik} A_{kj}$. The continuity equation (12a) does not contain a dissipative part. The pressure, and the different stress tensor contributions entering Eq. (12b), are [36]

$$p = \rho \frac{\delta \mathcal{F}_V}{\delta \rho} - f, \tag{14a}$$

$$\sigma_{ij}^D = 2\eta D_{ij} + \eta_b \delta_{ij} \nabla \cdot \mathbf{v}, \tag{14b}$$

$$\sigma_{ij}^E = -\frac{\partial f_{\rm int}}{\partial(\partial_j G_{kl})} \partial_i G_{kl} = -K_G \partial_i G_{kl} \partial_j G_{kl}, \tag{14c}$$

$$\sigma_{ij}^G = 2G_{jk} \frac{\delta \mathcal{F}_V}{\delta G_{ik}} - \delta_{ij} G_{kl} \frac{\delta \mathcal{F}_V}{\delta G_{kl}}. \tag{14d}$$

For the sake of simplicity, in Eq. (14b) we assumed that the dissipative part of the stress tensor, σ_{ij}^D, is that of an isotropic fluid with shear and bulk viscosity coefficients η and η_b, respectively. In uniaxial anisotropic situations, one expects five instead of just these two viscosities [38]. The contributions σ_{ij}^E and σ_{ij}^G are reactive and the former corresponds to Ericksen stresses in liquid crystals [38]. For homogeneous phases, Eq. (12c) should be compatible with the thermodynamic stability of Sect. 4.1. Thus, for zero average velocity, Eq. (12c) is a gradient system with the phases as homogeneous solutions. The coefficient tensor Γ_{ijkl} in Eq. (12d) is such that the dynamic stability of the homogeneous phases is the same as the thermodynamic stability [30]. For homogeneous solutions under a shear flow $v_y = 0$, $v_x = \dot{\gamma} y$, the coarse-grained equations are [30]

$$\begin{aligned} \dot{R} &= \dot{\gamma} R \tanh c \sin 2\theta - \gamma_1 \frac{\partial f_{\rm sc}}{\partial R} \\ &= \dot{\gamma} R \tanh c \sin 2\theta - \frac{\gamma_1}{2}\left[\frac{\kappa\mu}{\cosh c}\left(\frac{\mu R}{2\cosh c} - A_0\right) + \frac{\Lambda \nu}{\sqrt{R}} + \Gamma \nu^2\right], \end{aligned} \tag{15a}$$

$$\dot{c} = \dot{\gamma} \sin 2\theta - \gamma_2 \frac{\partial f_{\rm sc}}{\partial c} = \dot{\gamma} \sin 2\theta + \gamma_2 \frac{\kappa \mu R \sinh c}{2\cosh^2 c}\left(\frac{\mu R}{2\cosh c} - A_0\right), \tag{15b}$$

$$\dot{\theta} = \frac{\dot{\gamma}}{2}\left(1 + \frac{\cos 2\theta}{\tanh c}\right). \tag{15c}$$

Equation (15c) has a stationary solution with $c \le 0$ defined on $\frac{-\pi}{4} < \theta \le \frac{\pi}{4}$, and another with $c > 0$ on $\frac{\pi}{4} < \theta < \frac{3\pi}{4}$. Insertion of these values into Eqs. (15a) and (15b) yields the stationary solutions under shear flow. Their analysis shows that these solutions modify the bifurcation diagram of the homogeneous phases as depicted in Fig. 5: $\dot{\gamma}$ acts as a small imperfection of the pitchfork bifurcation diagram for $\dot{\gamma} = 0$. Performing a linear stability analysis of solutions under shear flow including space dependent terms in the governing equations, it tuns out that the stationary solutions with $\dot{\gamma} c > 0$ (corresponding to continuous solution branches existing for all values of the line tension Λ in Fig. 5) are always stable [30]. In contrast, the branches issuing from the turning points in Fig. 5 are unstable for wavenumbers on the interval $0 < k^2 < -\dot{\gamma}/(\gamma_3 K_G c)$. They will produce spatially non-homogeneous solutions when the size of the tissue exceeds a critical value. Here the kinetic coefficient γ_3 plays a stabilizing role [30].

5 Final Remarks

We have presented simulation results of the active vertex model under wound healing (formation of fingers) and antagonistic migration assays with a dynamic law that includes inertia [15]. Huang et al. [27] solved the vertex model with Lees-Edwards shear boundary conditions and found that there is a rigidity transition for shape indices corresponding to liquid-like behavior without shear. Later Fielding et al. [28] interpreted these simulation results using phenomenological continuum PDEs. Using space dependent target area and perimeter, the vertex model describes inhomogeneous patches with apical contractions of interest in morphogenesis [39].

To study epithelial tissues at macroscopic equations, we should derive coarse-grained continuum equations from microscopic models such as self-propelled and active vertex models. We have presented one such derivation using Poisson brackets and selected the Onsager coefficient tensor so that the stability of homogeneous phases under zero fluid flow coincides with their thermodynamic stability [30]. The resulting equations are the usual hydrodynamic equations coupled with equations for the trace of the shape tensor R, the anisotropic field c and the director θ. It would be interesting to analyze the rigidity transition under shear using these equations instead of the phenomenological model of Ref. [28]. In Eqs. (12), the deviatoric part of the shape tensor is similar to the nematic order parameter of liquid crystals [38]. In a relevant work in this regard, Hadjifrangiskou et al. have studied active nematics near the anisotropic transition in cells. They use hydrodynamic equations for order parameter and an equation for the shape distribution with a double well free energy. Their results include a linear stability analysis about the quiescent isotropic state, active turbulence from simulations, calculations of the average aspect ratio, and the number of topological defects depending on activity [40]. It would be worthwhile to study whether these phenomena can also be captured by including active forces in our coarse-grained equations (12).

Acknowledgements I am indebted to Ana Carpio, Carolina Trenado, Gloria Triguero and Falko Ziebert for useful discussions and collaborations in the topics of the present paper. This work has been supported by the FEDER/Ministerio de Ciencia, Innovación y Universidades—Agencia Estatal de Investigación grant PID2020–112796RB–C22, by the Madrid Government (Comunidad de Madrid, Spain) under the Multiannual Agreement with UC3M in the line of Excellence of University Professors (EPUC3M23), and in the context of the V PRICIT (Regional Programme of Research and Technological Innovation).

References

1. Gilbert, S.F., Barresi, M.J.F.: Developmental Biology, 11th edn. Sinauer Associates, Sunderland, MA (2018)
2. Hakim, V., Silberzan, P.: Collective cell migration: a physics perspective. Rep. Prog. Phys. **80**, 076601 (2017)

3. Trepat, X., Sahai, E.: Mesoscale physical principles of collective cell organization. Nat. Phys. **14**(7), 671–682 (2018)
4. Giavazzi, F., et al.: Flocking transition in confluent tissues. Soft Matter **14**, 3471–3477 (2018)
5. Alert, R., Trepat, X.: Physical models of collective cell migration. Ann. Rev. Cond. Matt. Phys. **11**, 77–101 (2020)
6. Poujade, M., et al.: Collective migration of an epithelial monolayer in response to a model wound. Proc. Natl. Acad. Sci. U. S. A. **104**, 15988–15993 (2007)
7. Hufnagel, L., et al.: On the mechanism of wing size determination in fly development. Proc. Natl. Acad. Sci. U. S. A. **104**, 3835–3840 (2007)
8. Lecuit, T., Lenne, P.F., Munro, E.: Force generation, transmission, and integration during cell and tissue morphogenesis. Annu. Rev. Cell Dev. Biol. **27**, 157–184 (2011)
9. Goodwin, K., Nelson, C.M.: Mechanics of Development. Dev. Cell **56**, 240–250 (2021)
10. Abercrombie, M., Ambrose, E.: Surface Properties of Cancer Cells. Cancer Res. **22**, 525–548 (1962)
11. Friedl, P., Gilmour, D.: Collective cell migration in morphogenesis, regeneration and cancer. Nature Rev. Mol. Cell Biol. **10**, 445–457 (2009)
12. Ben-Jacob, E., Coffey, D.S., Levine, H.: Bacterial survival strategies suggest rethinking cancer cooperativity. Trends Microbiol. **20**, 403–410 (2012)
13. Streitberger, K.-J., et al.: How tissue fluidity influences brain tumor progression. Proc. Natl. Acad. Sci. U. S. A. **117**, 128–134 (2020)
14. Moitrier, S., et al.: Collective stresses drive competition between monolayers of normal and Ras-transformed cells. Soft Matter **15**, 537–545 (2019)
15. Bonilla, L.L., Carpio, A., Trenado, C.: Tracking collective cell motion by topological data analysis. PLoS Comput. Biol. **16**(12), e1008407 (2020)
16. Flegg, J.A., Menon, S.N., Byrne, H.M., McElwain, D.L.S.: A current perspective on wound healing and tumour-induced angiogenesis. Bull. Math. Biol. **82**, 23 (2020)
17. Grosser, S., et al.: Cell and nucleus shape as an indicator of tissue fluidity in carcinoma. Phys. Rev. X **11**, 011033 (2021)
18. Gottheil, P., et al.: State of cell unjamming correlates with distant metastasis in cancer patients. Phys. Rev. X **13**, 031003 (2023)
19. Kechagia, Z., et al.: The laminin-keratin link shields the nucleus from mechanical deformation and signalling. Nat. Mater. **22**(11), 1409–1420 (2023)
20. Vega, R., Carretero, M., Bonilla, L.L.: Anomalous angiogenesis in retina. Biomedicines **9**, 224 (2021)
21. Mayor, R., Etienne-Manneville, S.: The front and rear of collective cell migration. Nat. Rev. Mol. Cell Biol. **17**, 97–109 (2016)
22. Stramer, B., Mayor, R.: Mechanisms and in vivo functions of contact inhibition of locomotion. Nat. Rev. Mol. Cell Biol. **18**, 43–55 (2017)
23. Ladoux, B., Mège, R.-M.: Mechanobiology of collective cell behaviours. Nat. Rev. Mol. Cell Biol. **18**, 743–757 (2017)
24. Xi, W., et al.: Material approaches to active tissue mechanics. Nat. Rev. Mater. **4**, 23–44 (2019)
25. Löber, J., Ziebert, F., Aranson, I.S.: Collisions of deformable cells lead to collective migration. Sci. Rep. **5**, 9172 (2015)
26. Bi, D., Yang, X., Marchetti, M.C., Manning, L.: Motility-driven glass and jamming transitions in biological tissues' Phys. Rev. X **6**, 021011 (2016)
27. Huang, J., et al.: Shear-driven solidification and nonlinear elasticity in epithelial tissues. Phys. Rev. Lett. **128**, 178001 (2022)
28. Fielding, S.M., et al.: Constitutive model for the rheology of biological tissue. Phys. Rev. E **108**, L042602 (2023)
29. F. Pérez-Verdugo, S. Banks, S. Banerjee, Excitable dynamics driven by mechanical feedback in biological tissues, Commun. Phys. **7**, 167 (2024)
30. Triguero-Platero, G., Ziebert, F., Bonilla, L.L.: Coarse-graining the vertex model and its response to shear. Phys. Rev. E **108**, 044118 (2023)

31. Barton, D.L., Henkes, S., Weijer, C.J., Sknepnek, R.: Active Vertex Model for Cell-Resolution Description of Epithelial Tissue Mechanics. PLoS Comput. Biol. **13**, e1005569 (2017)
32. Weaire, D.L., Hutzler, S.: The Physics of Foams. Oxford U.P, Oxford UK (1999)
33. Bi, D., Lopez, J.H., Schwarz, J.M., Manning, M.L.: A density-independent rigidity transition in biological tissues. Nat. Phys. **11**, 1074–1079 (2015)
34. Lv, J.-Q., Chen, P.-C., Góźdź, W.T., Li, B.: Mechanical adaptions of collective cells nearby free tissue boundaries. J. Biomech. **104**, 109763 (2020)
35. Staddon, M.F., et al.: Mechanosensitive junction remodeling promotes robust epithelial morphogenesis. Biophysical Journal **117**, 1739–1750 (2019)
36. Hernandez, A., Marchetti, M.C.: Poisson-bracket formulation of dynamics of fluids of deformable particles. Phys. Rev. E **103**, 032612 (2021)
37. De Groot, S.R., Mazur, P.: Non-Equilibrium Thermodynamics. Dover, N.Y. (1984)
38. De Gennes, P.G., Prost, J.: The Physics of Liquid Crystals. Oxford University Press, Oxford UK (1993)
39. Pérez-Verdugo, F., Soto, R.: Continuum description of confluent tissues with spatial heterogeneous activity. Soft Matter **19**, 6501–6512 (2023)
40. Hadjifrangiskou, I., Ruske, L.J., Yeomans, J.M.: Active nematics with deformable particles. Soft Matter **19**, 6664–6670 (2023)

Structure of Fields in Extremal 2D Conducting Multimaterial Composites

Andrej Cherkaev

Abstract The paper discusses a long-standing problem of exact bounds for the effective properties of multimaterial composites. The suggested bounds refine Hashin-Shtrikman's bounds in the region of parameters where the last ones are loose. They are derived by the description of gradient fields in the components of optimal structures. We show that the gradient fields vary in restricted domains; taking this into account, we modify the Translation method to obtain new exact bounds. These bounds are multifaced: the active or passive inequality constraints lead to algebraically different expressions and topologically different optimal structures. The bounds for the energy and optimal structures are explicitly computed for three-material conducting and elastic composites (one material is void). However, a unified approach to deriving a set of constraints in the supporting fields is yet to be developed.

Keywords Optimal composites · Multimaterial composites · Quasiconvex envelope · Compatibility

1 Introduction

The problem of the best composite structure is formulated as a problem of the quasiconvex envelope for a multiwell Lagrangian. While that envelope is explicitly found for many cases of two-well Lagrangians, see [7, 23], the technique for multiwell quasiconvex envelopes is still in its infancy. The paper discusses a method for constructing the envelopes for piece-wise quadratic Lagrangians that exploits inequalities on the supporting points (gradient fields in the mixed materials) that define the envelope. The inequalities require boundedness and the rank-one connectivity of the set of supports or gradient fields in an optimal structure.

A. Cherkaev (✉)
Department of Mathematics, University of Utah, Salt Lake City, UT, USA
e-mail: cherk@math.utah.edu

F. Willot et al. (eds.), *Continuum Models and Discrete Systems*,
Springer Proceedings in Mathematics & Statistics 457,
https://doi.org/10.1007/978-3-031-58665-1_6

The problem of bounds can be reduced to finite-dimensional optimization. Indeed, the optimization problem *compares all structures*; therefore, the minimizers—the gradient fields are determined by only a finite number of parameters: resistivity of the mixed materials, their volume fractions, and an applied field; this implies that some components of the fields are constant in all points of the subdomain. The optimal energy and the effective moduli are functions of the same prescribed parameters; they can be found by an algebraic procedure.

The Hashin-Shtrikman [16, 17] and translation bounds [19, 24, 29] for two-material mixtures are reduced to a finite-dimensional optimization: the gradient field in one of the materials is constant, some components of the field in the second material are constant too, the remaining components freely vary without affecting the optimal energy because they belong to zero subspace of the "translated energy." The bounds are attainable by special structures: coated spheres by Hashin and Shtrikman [16], second-rank lamination by Lurie and Cherkaev [19], ellipses assembly by Murat and Tartar [26], Benveniste and Milton [4], special periodic structures by Vigdergauz [30], and by Milton [23].

For multicomponent mixtures, the translation bounds tend to the incorrect asymptotic when the fraction of the "best" material tends to zero. For a larger fraction, these bounds are achievable by structures found by Milton [22, 23], Lurie and Cherkaev [20], Milton and Kohn [25], Gibiansky and Sigmund [15], Lui [18]. Such behavior suggests that the Translation Bound misses some inequalities on the gradient fields in optimal structures that may or may not be active depending on the parameters.

An extended literature exists on the optimal design of multimaterial composites. The problem was discussed by Burns and Cherkaev [6], Pedregal and Zhang [28], Briggs et al. [5]. Cherkaev and Kucuk [13], Albin and Cherkaev [1], Albin et al. [2] investigated the fields in optimal composites, see also [7]. Alessandrino and Nesi [3] noticed the inequalities on the gradient fields in all (not only optimal) structures; based on them, Nesi suggested improved bounds [27] that account for these inequalities. The paper [8] described constraints on the field in optimal structures that realize the quasiconvex envelope of the energy. Based on them, the optimal bounds and matching structures for isotropic three-material composites (one material is ideal) were found in [9]. Cherkaev and Zhang described the whole collection of anisotropic bounds and optimal structures for the optimally conducting composite [14] from the two materials and void. Cherkaev and Dzierżanowski [11] found the set of optimal elastic structures from two materials and void.

Optimal microstructures of multi-material composites depend on the active constraints. The two-material structures of minimal conductivity have a steady and intuitively expected topology: A resistive material surrounds conducting inclusions, as in Hashin-Shtrikman coated circles and second-rank laminates. The topology of multicomponent composite depends on the volume fractions. The best resistor either envelopes the inclusions or penetrates them; the layers of the best and worst conductors direct the currents, etc.

This paper provides informal arguments for the constraints on fields in optimal structures and the algebraic structure of the constrained bounds.

2 Preliminaries

2.1 The Problem. Notations

The periodicity cell Ω of a conducting composite consists of n subdomains $\Omega_1, < \Omega_2, \dots \Omega_n$ occupied with different materials with properties $k_1, \dots k_n$, $k_1 < k_2 < \dots < k_n$. The variable conductivity $k(x)$ in the cell is

$$k(x) = \sum_i k_i \chi_i(x), \quad \chi_i(x) = \begin{cases} 1 & x \in \Omega_i \\ 0 & x \notin \Omega_i \end{cases}. \tag{1}$$

where $\chi_i(x)$ is the index function of the domain Ω_i. The structure conducts the fields $e_p = \nabla u_p$, $p = 1, 2$; the prescribed average fields e_{01} and e_{02} are orthogonal, $e_{01}^T e_{02} = 0$. The conductivity equations are

$$\nabla \cdot k(x) \nabla u_p = 0, \quad \int_\Omega \nabla u_p dx = e_{0p}, \quad \nabla u_p \text{ are } \Omega\text{-periodic}, \quad p = 1, 2. \tag{2}$$

These equations are the stationarity conditions to the variational problems for vector potential $u = (u_1, u_2)$ and its gradient, the 2×2 matrix $e = \nabla u = (\nabla u_1 | \nabla u_2)$

$$J(\chi) = \min_u \int_\Omega W(\chi, e) dx, \quad \text{subject to} \int_\Omega e\, dx = (e_{01} | e_{02}) \tag{3}$$

where $\chi = (\chi_1, \dots \chi_n)$ and $W(k, e) = \frac{1}{2} k(\chi) e^T e$ is the energy density.

We assume that the areas m_i of the domains Ω_i are fixed:

$$\|\Omega_i\| = \int_\Omega \chi_i\, dx = m_i, \quad \sum_{i=1}^n m_i = 1. \tag{4}$$

The optimization problem asks for the shape of Ω_i (or the values of $\chi_i(x)$) that minimizes the sum of conductivity energies caused by the two orthogonal excitations e_{p0}, $p = 1, 2$.

$$I(e_0) = \inf_\chi \min_e \int_\Omega \left(W(\chi, e) + \sum_k \gamma_k \chi_k(x) \right) \text{ subject to (3), (4)} \tag{5}$$

where $\gamma_1, \dots \gamma_n$ are Lagrange multipliers by the constraints (4).

Interchanging the minimization operations and choosing the optimal χ_i, see (1), we transform the problem to the nonconvex problem for e

$$I(e_0) = \inf_e \int_\Omega F(e, \gamma) dx, \quad F(e, \gamma) = \min_i \{W(\chi_i, e) + \gamma_i\} \tag{6}$$

and observe that $F(e, .)$ is the multiwell piece-wise quadratic Lagrangian. The solution of (6) is the quasiconvex $\mathcal{Q}F(e)$ envelope of $F(e)$. The quasiconvex envelope of a multiwell Lagrangian is characterized by supporting points in each well

$$\mathcal{Q}F(e) = \inf \sum_i \int_{\Omega_i} \mu_i W(e_i) dx, \quad \int_{\Omega_i} \mu_i = m_i, \quad \mu_i \geq 0, \tag{7}$$

$$e_i \in \mathcal{E}_i, \quad \nabla \times e = 0, \quad \int_\Omega e\, dx = \sum_i \int_{\Omega_i} e_i\, dx = e_0. \tag{8}$$

The energy in a material plus γ is called the well. The wells $W(k_i, e)$ are convex functions of e; without the differential constraint $\nabla \times e = 0$, the support e_i in each well is constant of x. The constraint requires non-constant support e_i to satisfy the continuity of tangent component $[e] \cdot \tau = 0$ of minimizers-gradients at the boundary of Ω_i. The variable within a convex well support increases the cost of the problem.

2.2 Fields in Optimal Structures According to Translation Method

It is impossible to account directly for the differential constrain $e = \nabla u$ because the domains Ω_i are unknown. However, it is possible to account for weaker integral identities (polyconvexity).

$$\int_\Omega \det e\, dx = \det e_0. \tag{9}$$

Determinant $\det e$ and the square of norm $\|e\|^2$ are rotation-invariant quadratic functions of the entries of e. It is convenient to represent them through the half sum S and half difference D of the eigenvalues λ_1, λ_2 of e

$$S = \frac{1}{2}(\lambda_1 + \lambda_2), \quad D = \frac{1}{2}(\lambda_2 - \lambda_1). \tag{10}$$

We have

$$W(k_i, e) = k_i(\lambda_1^2 + \lambda_2^2) = k_i(S^2 + D^2), \quad \det e = \lambda_1 \lambda_2 = S^2 - D^2$$

Adding the constraint (9) with Lagrange multiplier t, we define the *translated energy* $W_i^t(e)$) of ith material

$$W^t(k_i, e) = W(k_i, e) + t\det(e) = (k_i + t)S^2 + (k_i - t)D^2 \tag{11}$$

Then, ignoring other differential and pointwise constraints on e and treating e as an arbitrary matrix, we minimize the whole energy of the cell. The minimizer e is constant in Ω_i. The energy $W(k_0, e_0)$ of the cell is the solution to the problem

$$W(k_0, e_0) = \max_t \left(\min_{e_1,\dots,e_n} \sum_i^n m_i \langle W_i^t(e)\rangle_i - tT(e_0) \right) \tag{12}$$

subject to

$$\sum_i m_i \langle S\rangle_i = 1, \quad \sum_i m_i \langle D\rangle_i = D_0. \tag{13}$$

where $\langle . \rangle_i$ denotes the average in ith well as follows

$$\langle e\rangle_i = \frac{1}{|\Omega_i|}\int_{\Omega_i} e_i\, dx, \quad \langle e\rangle_i \in \mathcal{C}\partial\mathcal{E}_i. \tag{14}$$

Solving this problem, we find the following:

- When $|t| < k_i$, the translated energy $W_i^t(e)$) is convex, and optimal fields S_i and D_i are constant, $S_i = \langle S\rangle_i$ and $D_i = \langle D\rangle_i$,

$$S_i = \frac{1}{k_i + t}\left(\sum_i^n \frac{m_i}{k_i + t}\right)^{-1} S_0, \quad D_i = \frac{1}{k_i - t}\left(\sum_i^n \frac{m_i}{k_i - t}\right)^{-1} D_0. \tag{15}$$

- (b) When $|t| = k_1$, $S_i = \langle S\rangle_i$, D_1 is not uniquely defined pointwise,

$$S_i = \frac{1}{k_i + k_1}\left(\sum_i^n \frac{m_i}{k_i + k_1}\right)^{-1} S_0, \quad D_i = 0,\ i = 2, \dots, n, \quad \langle D\rangle_1 = \frac{1}{m_1} D_0. \tag{16}$$

  These fields correspond to the Hashin-Shtrikman and Translation bounds. The last relation (16) shows that the constraint (13) is satisfied due to the choice of D_1. The optimal values $D_i = 0,\ i = 2, \dots, n$ in other materials are zero independently of D_0. Notice, that $\langle D\rangle_1 \to \infty$ when $m_1 \to 0$.
- When $|t| > k_1$, $W_i^t(e)$ is a saddle function, and the bound is trivial: $W_0^t = -\infty$ because the variable optimal field D_1 becomes improper, $(D_i^2)_{opt} \to \infty$ while $\langle D\rangle_1$ remains as in (16). This feature hints that accounting for the boundedness of D in the optimal composite makes this bound nontrivial even in that case.

Notice that the fields (16) in all subdomains depend on k_1 no matter how small the volume fraction $m_1 > 0$ is. But when $m_1 = 0$, the positivity of $W_1^t(e)$ is not required, and the translation parameter varies in the interval ($|t| \leq k_2$). That explains the discontinuity in the Translation bound and its non-exactness for small values of m_1.

3 Translation Bounds with Constraints

3.1 Finite-Dimensional Problem. Arguments

Because all partitions Ω_i are compared in the optimization problem (5), the points in Ω_i are not distinguishable. The gradient field $e = \nabla u$ in each point of an optimal domain Ω_i is independent of its shape. The optimal fields may be subject only to uniform integral constraints or to pointwise equalities or inequalities constraints. Thus, the problem of bounds can be reduced to finite-dimensional nonlinear programming. The following parameters define the solution

- Uniformly bounded pointwise values of the supporting points in each well. The supporting points e_i are the values of optimal fields in the subdomains Ω_i of the structure (or limits of these values). They vary in the sets $\mathcal{E}_1, ...\mathcal{E}_n$ respectively, $e_i \in \mathcal{E}_i,\ i = 1, \dots, n$. Due to the definition of F, the supports are ordered:

$$\|e_n\| < \|e_2\| < \cdots < \|e_1\| < M, \quad \|e_n\| < \|e_0\| < \|e_1\|. \tag{17}$$

  Here, $\|e_n\|^2 = S^2 + D^2$.
- The field e_1 with the largest norm is also bounded in an optimal composite. In the region of large applied fields $\|e_0\| > M$, Lagrangian $F(e)$ is convex, because $F(e)$ for larger $|e|$ coincides with $W_1(e) + \gamma_1$.

$$F(e) = W_1(e) + \gamma_1 \quad \forall e : \ |e| > M$$

  In that region, $F(e)$ grows quadratically; therefore, the solution to (7) and (8) is constant: the composite degenerates into the pure material k_1. All supporting fields are bounded, and the quasiconvex envelope $\mathcal{Q}F$ of Lagrangian is smaller than it, $\mathcal{Q}F < F$. Notice that convexity implies quasiconvexity; therefore, the field e_1 in a multiphase composite certainly belongs to the set of non-convexity of F.
- Solution depends on averages $\langle e \rangle_i$ of the supports in each well. The subdomain Ω_i where $k = k_i$ may consist of several disjoint pieces $\Omega_i^{(p)}$. The fields in these domains may belong to the boundary $\partial\mathcal{E}_i$ of the domain of variation $\mathcal{E}_i$ of e_i. They have the same eigenvalues and translated energy $W_t(e_i)$ but differ by the orientation of the eigenvectors. When the relative volume fractions of $\Omega_i^{(p)}$ varies, the average field $\langle e \rangle_i$ in Ω_i varies in the convex envelope $\mathcal{C}\partial\mathcal{E}_i$ stretched at the optimal fields $e_i \in \partial\mathcal{E}_i$. These averages are constrained by (13)

Remark 1 The boundedness of the fields e_i excludes the isolated points in an optimal structure where more than two materials meet because the fields are singular at these points. The materials in an optimal composite could either occupy disjunctive areas in multiple connected subdomains (islands) of one of the materials, or their meeting points could be dense. For example, a laminate from two materials can be adjacent to the domain of the third one if the layers are perpendicular to the boundary with the third material.

3.2 *Rank-One Connectivity*

The field e in an optimal composite alternate values $e^{(i)}$ in subdomains Ω_i.

$$e = e^{(i)} \in \mathcal{E}_i \quad \text{if } x \in \Omega_i$$

where $\mathcal{E}_i$ is the set of the fields (supporting points) in the ith subdomain. The quasiconvex envelope $\mathcal{Q}F(e_0)$ (energy of an optimal composite) is the limit of the averaged energy computed on these minimizers alternating between $\mathcal{E}_i$.

Compatibility requirement, see [8, 10]. Due to the continuity of potentials $u = (u_1, u_2)$, the tangent component of gradient $e \cdot \tau = \nabla u \cdot \tau$ is continuous at the boundaries Ω_i, which implies that $\text{Rank}\,[e]_-^+ = 1$ at any boundary. Thus, the fields in subdomains Ω_i are connected. For example, the isotropic fields $e_i = \beta_i Id$ cannot be joined in a composite structure.

However, the compatibility does not imply that all sets $\mathcal{E}_i$ are rank-one connected. If the composite is assembled from more than two materials, the exterior of subdomain Ω_i can be a smaller-scale mixture of several other materials. The continuity on the boundary of Ω_i applies to the average field in the mixture but not the field in any mixed in smaller scale materials. The gradient field $e|_{\text{ext}}$, averaged on a smaller scale in the exterior, is in rank-one contact with the interior field $e_i \in \mathcal{E}_i$, $(e|_{\Omega_i} - e|_{\text{ext}}) \cdot \tau = 0$.

The field $e|_{\text{ext}}$ in a mixture belongs to the convex envelope $\mathcal{C}\mathcal{E}_{ext}$ of the fields of involved,

$$e|_{\text{ext}} \in \mathcal{C}\mathcal{E}_{ext}$$

where $\mathcal{E}_{ext} = \cup \mathcal{E}_k$ is the set of optimal fields in involved subdomains.

Rank-one connectivity of supports $\mathcal{E}_i$ Each supporting point e of the quasiconvex envelope is in the rank-one contact with the gradient field e_α belonging to the convex envelope $\mathcal{C}\mathcal{E}_\alpha$ of the other supporting points, $\mathcal{E}_\alpha = \mathcal{E} - e_\alpha$:

$$\forall e \in \mathcal{E} \;\; \exists s_\alpha \in \mathcal{C}E_\alpha : \; \text{Rank}\,(e - e_\alpha) = 1. \tag{18}$$

Notice that (18) is only a necessary condition for s_α to be a supporting point because not all points of the convex envelope $\mathcal{C}\mathcal{E}_A$ may correspond to a mixture in a smaller scale structure (some points in $\mathcal{C}\mathcal{E}_A$ may be incompatible).

Example Assume that the fields in materials are as predicted by the Translation bound (16). The fields in materials $k_2, \ldots k_n$ are isotropic $D_i = 0$, i ≥ 2 and not rank-one connected; their normal components $e_i \cdot \tau = S_i$ are different from each other. The field in k_1 is connected to all others because $e_1 \cdot \tau = S_1 + D_1$ and D_1 is arbitrary. The rank-one connectivity restricts the values of the D_1 component of e_1 so that

$$0 \leq S_1 + D_1 \leq S_n. \tag{19}$$

because S_n is the maximal of the ordered S components. Inequality (19) becomes an equality if the materials k_1 and k_n form a laminate in an optimal structure.

Remark 2 In particular, if $k_n = \infty$, $S_n = 0$. When e_0 are positively defined, all eigenvalues of the supporting points are nonnegative, $e_i \geq 0$, or $D_1 \leq S_1$. Notice that this inequality is valid for all, not only optimal structures; see also arguments in Alessandrini and Nesi [3] and Nesi [27].

3.3 Inequality Between e_0 and e_1

The optimal field e_1 is extremal; it belongs to the line $\mathcal{G}L$ where the quasiconvex envelope $\mathcal{Q}F(e)$ of the Lagrangian $F(e)$ departs from the Lagrangian. For all values e outside of $\mathcal{Q}L$, the optimal composite degenerates into pure material k_1.

The average field e_0 is a point in the quasiconvex envelope; at least one one eigenvalue of e_0 belongs to interior of the set bounded by $\mathcal{Q}L$

$$\det(e_0 - e_1) \leq 0. \tag{20}$$

The field e_1 is optimal if at least one eigenvalue of e_0 lies on $\mathcal{Q}L$, or when (20) is satisfied as equality.

The composite that realizes equality in (20) contains a layer from k_1 that propagates across the whole structure. The compatibility conditions for the tangent to the layer $e \cdot \tau$ and the normal current $j \cdot n$ becomes

$$(e_1 - e_0) \cdot \tau = 0, \quad (j_1 - j_0) \cdot n = 0. \tag{21}$$

This inequality (20) explains, in particular, why the best material in an optimal structure tends to form an envelope around the core of other materials or substructures: In this case, the normal current in the penetrating layer is equal to the normal averaged current.

4 Examples

Constrained translation bounds The constrained bounds require minimization of the translated energy (12) subject to the additional inequalities (18) and (21). The algebraic expression of the bound depends on the anisotropy of e_0 and the optimal value of the translation parameter t.

4.1 *Two-Material Composites: Optimal Second-Rank Laminate*

An optimal two-material layered structure is a second-rank laminate. Materials k_1 and k_2 form a laminate on a smaller scale (first-rank laminate); this laminate forms an orthogonal layered structure with k_1 (second-rank laminate). These structures degenerate into a simple laminate; see [7, 13, 19].

In the optimal second-rank laminate, the field e_2 is constant and isotropic, $D_2 = 0$. Field e_1 is piece-wise constant. In the first-rank layer $e_1 = e_{11} = S_1 + D_{11}$ and in the second-rank layer $e_1 = e_{12} = S_1 + D_{12}$. The translation optimality conditions (16) are satisfied. Translation parameter $t = k_1$ allows a variable D-component in the material k_1. The component S_1 is constant in the first- and second-rank layers due to adjusting the relative amount of k_1 in the layers.

Connectivity: Field e_{11} is in rank-one contact with e_2: constraint (18) is active; e_{12} is in rank-one contact with e_0: constraint (21) is active,

$$\det(e_{11} - e_2) = 0, \quad \det(e_{12} - e_0) = 0. \tag{22}$$

This case corresponds to the moderate anisotropy of external excitation, $m_1 D_0 < S_1 - S_2$.

The simple laminate is optimal in the "strongly anisotropic" external field: $m_1 D_0 \geq S_1 - S_2$ and $t < k_1$. In the laminate, the fields e_1 and e_2 are in rank-one contact with themselves and the average field e_0.

In other known optimal assemblages, the Hashin-Shtrikman coated circles [16], Vigdergaus structures [30], Milton [23, 24], Milton and Benveniste [4] structures the fields satisfy the translation optimality conditions (16) and the same inequalities

$$\det(e_1 - e_2) \geq 0; \quad \det(e_1 - e_0) \leq 0; \quad \forall e_1 \in \mathcal{E}_1. \tag{23}$$

The fields are rank-one connected at the boundary between two materials, and the first inequality becomes equality. At the outer boundary of the Hashin-Shtrikman coated circles, the field in the annulus is rank-one connected with the surrounding effective medium, and the second inequality becomes equality.

4.2 Multimaterial Composites. Optimal Fields

The rank-one connected set $\mathcal{E}_A$ of admissible fields in materials is as follows.

Optimal fields in materials are ordered, $\det(e_i - e_n) \geq 0$. The rank-one connectedness constraint restricts the range of variation of S-components,

$$\mathcal{E}_A : (S_i - D_i) \geq S_n - D_n \tag{24}$$

and allows for variation of the translation parameter t above k_1. The translation-optimal D-fields that minimize translated energy (11) are as follows

$$\begin{array}{ll} D_i = 0 & \text{if } t < k_i \\ D_i \in [0, S_i - S_n - D_n] & \text{if } t = k_i \\ D_i = S_i - S_n & \text{if } t > k_i \end{array} . \tag{25}$$

Here, we assume that e_0 is "close to isotropy," that is, the constraint (13) or

$$\sum_{i=1}^{n-1} m_i \langle D_i \rangle = D_0 \tag{26}$$

can be satisfied by the variable orientation of D_i without varying the average $\langle D_i \rangle$ due to relation $\langle D_i \rangle < D_i$, if $k_i \leq t$, see (14).

Translated energy in the materials: If $t < k_i$, the translated energy (11) is strictly convex, and its minimal value corresponds to $D_i = 0$ (isotropic field). The translated energy is

$$W_i^t(t, e) = (k_i + t)S_i^2. \tag{27}$$

These translation-optimal fields e_i are not rank-one connected.

If $t = k_i$, the translated energy (11) is convex but has a null space: the value of D_i does not affect it. D_i can vary within Ω_i in the range $D_i \in [0, S_i - S_n - D_n]$ that is allowed by the rank-one connectivity. The mean value $\langle D_i \rangle$ varies in the same interval $\langle D_i \rangle \in [0, S_i - S_n - D_n]$. The translated energy is as in (27).

If $t > k_i$, the translated energy is nonconvex; its minimal value corresponds to maximal value $D_{\max}^2 = (S_i - S_n - D_n)^2$ on the boundary of the region (24) of rank-one connectedness. The translated energy is

$$W_i = (k_i + t)S_i^2 + (k_i - t)(S_i - S_n - D_n)^2 \quad \text{if } k_i < t. \tag{28}$$

The domain Ω_i may be multi-connected. The orientation of the pieces of optimal fields can be different in different subdomains of Ω_i, but the eigenvalues in all such pieces are constant. The trace S of the averaged field is invariant to rotation; the average values $\langle D_i \rangle$ of D_i vary in the interval:

$$\langle D_i \rangle \in [0, D_i].$$

The values of $\langle D_i \rangle$ do not affect the translated energy because $D = D_{\max}$ pointwise.

If the excitation field e_0 is highly anisotropic and (26) is not satisfied, e_1 is constant; it belongs to the boundary of rank-one connected region, $D_i = (S_1 - S_n), \langle D_i \rangle = D_i$. In this case, we check if the constraint

$$\sum_{i=2}^{n-1} m_i \langle D_i \rangle = D_0 - m_1 D_1 \quad \text{and} \quad \sum_{i=2}^{n-1} m_i D_i \geq D_0 - m_1 D_1 \tag{29}$$

for D_2 is satisfied. If it is not, then e_2 is a constant, and so on. The formulae change accordingly for "very anisotropic" external fields when $D_n > 0$.

The constraint (21) becomes active when m_1 is larger than a threshold, and the external field is close to unidirectional, $\det e_0 < \alpha < 1$. The bound corresponds to $t > k_1$. The field in Ω_1 alternates between the two e_{11} and e_{12} of equal values of the translated energy, such that $\det(e_{11} - e_3) = 0$ and $\det(e_{12} - e_0) = 0$.

4.3 *Optimal Structures of Three-Material Composites, One Material Is Ideal*

The described theory [8] has been checked [9, 14] for a particular case of three materials, one of them is a superconductor, $k_3 = \infty$, and for 2D elastic composites of maximal stiffness where one of three materials is void [9, 11]. In these cases, it is possible to explicitly compute the parameters of optimal fields and structures, solving the corresponding finite-dimensional optimization problems. The optimal anisotropic structures are shown in Fig. 1.

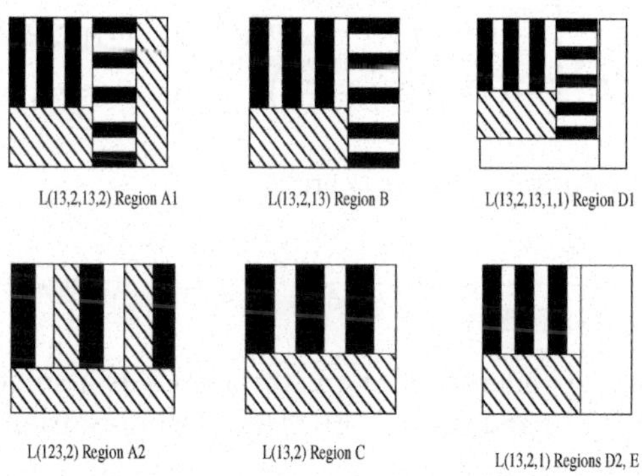

Fig. 1 Cartoon of optimal multi-rank laminates that minimize the conductance energy of a three-material composite [11, 14]. Black fields denote a superconductor, $k_3 = \infty$, striped fields denote a material of intermediate conductance k_2, and white fields indicate the least conductive material k_1. The notation like L(13, 2, 13) shows the order of laminating as follows: materials k_1 and k_3 are laminated first, then they are laminated with material k_2 in an orthogonal direction, then again laminated with $k_1 - k_3$ laminate

If $k_1 = t$, material k_1 occupies inclusions (regions D1, C); the trace S_1 of the field in k_1 is constant in both regions and D_1 varies $D_1 \in [0, S_1]$.

If $k_1 < t < k_2$, the field in k_1 on the boundary $\partial\mathcal{E}_1$; material k_1 is in rank-one contact with k_3 forming laminate, the (k_1, k_3)-layers may be arbitrarily oriented, (region B); also k_1 may form a layer across the whole structure, (regions D2, E).

If $k_2 = t$, the material k_1 is in rank-one contact with k_3, S_2 is constant and $D_2 \in [0, S_1]$ is variable (regions A1, A2).

Remark 3 The structure L(13,2,1) in Fig. 1 (regions D2, E) consists of laminates from k_1 and k_3 where $\det(e_{11} - e_3) = 0$ and the layer of k_1 across the structure where $\det(e_{12} - e_0) = 0$. The corresponding structure, see [12], does not degenerate into simple layers when the external field tends to unidirectional $\det(e_0) \to 0$. By properly choosing volume fractions, the conductivity of the k_1, k_3 layers in the horizontal direction can be made equal to k_2. Then, the horizontal current is constant everywhere in the structure, and the horizontal conductivity equals the harmonic mean of the components' conductivities. However, the conductivity in the vertical directions is less than their arithmetic mean.

4.4 *Bound for Effective Conductivity of an Isotropic Three-Material Composite, One Material Is Ideal*

The optimal effective conductivity tensor was explicitly computed for this example; see [11, 14]. Here, we show the extremal isotropic effective compliance k_* [9].

$$\kappa_*(m_1, m_2) \geq \begin{cases} -\kappa_1 + \left(\frac{m_1}{2\kappa_1} + \frac{m_2}{\kappa_1+\kappa_2}\right)^{-1} & \text{if } m_{11} \leq m_1 \leq 1, \\ \kappa_2 + 2\frac{\kappa_1}{m_1}(1 - \sqrt{m_2})^2 & \text{if } m_{11} \leq m_1 \leq m_{12}, \\ -\kappa_2 + \left(\frac{m_1}{2\kappa_1} + \frac{m_2}{2\kappa_2}\right)^{-1}. & \text{if } 0 \leq m_1 \leq m_{12}. \end{cases} \tag{30}$$

where the threshold values m_{11} and m_{12} are

$$m_{11} = \frac{2k_1}{k_2 + k_1}(\sqrt{m_2} - m_2), \quad m_{12} = \frac{k_1}{k_2}(\sqrt{m_2} - m_2). \tag{31}$$

Notice the irrational dependence of m_2; $k_*(m_1) \in C_1$ is a smooth function of m_1. The first line on the right-hand side of (30) is the Hashin-Shtrikman bound [17], and the third line is Nesi bound [27].

5 Conclusion

The translation bound for multicomponent composite needs to be complemented by pointwise constraints of the supporting fields. There is yet to be a test for the completeness of the system of constraints. We can demonstrate the completeness of the set of constraints only when we can explicitly show optimal structures that realize the constrained bound for any given excitation e_0 and any parameters m_i, k_i. This procedure involves the solution of a polynomial equation of high order.

References

1. Albin, N., Cherkaev, A.: Optimality conditions on fields in microstructures and controllable differential schemes. In: H. Ammari, H. Kang (eds.) Inverse Problems, Multi-scale Analysis, and Effective Medium Theory. Series: Contemporary Mathematics, vol. 408, pp. 137–150 (2006)
2. Albin, N., Cherkaev, A., Nesi, V.: Optimal structures of multimaterial composites. J. Mech. Phys. Solids **55**(7), 1513–1553 (2007)
3. Alessandrini, G., Nesi, V.: Univalent σ-harmonic mappings: applications to composites. ESAIM: Control, Optim. Calculus Var. **7**, 379 (2002)
4. Benveniste, Y., Milton, G.W.: The effective medium and the average field approximations vis-à-vis the Hashin-Shtrikman bounds. II. The generalized self-consistent scheme in matrix-based composites. J. Mech. Phys. Solids **58**, 1039–1056 (2010)
5. Briggs, N., Cherkaev, A., Dzierżanowski, G.: Note of Optimal design from three materials. J. Struct. Optim. Int. (2014). https://doi.org/10.1007/s00158-014-1156-1
6. Burns, T., Cherkaev, A.: Optimal distribution of multimaterial composites for torsion beams. Struct. Optim. **13**(1), 1–4 (1997)
7. Cherkaev, A.: Variational Methods for Structural Optimization. Springer, NY (2000)
8. Cherkaev, A.: Bounds for effective properties of multimaterial two-dimensional conducting composites and fields in optimal composites. Mech. Mater. **41**, 411–433 (2009)
9. Cherkaev, A.: Optimal three-material wheel assemblage of conducting and elastic composites. Int. J. Eng. Sci. **59**, 27–39 (2011)
10. Cherkaev, A.: Variational method for optimal multimaterial composites and optimal design. Int. J. Eng. Sci. **83**, 162–173 (2014)
11. Cherkaev, A., Dzierżanowski, G.: Three-phase plane composites of minimal elastic stress energy: high-porosity structures. Int. J. Solids Struct. **50**, 4145–4160 (2013)
12. Cherkaev, A.V., Gibiansky, L.V.: Extremal structures of multiphase heat conducting composites. Int. J. Solids Struct. **33**(18), 2609–2623 (1996)
13. Cherkaev, A., Kucuk, I.: Detecting stress fields in an optimal structure I: Two-dimensional case and analyzer. Int. J Struct. Opt. 1–15 (2004); II: Three-dimensional case. Int. J Struct. Opt. 15–24 (2004)
14. Cherkaev, A., Zhang, Y.: Optimal anisotropic three-phase conducting composites: plane problem. Int. J. Solids Struct. **48**(20), 2800–2813 (2011)
15. Gibiansky, L.V., Sigmund, O.: Multiphase composites with extremal bulk modulus. J. Mech. Phys. Solids **48**(3), 461–498 (2000)
16. Hashin, Z., Shtrikman, S.: A variational approach to the theory of the elastic behavior of multiphase materials. J. Mech. Phys. Solids **11**, 127–140 (1963)
17. Hashin, Z., Shtrikman, S.: A variational approach to the theory of the effective magnetic permeability of multiphase materials. J. Appl. Phys. **33**, 3125–3131 (1963)

18. Liu, L.P.: New optimal microstructures and restrictions on the attainable Hashin-Shtrikman bounds for multiphase composite materials. Philso. Mag. Lett. **91**, 473–482 (2011)
19. Lurie, K., Cherkaev, A.: Exact estimates of conductivity of mixtures composed of two isotropic media taken in prescribed proportion. Proc. R. Soc. Edinburgh, sect. A, 1984, **99**(P1-2), 71-87. * First has been published in Russian in 1982: K. Lurie, A. Cherkaev. Report 783, Physical Technical. Inst. Acad. Sci. USSR, 1982, p. 32 (in Russian)
20. Lurie, K., Cherkaev, A.: Optimization of properties of multicomponent isotropic composites. J. Optim. Theory Appl. **46**(4), 571–580 (1985)
21. Lurie, K., Cherkaev, A.: Optimization of properties of multicomponent isotropic composites. J. Optim. Theory Appl. **46**(4), 571–580 (1985)
22. Milton, G.W.: Concerning bounds on the transport and mechanical properties of multicomponent composite materials. Appl. Phys. A **26**, 125 (1981)
23. Milton, G.W.: The Theory of Composites. Cambridge University Press (2001)
24. Milton, G.W.: On characterizing the set of possible effective tensors of composites: the variational method and the translation method. Communications (1990)
25. Milton, G.W., Kohn, R.: Variational bounds on the effective moduli of anisotropic composites. J. Mech. Phys. Solids **36**(6), 597–629 (1988)
26. Murat, F., Tartar, L.: On the control of coefficients in partial differential equations. In: Cherkaev, Kohn (ed.) Topics in the Mathematical Modeling of Composite Materials, pp. 1–8. Birkhäuser, Boston, MA (1997)
27. Nesi, V.: Bounds on the effective conductivity of two-dimensional composites made of n greater than or equals $3n \geq 3$ isotropic phases in prescribed volume fraction: the weighted translation method. Proc. R. Soc. Edinburgh. Sect. A, Math. Phys. Sci. **125**(6), 1219–1239 (1995)
28. Pedregal, P., Zhang, Y.: Optimal design for multimaterials. Analy. Appl. **10**(4), 413–438 (2012)
29. Tartar, L.: Estimation fines des coefficients homogénéisés. In: Kree, P. (ed.) E. De Giorgi Colloquium (Paris, 1983), pp. 168–187. London (1985)
30. Vigdergauz, S.B.: Effective elastic parameters of a plate with a regular system of equal-strength holes. MTT **21**(2), 165–169 (1986)

Effect of Crystallographic Twins on the Elastoplastic Response of Polycrystals

Lucas Monteiro Fernandes, Philipp Rieder, Matthias Neumann, Aude Mulard, Henry Proudhon, Volker Schmidt, and François Willot

Abstract We investigate the influence of crystallographic twins on the elastoplastic response of γ-TiAl intermetallics via full-field FFT-based computations. We first introduce a hierarchical stochastic model, which is used to simulate synthetic polycrystalline microstructures containing twin grains with certain morphologies, and apply it to generate representative volume elements. Second, we develop a Fourier-based method with regularization for solving the effective and local mechanical response of polycrystalline media using the Méric-Cailletaud crystal plasticity constitutive law. Numerical results show that, across configurations of twinning, the corresponding average effective response is similar. Although differences were quantified, the effect of twins on the yield stress is negligible in practice (less than 1%).

L. Monteiro Fernandes (✉) · F. Willot
Center for Mathematical Morphology, Mines Paris PSL, 35 rue Saint Honoré, 77305 Fontainebleau, France
e-mail: lucas.monteiro_fernandes@minesparis.psl.eu

F. Willot
e-mail: francois.willot@minesparis.psl.eu

P. Rieder · M. Neumann · V. Schmidt
Institute for Stochastics, Ulm University, Helmholtzstr. 18, 89069 Ulm, Germany
e-mail: philipp.rieder@uni-ulm.de

M. Neumann
e-mail: matthias.neumann@uni-ulm.de

V. Schmidt
e-mail: volker.schmidt@uni-ulm.de

A. Mulard · H. Proudhon
Center for Materials Science, Mines Paris PSL, 63-65 Rue Henri-Auguste Desbruères, 91100 Corbeil-Essonnes, France
e-mail: aude.mulard@minesparis.psl.eu

H. Proudhon
e-mail: henry.proudhon@minesparis.psl.eu

F. Willot et al. (eds.), *Continuum Models and Discrete Systems*,
Springer Proceedings in Mathematics & Statistics 457,
https://doi.org/10.1007/978-3-031-58665-1_7

Keywords Polycrystals · Twin boundaries · Representative volume element · Stochastic microstructures · Crystal plasticity FFT simulations

1 Introduction

Crystallographic twins are adjacent grains whose crystalline lattices are symmetric in a sense that they share the points lying on their joint grain boundary. They may originate during certain thermodynamic conditions of solidification, during lattice symmetry changes at the solid state, or during the response to shear stress. Several authors have proposed to model twinning as a mechanism of plastic straining, generally via a crystal plasticity framework. Clausen et al. [1] studied twinning in magnesium alloys and accurately modeled twin activation and the subsequent textural evolution. They notably discussed stress relaxation mechanisms due to twinning. Juan et al. [2] proposed a double-inclusion elastoplastic self-consistent scheme to model polycrystals in twinning relations. This approach allows one to take into account relevant phenomena such as the initial stresses experienced by grains with respect to the stress state of their twin domain. The model is based on the assumption of ellipsoidal inclusions embedded in a homogeneous medium, common in homogenization schemes. Local effects that arise from the interaction between grain boundary geometries and crystalline orientations are not explicitly represented. Zhang et al. [3] proposed a coupled FEM-FFT model to simulate dislocation slip and twinning propagation during the compression of magnesium micro-pillars, and compared the predictions of the numerical model to experimental data. Numerical predictions show that for a 2D polycrystalline microstructure, in a scenario where the grains are in a twin relation, the ultimate stresses are lower compared to a scenario where no twins are modeled. Numerical results and analytical estimates obtained from self-consistent homogenization theories also indicate a stiffer response for polycrystals without twins [4].

The present work is concerned with the elastoplastic response of twinned structures, derived from a stochastic 3D microstructure model. In particular, we investigate by numerical tools how twin grains influence the effective response of polycrystals at a macroscopic scale. The microstructure model, presented in Sect. 2, is used to generate grain architectures of polycrystaline materials and introduces the grain-wise crystallographic orientations that model twinning relations. We consider a crystal plasticity law based on small strains explicitly integrated locally, while taking into account not only dislocation slip systems but also pseudo-slip systems modeling micro-twinning. The constitutive law coupled to a Green operator-based FFT solver used to solve the mechanical problem are described in Sects. 3 and 4. The obtained results are presented and discussed in Sect. 5. Section 6 concludes.

2 Stochastic 3D Modeling of Polycrystalline Grain Architectures

In this section we outline the stochastic 3D microstructure model whose realizations serve as geometric input for the numerical simulations described in Sect. 5. The model is realized within a cubic sampling window $W \subset \mathbb{R}^3$ and implemented with periodic boundary conditions to avoid edge effects. For a more comprehensive description of this model, we refer to [4].

2.1 *Space Tessellation Model*

In particular, in the present work, a Voronoi tessellation is employed, which subdivides the Euclidean space $\mathbb{R}^3$ into subsets with disjoint interiors, called cells in stochastic geometry [5] and grains in the context of crystallography. The n-th grain $G_n \subset \mathbb{R}^3$ of the tessellation is given by the set of those points $x \in \mathbb{R}^3$, which are not further away (with respect to the Euclidean distance) from some generator point $s_n \in \mathbb{R}^3$ than from all other generator points $s_m \in \mathbb{R}^3$ $(m \neq n)$:

$$G_n = \left\{x \in \mathbb{R}^3 : |x - s_n| \leq |x - s_m| \text{ for all } m \neq n\right\}, \tag{1}$$

where $|\cdot|$ denotes the Euclidean distance. Note that the stochastic nature of random Voronoi tessellations arises from the randomness of the underlying point process of generator points. An inherent property of Voronoi tessellations is that the grains are convex polyhedrons, in particular their boundaries are given by planar facets. Furthermore, a Voronoi tessellation is uniquely characterized by its generating point pattern. In this study, the generating point pattern $\{s_n\}$ in $\mathbb{R}^3$ is sampled from a Matérn-hardcore point process with intensity $\lambda > 0$ and hardcore radius $r \geq 0$. The Matérn-hardcore process is obtained by a thinning of a homogeneous Poisson process [6], such that the minimum pairwise distance between two points of the thinned point pattern is larger than r. This choice of a point-process model allows us to control the grain size of the corresponding Voronoi tessellation via the hardcore radius r. For a formal definition of Matérn-hardcore point processes, we refer to [5]. Once a Matérn-Voronoi tessellation has been generated, one must define the crystallographic orientation within each grain which will be addressed in the following section.

2.2 *Modeling of Crystallographic Orientations and Twinning Effects*

A twin relation between two grains depends not only on the grains respective crystallographic orientations, but also on the spatial orientation of their joint grain boundary.

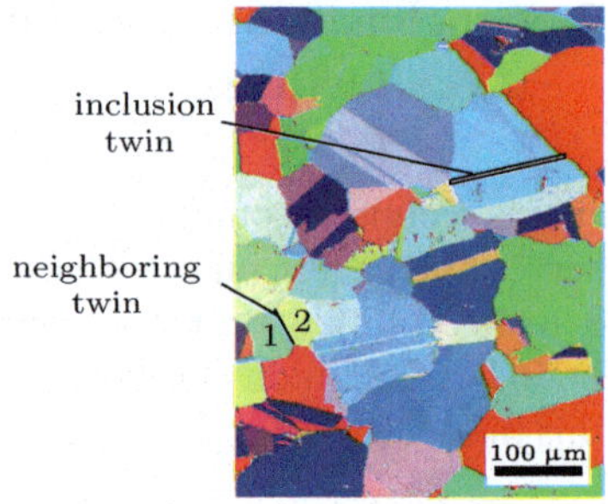

Fig. 1 Electron backscatter diffraction (EBSD) scan of a γ-TiAl sample. Pairs of twin grains with different morphologies can be observed

Crystallographic twins are pairs of neighboring grains, which are in a certain relationship to each other. On the one hand, the crystallographic misorientation between such neighboring grains, as well as the spatial orientation of their joint grain boundary have to fulfill two main criteria: the misorientation must coincide with a predefined angle and the crystallographic habit plane has to be aligned parallel to the joint grain boundary of the neighboring grains in a twin relation. In general, these properties of twin relations depend on the type of crystal symmetry. For the tetragonal crystal symmetry of γ-TiAl, considered in this paper, the misorientation angle θ has to be equal to $2\tan^{-1}(a/c) \approx 69.79°$, where a and c denote the lattice constants. Furthermore, the habit plane is either the (011) or (101) plane of the tetragonal lattice. The model used in the present study incorporates two different types of twins, which we treat separately. These two types of twins have been observed on EBSD maps of γ-TiAl polycrystalline aggregates, see Fig. 1. The first type, referred to in this study as "neighboring twins", are modeled by randomly choosing contiguous pairs of grains. For each of these pairs, the crystallographic orientations of the adjacent grains are determined such that they are in a twin relation. For the second type of twins, which are labeled as "inclusion twins", a crystallographic orientation is attached to randomly chosen grains from the initial tessellation, the so-called parent grains. For each of them, depending on their crystallographic orientation, another grain, the child grain, is inserted in the interior of the parent grain, such that they are in a twin relation, as specified hereafter.

2.2.1 Neighboring Twins

To determine the number of neighboring twins within the sampling window $W \subset \mathbb{R}^3$, we realize a Poisson-distributed random variable N_{neigh} with expectation equal to $p_{\text{neigh}} \cdot N_W/2$, where $N_W \in \{1, 2, \ldots\}$ denotes the number of grains contained in W, and $p_{\text{neigh}} \in [0, 1]$ a neighboring twins parameter. Note that the realization of N_{neigh} is rejected until $N_{\text{neigh}} \leq N_W/2$. The following procedure randomly generates pairs of twin grains, by assigning them with a crystallographic orientation. The procedure is repeated until N_{neigh} pairs are created, or no grains with free neighbors are left. It is important to note that once a grain is chosen and assigned a crystallographic orientation, it cannot be chosen a second time. Consequently, the structure contains no so-called twin related domains, where twin grains form a "chain" with several grains.

In order to model the crystallographic orientations of neighboring twins, we proceed as follows. At first, a grain G_k and one of its neighbours, denoted by G_ℓ, are chosen at random. Let $\partial G_{k,\ell} = \partial G_{\ell,k} = G_k \cap G_\ell$ denote the joint grain boundary of G_k and G_ℓ. Recall that the Voronoi tessellations form polyhedral grains, which implies that $\partial G_{k,\ell}$ is a two-dimensional facet. Furthermore, the crystallographic orientation O_k of grain G_k is sampled from a uniform distribution on the space of rotations SO_3, under the condition that one of its habit planes aligns with the joint grain boundary $\partial G_{k,\ell}$. Subsequently, the orientation of O_ℓ of G_ℓ is determined by rotating O_k by the misorientation angle θ around [100] or [010], respectively.

2.2.2 Inclusion Twins

So far, only those grains involved in neighboring twin relations are assigned with a crystallographic orientation, in general some grains are left without an orientation. These remaining orientations are sampled from a uniform distribution on the space of rotations SO_3.

The generation of inclusion twins is controlled by a probability $p_{\text{incl}} \in [0, 1]$ such that a grain G is equipped with an inclusion $I \subset G$, which splits G into two parts. If G is selected to get an inclusion, the following procedure is executed, otherwise the algorithm continues with the next grain.

First, the crystallographic orientation O_I of I is determined by rotating O, the orientation of G, by the misorientation angle θ around a randomly chosen [100] or [010] direction. Note that the choice of the rotation axis uniquely determines the spatial orientation of the habit plane (up to translations).

Second, we describe how the inclusion twin I is defined, namely as dilated plane, parallel to the habit plane. Let ∂G denote the grain boundary of G and $x \in \partial G$ a point on it chosen at random. Furthermore, we define $\mathcal{H}$ as plane, that is parallel to the habit plane and contains x. Additionally, we introduce $\mathcal{H}_\delta$ as $\mathcal{H}$, which is dilated to a thickness of $\delta > 0$ and define I as intersection of G and $\mathcal{H}_\delta$. The value of δ is sampled at random on $[\delta_- \cdot \text{ved}(G), \delta_+ \cdot \text{ved}(G)] \subset [0, \text{ved}(G)]$, where ved$(G)$ denotes the volume equivalent diameter of G with $0 \leq \delta_- \leq \delta_+ \leq 1$, the bounds of the relative thickness of the inclusion.

For a visualization of the generation of inclusion twins, we refer to [4].

2.3 *Specification of Model Parameters*

We draw a total of 15 realizations from the stochastic model described above for each of the following twinning configurations: (i) neighboring twins ($p_{\text{neigh}} = 1$, $p_{\text{incl}} = 0$), (ii) inclusion twins ($p_{\text{neigh}} = 0$, $p_{\text{incl}} = 1$) and (iii) polycrys-

Fig. 2 Examples of inverse pole figure (IPF) colored 2D cross sections of the generated 3D microstructures: **a** Realization without twins. **b** The same realization following the insertion of local correlations between orientations for modeling neighboring twins. **c** The same realization following the insertion of inclusion twins, with both morphological and crystallographic alteration

tals without twins ($p_{\text{neigh}} = p_{\text{incl}} = 0$). Note that for neighboring twins, although p_{neigh} is equal to 1, not every grain necessarily participates in a twin relationship. The sampling window W contains 256^3 voxels. The parameters of the Matérn-hardcore point process are given by a hardcore radius r of 5 voxels and an intensity $\lambda = 1920/(256\ \text{voxels}^3)\ \approx 1.14 \cdot 10^{-4}/\text{voxels}^3$, which corresponds to 1920 expected grains within the sampling window W. To reduce edge effects, which may influence the results of the mechanical simulations described in Sect. 5, realizations of the stochastic model are generated with periodic boundary conditions.

3 Méric-Cailletaud Crystal Plasticity

The single-crystal plasticity model presented in this article is inspired by the one described by Méric and Cailletaud in [7] and experimentally fitted for a γ-TiAL sample by Mulard in [8], from which the relevant parameters were used in the following. In the present study, we consider a crystalline material that possesses 12 slip systems relating to movement of dislocations and superdislocations and 4 pseudo-slip systems relating to micro-twinning mechanisms. Assuming the validity of small strain theory, one can separate the strain tensor ε into elastic ε^e and plastic ε^p contributions by $\varepsilon = \varepsilon^e + \varepsilon^p$. ε^e is linked to the stress tensor σ via the Hooke's law with an anisotropic stiffness tensor $\mathbb{C}$, i.e. $\sigma = \mathbb{C} : \varepsilon^e$. The plastic contribution on strain is determined by integrating the plastic strain rate $\dot{\varepsilon}^v$, i.e.,

$$\dot{\varepsilon}^v = \left(1 - \sum_{\beta=1}^{N_\beta} f^\beta\right) \sum_{s}^{N_s} \dot{\gamma}^s \mathbf{m}^s + \sum_{\beta}^{N_\beta} \dot{\gamma}^\beta \mathbf{m}^\beta\,, \tag{2}$$

where N_s and N_β denote respectively the number of dislocation and superdislocation s slip systems and micro-twinning β pseudo-slip systems. Further, f^β denotes the volume fraction of micro-twinnning at the considered pseudo-slip system β, $\dot{\gamma}^s$ and $\dot{\gamma}^\beta$ represent the slip rate and $\mathbf{m}^{s,\beta}$ is the Schmid tensor. The tensor $\mathbf{m}^{s,\beta}$ compactly describes a slip system by condensing slip planes and directions into the tensor

$$\mathbf{m}^{s,\beta} = \frac{1}{2}\left(\mathbf{n}^{s,\beta} \otimes \mathbf{l}^{s,\beta} + \mathbf{l}^{s,\beta} \otimes \mathbf{n}^{s,\beta}\right), \tag{3}$$

where $\otimes$ denotes the outer product. The driving force of slip is the resolved shear stress $\tau^{s,\beta}$, obtained by projecting the stress tensor at both a slip plane and a slip direction. Doing so, we obtain

$$\tau^{s,\beta} = \sigma : \mathbf{m}^{s,\beta}. \tag{4}$$

Increase in slip depends on whether the driving force has reached a critical value, called the yield stress $r^{s,\beta}$ for the slip system

$$r^s = \tau_0^s + Q_s \sum_{s'=1}^{N_s} q_{ss'} \left(1 - \mathrm{e}^{-b_s v^{s'}}\right), \quad r^\beta = \tau_0^\beta + Q_\beta \sum_{\beta'=1}^{N_\beta} q_{\beta\beta'} \left(1 - \mathrm{e}^{-b_\beta v^{\beta'}}\right) \tag{5}$$

under consideration, where $\tau_0^{s,\beta}$ denotes the initial resistance to slip, $Q_{s,\beta}$ is the magnitude of hardening interaction, $q_{ss',\beta\beta'}$ is the hardening interaction matrix. Here, s' and β' denote slip and pseudo-slip system indices respectively, $b_{s,\beta}$ is the hardening saturation and $v^{s',\beta'}$ is the accumulated slip, which is obtained by integrating the absolute value of the slip rate at each slip and pseudo-slip system: $\dot{v}^{s',\beta'} = |\dot{\gamma}^{s',\beta'}|$. In order to determine the slip rates, not only the yield stresses are taken into account but also some viscous parameters K_s and n from Norton's law. For dislocation slip systems, this reads as:

$$\dot{\gamma}^s = \left\langle \frac{|\tau^s| - r^s}{K_s} \right\rangle^n \operatorname{sign}(\tau^s). \tag{6}$$

For micro-twinning pseudo-slip systems, the relation that exists between slip and the volume fraction of twins must be accounted for, namely $\dot{\gamma}^\beta = \dot{f}^\beta/\sqrt{2}$. Indeed, a certain volume becomes a micro-twin if the required shear displacement is a fraction of the interatomic distance, which is determined by the geometry of the lattice and called twinning shear γ^T. For the material under consideration, $\gamma^T = 1/\sqrt{2}$. One then derives:

$$\dot{f}^\beta = K_c \left(f_m - \sum_\beta f^\beta\right) \left\langle \frac{\tau^\beta - r^\beta}{K_t} \right\rangle^{n_t}, \tag{7}$$

where K_c is the material-dependent magnitude of twinning and f_m is a parameter modeling the maximum twinned volume fraction observed experimentally in a grain.

Table 1 Left: planes $\mathbf{n}^s$ and directions $\mathbf{l}^s$ of dislocation slip systems of the γ-TiAl single-crystal; "o" and "s" denote respectively dislocations and superdislocations. Right: planes $\mathbf{n}^\beta$ and directions $\mathbf{l}^\beta$ of the pseudo-slip systems associated to the twinning of the γ-TiAl single-crystal

s	1	2	3	4	5	6	7	8	9	10	11	12
$\mathbf{n}^s$		$(11\bar{1})$			$(\bar{1}11)$			$(1\bar{1}1)$			(111)	
$\mathbf{l}^s$	$[011]$	$[101]$	$[1\bar{1}0]$	$[0\bar{1}1]$	$[\bar{1}0\bar{1}]$	$[\bar{1}\bar{1}0]$	$[011]$	$[10\bar{1}]$	$[011]$	$[01\bar{1}]$	$[10\bar{1}]$	$[1\bar{1}0]$
type	s	s	o	s	s	o	s	s	o	s	s	o

β	1	2	3	4
$\mathbf{n}^\beta$	$(11\bar{1})$	$(\bar{1}11)$	$(1\bar{1}1)$	(111)
$\mathbf{l}^\beta$	$[112]$	$[\bar{1}1\bar{2}]$	$[1\bar{1}\bar{2}]$	$[11\bar{2}]$

K_t and n_t are parameters from Norton's law in case of twinning pseudo-slip systems. It is important to highlight that, contrarily to dislocation and superdislocation slip, twinning is unidirectional. The formulation was implemented with an explicit discrete integration scheme (Runge-Kutta 4th order method) on a spectral solver based on the Green's operator (commonly known as FFT solver). As for elasticity, the following moduli have been measured experimentally [9] at 298K:

$$c_{11} = 183, \quad c_{33} = 183, \quad c_{12} = c_{13} = 74, \quad c_{44} = 105, \quad c_{66} = 78 \text{ GPa}. \tag{8}$$

In addition, we follow [8] (Table 5.7, p. 79) and use the following single-crystal plasticity parameters found by a coupled numerical and experimental identification:

$$K_s = K_t = 104 \,\text{MPa}\, \text{s}^{1/n}, \quad K_c = 1.45, \quad n = n_t = 6.42, \quad Q_s = 9.57 \,\text{MPa}, \tag{9a}$$
$$Q_\beta = 10.5 \,\text{MPa}, \quad f_m = 0.8, \quad \tau_0^{s,\mathcal{O}} = \tau_0^\beta = 130 \,\text{MPa}, \quad \tau_0^{s,\mathcal{S}} = 200 \,\text{MPa}. \tag{9b}$$

Here $\tau_0^{s,\mathcal{O}}$, $\tau_0^{s,\mathcal{S}}$ and τ_0^β refer to the initial resistance to slip respectively for dislocation and superdislocation slip systems and for pseudo-slip systems associated to micro-twinning (Table 1).

4 Regularization by Viscous Hardening

When the norm of the local stress tensor is larger than $E\delta\bar{\varepsilon}$, where E is the Young's modulus and $\delta\bar{\varepsilon}$ the strain increment, mechanical equilibrium is not enforced in certain voxels. Algorithms such as Newton's method [10] can be used to deal with this problem, which is commonly encountered in the case of finite strains [11]. In the present case, we make use of an explicit scheme with a regularization technique inspired by viscous hardening, as a way to obtain convergence even with large strain increments in the elastoplastic regime. Namely, we introduce a parameter η which instantaneously increases the yield stresses for each slip and pseudo-slip system by a term proportional to the absolute value of the slip rate at the considered slip and pseudo-slip system:

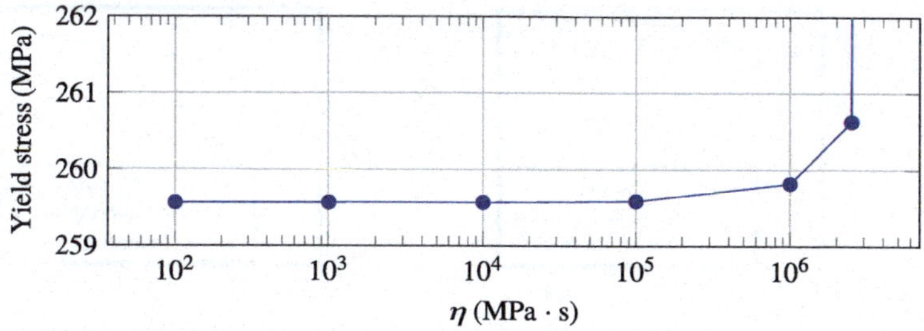

Fig. 3 Influence of the linear viscous hardening parameter η on the yield stresses for the integration of the constitutive behavior and arbitrary imposed strain. It can assume a wide range of values provided it is high enough to avoid divergence of the simulation and low enough to prevent unrealistically high hardening

$$r'^{s,\beta} = r^{s,\beta} + \eta|\dot{\gamma}^{s,\beta}|. \tag{10}$$

This regularization term can be interpreted as viscosity that penalizes sudden and high variation in slip, and privileges elastic deformation instead of plastic deformation. By regularizing, elastoplastic strain steps alternates with purely elastic strain steps in the elastoplastic regime. When η takes values inside a certain range, the regularized solution varies little enough to be consistent. On the one hand, if η is too low, convergence is not obtained. On the other hand, if it is too high, purely elastic behavior is recovered, which is not physically valid. In the present work, we set $\eta = 10^4$ MPa $\cdot$ s (Fig. 3).

5 Elastoplastic Response of Twinned Polycrystals

In order to compare the configurations of crystallographic twins, simulations were carried out for prescribed macroscopic uniaxial shortening strain of 0.5 %. Accordingly, we consider both the equivalent stress of the spatial average of the stress tensor field $\langle\sigma\rangle_{eq}$ and the spatial average of the equivalent stress field $\langle\sigma_{eq}\rangle$ as measures of the macroscopic response. Then, realization-wise averages of these quantities can be computed for each configuration of twinning.

For each configuration, fifteen realizations of the stochastic model described in Sect. 2 with a size of 256^3 voxels were simulated. Figure 4 shows averaged macroscopic stress-strain curves, more precisely the realization-wise average of Von Mises equivalent stresses of the RVE-wise average stress tensors as a function of the prescribed macroscopic strain at each strain step of the quasi-static simulation (1000 steps in total). The observed average response is rather similar across the three configurations of twin grains (Fig. 5).

Due to the isotropy of the Voronoi tessellation and the fact that crystalline orientations were sampled from a uniform distribution on the space of rotation matrices, the

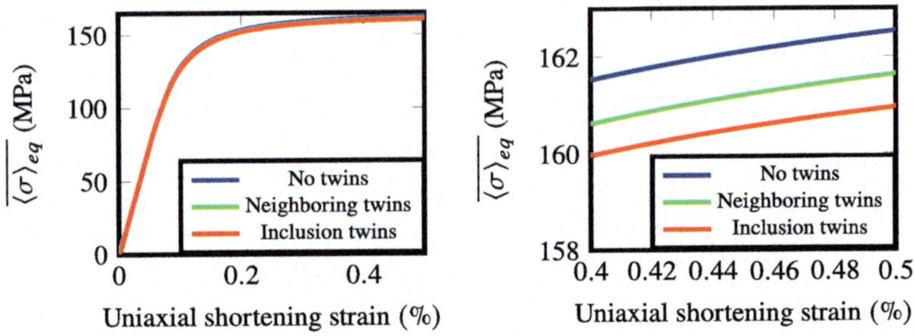

Fig. 4 Von Mises stresses of the RVE-wise average of the stress tensors vs. strain curves of polycrystalline RVE realizations representing twin configurations. Prescribed macroscopic uniaxial shortening strain along the x axis. In the elastoplastic regime, small differences are observed between the three configurations

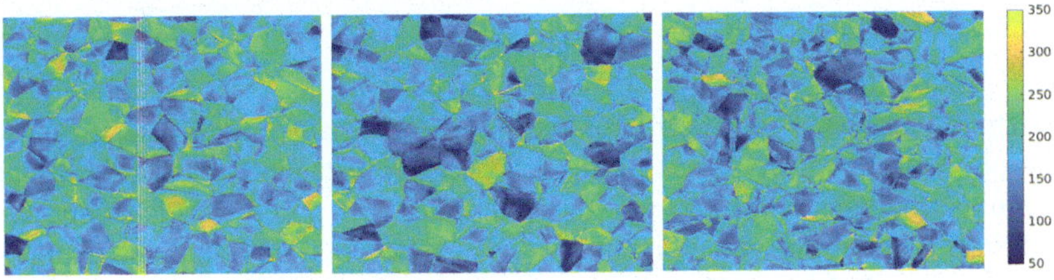

(a) (b) (c)

Fig. 5 Examples of Von Mises stress fields obtained with the FFT computations (Color map values in MPa), corresponding to the images in Fig. 2: **a** Realization without twins. **b** The same realization following the insertion of neighboring twins. **c** The same realization following the insertion of inclusion twins

Table 2 Realization-wise averages and standard deviations (SD) of grain numbers and sizes for each configuration of twinning. Unit of grain size: voxels or μm^3

	Average nb. of grains	Average grain size	SD (# grains)	SD (grain size)
Neighboring twins	1932	8689	44	201
Inclusion twins	5589	3003	132	71
No twins	1932	8689	44	201

results are statistically independent of the direction of the prescribed load. In order to verify that, the responses to loads along x, y and z axes were analyzed (Table 3).

Our results show that although slightly, structures without twins are more resistant than structures containing neighboring twins, whereas polycrystals with inclusion twins provide the least resistant response. Additionally, the difference in dispersion regarding yield stresses is partly explained by the difference in dispersion concerning average grain sizes (Table 2). While the average yield stresses differ little, their

Table 3 Realization-wise averages and standard deviations of the yield stresses (in MPa) for each configuration of twinning and each load direction

		Load along x	Load along y	Load along z
Neighboring twins	$\overline{\langle\sigma\rangle_{eq}}$	161.65	161.44	161.91
	$\overline{\langle\sigma_{eq}\rangle}$	186.06	185.89	186.29
	SD($\langle\sigma\rangle_{eq}$)	1.36	1.28	1.01
	SD($\langle\sigma_{eq}\rangle$)	1.30	1.15	1.00
Inclusion twins	$\overline{\langle\sigma\rangle_{eq}}$	160.97	160.33	160.92
	$\overline{\langle\sigma_{eq}\rangle}$	185.20	184.59	185.14
	SD($\langle\sigma\rangle_{eq}$)	0.78	1.22	0.67
	SD($\langle\sigma_{eq}\rangle$)	0.73	1.05	0.63
No twins	$\overline{\langle\sigma\rangle_{eq}}$	162.54	162.26	162.29
	$\overline{\langle\sigma_{eq}\rangle}$	187.04	186.69	186.84
	SD($\langle\sigma\rangle_{eq}$)	1.11	1.43	0.92
	SD($\langle\sigma_{eq}\rangle$)	1.06	1.29	0.75

high dispersion requires careful consideration. Statistical significance tests should be appropriately performed to assess the representativity of these values. For that purpose, Welch's t-test was chosen, due to it being applicable to populations with unequal variances. Additionally, it assumes that both populations are normally distributed. This condition is asymptotically satisfied when the domain is large, for certain diffusion problems with elliptic PDEs [12]. The tested null hypothesis is that the two populations have equal mean, in which a two-tailed test is applied. Table 4 shows that for 5% significance level, the null hypothesis of equal means in only rejected for all scenarios when comparing inclusion twins and no twins populations.

The probability density function of σ_{eq} over all realizations is visualized in Fig. 6. The graph shows that the distributions are multimodal. There are essentially two highest peaks likely associated to the initial resistance to slip thresholds from Eq. (9). In order to interpret tehse results, the averages of twin volume fraction f_β and accumulated slip v^s fields were estimated for each twinning configuration (Table 5), by averaging over slip systems, voxels and realizations. On the one hand, the averages for neighboring twins and no twins are the same. On the other hand, RVEs with inclusion twins present systematically higher averages for $\langle f_\beta \rangle$ and lower averages for $\langle v^s \rangle$. While the absence of difference between neighboring twins and no twins confirms the lack of influence of crystallographic twins on the macroscopic elastoplastic response, the values obtained for inclusion twins hints at an interaction between morphology and texture.

A final remark concerns the Hall-Petch effect, which states that the yield strength of a polycrystalline materials decreases as $d^{-1/2}$, where d is the average grain size, for values of d larger than a certain critical size. Indeed, materials with larger grains in average tends to present a higher number of dislocations piling up at grain boundaries, favoring the transmission of dislocations to other grains. When the grain size becomes

Table 4 Results of the statistical significance tests for both yield stress measures and all three load directions. Two-tailed Welch's t-test was performed for the null hypothesis of equal means. "NeT", "InT" and "NoT" denote respectively "neighboring twins", "inclusion twins" and "no twins". NH denotes "null hypothesis", while "NR" and "R" denote respectively "not rejected" and "rejected"

	Test for $\langle \sigma \rangle_{eq}$						Test for $\overline{\langle \sigma_{eq} \rangle}$					
	x		y		z		x		y		z	
	p-value	NH	p-value	NH	p-value	NH	p-value	NH	p-value	NH	p-value	NH
NeT versus InT	$1.08 \cdot 10^{-1}$	NR	$2.14 \cdot 10^{-2}$	R	$4.31 \cdot 10^{-3}$	R	$3.68 \cdot 10^{-2}$	R	$3.02 \cdot 10^{-3}$	R	$9.77 \cdot 10^{-4}$	R
NeT versus NoT	$5.98 \cdot 10^{-2}$	NR	$1.11 \cdot 10^{-1}$	NR	$2.96 \cdot 10^{-1}$	NR	$3.12 \cdot 10^{-2}$	R	$8.70 \cdot 10^{-2}$	NR	$9.77 \cdot 10^{-2}$	NR
InT versus NoT	$1.42 \cdot 10^{-4}$	R	$4.62 \cdot 10^{-4}$	R	$8.67 \cdot 10^{-5}$	R	$9.08 \cdot 10^{-6}$	R	$4.10 \cdot 10^{-5}$	R	$3.37 \cdot 10^{-7}$	R

Table 5 Realization-wise averages and standard deviations of the slip-system-wise and voxel-wise averages of the elastoplastic variables f_β and υ^s (dimensionless) for each configuration of twinning and each load direction

		Load along x	Load along y	Load along z
Neighboring twins	$\overline{\langle f_\beta \rangle}$	$2.24 \cdot 10^{-4}$	$2.26 \cdot 10^{-4}$	$2.26 \cdot 10^{-4}$
	$\overline{\langle \upsilon^s \rangle}$	$1.55 \cdot 10^{-4}$	$1.55 \cdot 10^{-4}$	$1.55 \cdot 10^{-4}$
	SD ($\langle f_\beta \rangle$)	$6.27 \cdot 10^{-6}$	$5.73 \cdot 10^{-6}$	$3.60 \cdot 10^{-6}$
	SD ($\langle \upsilon^s \rangle$)	$2.06 \cdot 10^{-6}$	$2.60 \cdot 10^{-6}$	$1.44 \cdot 10^{-6}$
Inclusion twins	$\overline{\langle f_\beta \rangle}$	$2.41 \cdot 10^{-4}$	$2.42 \cdot 10^{-4}$	$2.41 \cdot 10^{-4}$
	$\overline{\langle \upsilon^s \rangle}$	$1.51 \cdot 10^{-4}$	$1.51 \cdot 10^{-4}$	$1.52 \cdot 10^{-4}$
	SD ($\langle f_\beta \rangle$)	$4.56 \cdot 10^{-6}$	$5.17 \cdot 10^{-6}$	$5.11 \cdot 10^{-6}$
	SD ($\langle \upsilon^s \rangle$)	$1.28 \cdot 10^{-6}$	$2.18 \cdot 10^{-6}$	$1.73 \cdot 10^{-6}$
No twins	$\overline{\langle f_\beta \rangle}$	$2.24 \cdot 10^{-4}$	$2.24 \cdot 10^{-4}$	$2.26 \cdot 10^{-4}$
	$\overline{\langle \upsilon^s \rangle}$	$1.56 \cdot 10^{-4}$	$1.56 \cdot 10^{-4}$	$1.55 \cdot 10^{-4}$
	SD ($\langle f_\beta \rangle$)	$7.13 \cdot 10^{-6}$	$8.56 \cdot 10^{-6}$	$6.68 \cdot 10^{-6}$
	SD ($\langle \upsilon^s \rangle$)	$2.37 \cdot 10^{-6}$	$3.13 \cdot 10^{-6}$	$2.18 \cdot 10^{-6}$

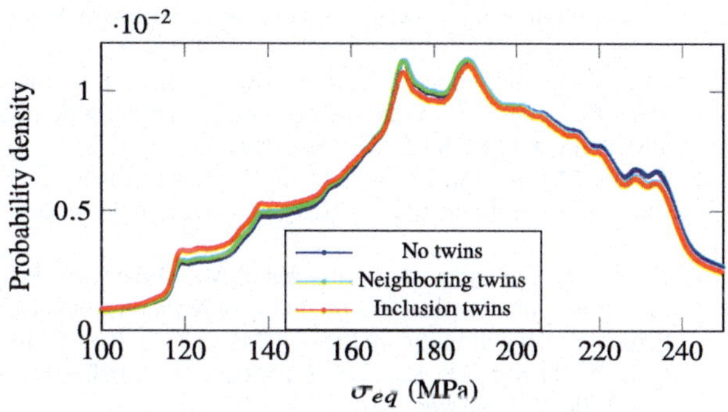

Fig. 6 Probability density function (PDF) of the von Mises stress field σ_{eq} estimated for the set of 15 realizations for each twinning configuration subjected to a load along the x axis (251.7 million voxels used for computing each curve)

smaller than the critical size, on the contrary, experimental observations [13] have shown that the yield strength actually decreases when d decreases. In this range of grain refinement, dislocation slip dominates other mechanisms of plastic flow, such as grain boundary sliding. In any case, we emphasize that the Hall-Petch relationship, while being a significant phenomenon to consider, is not taken into account in the present work.

6 Conclusion

The contribution of this study is twofold. First, we proposed a regularization strategy for accelerating an explicit integration scheme of a rate-independent crystal plasticity law implemented on an FFT-based solver. Second, the influence of twin grains with different morphologies on the macroscopic yield stress has been investigated, making use of stochastically-generated polycrystalline RVEs of a γ-TiAl intermetallics material with tetragonal crystal symmetry. Our result indicates that the presence of twins has no significant effect on the average elastoplastic behavior of polycrystals, provided untextured crystallographic orientations are considered. However, the interaction of grain morphology and size with local correlations arising from twin relations has to be further analyzed.

Acknowledgements We are indebted to S. Forest for useful suggestions made on a previous version of this paper. The authors acknowledge the financial support of the French Agence Nationale de la Recherche (ANR, ANR-21-FAI1-0003) and the Bundesministerium für Bildung und Forschung (BMBF, 01IS2109) within the French-German research project SMILE.

References

1. Clausen, B., Tomé, C., Brown, D., Agnew, S.: Reorientation and stress relaxation due to twinning: modeling and experimental characterization for Mg. Acta Materialia **56**(11), 2456–2468 (2008)
2. Juan, P.-A., Berbenni, S., Barnett, M., Tomé, C., Capolungo, L.: A double inclusion homogenization scheme for polycrystals with hierarchal topologies: application to twinning in Mg alloys. Int. J. Plast. **60**, 182–196 (2014)
3. Zhang, M., Ma, A., LLorca, J.: Simulation of dislocation slip and twin propagation in Mg through coupling crystal plasticity and phase field models (2023). arXiv:2302.13612 [cond-mat]
4. Rieder, P., Neumann, M., Fernandes, L.M., Mulard, A., Proudhon, H., Willot, F., Schmidt, V.: Stochastic 3d microstructure modeling of twinned polycrystals for investigating the mechanical behavior of γ-tial intermetallics (2024). arXiv:2401.08349 [cond-mat.mtrl-sci]
5. Chiu, S., Stoyan, D., Kendall, W., Mecke, J.: Stochastic Geometry and Its Applications, 3rd edn. Wiley, Chichester (2013)
6. Last, G., Penrose, M.: Lectures on the Poisson Process. Cambridge University Press, Cambridge (2017)
7. Méric, L., Poubanne, P., Cailletaud, G.: Single crystal modeling for structural calculations: Part 1-model presentation. J. Eng. Mater. Technol. **113**(1), 162–170 (1991)
8. Mulard, A.: Micro-mechanical modeling of γ-TiAl alloys using real 3D microstructures. Technical report, Mechanical and Materials Engineering, École des Ponts ParisTech (2018)
9. Tanaka, K.: Single-crystal elastic constants of γ-TiAl. Philoso. Mag. Lett. **73**(2), 71–78 (1996)
10. Besson, J., Cailletaud, G., Chaboche, J.-L., Forest, S.: Non-linear Mechanics of Materials, Springer Series on Solid Mechanics and Its Applications, vol. 167. Springer, Dortrecht (2010)
11. Ling, C., Forest, S., Besson, J., Tanguy, B., Latourte, F.: A reduced micromorphic single crystal plasticity model at finite deformations. application to strain localization and void growth in ductile metals. Int. J. Solids Struct. **134**, 43–69 (2018)

12. Nolen, J.: Normal approximation for a random elliptic equation. Probab. Theory Related Fields **159**(3–4), 661–700 (2014)
13. Conrad, H., Narayan, J.: On the grain size softening in nanocrystalline materials. Scripta Materialia **42**(11), 1025–1030 (2000)

Continuum Modelling of Orthotropic Hexatruss Lattice Materials: Effective Stiffness and Experimental Validation

Federica Ongaro, Kévin Mathis, Frédéric Masson, and Justin Dirrenberger

Abstract Architected materials are designed with specific configurations that offer enhanced properties, making them ideal for addressing various challenges in materials science, architecture, aerodynamics, and mechanical engineering. Their unique quality, coupled with the ability to tailor mechanical properties in every direction, renders them highly suitable for industries like aerospace, automotive, marine, and construction. However, the application of architected materials depends on the development of accurate models to understand the complex relationship between microstructure characteristics and macroscopic behavior. Despite the proposal and discussion of numerous analytical and numerical methods in recent years, very few studies have derived explicit formulas for effective mechanical properties. This paper contributes to this underexplored area by presenting a mathematical formulation and modeling technique for the effective elastic moduli of a three-dimensional orthotropic hexatruss lattice. The analytical relations we have established, validated through comprehensive experimental tests on a 3D-printed lattice, demonstrate the significant impact of lattice parameters on macroscopic properties. Practically, our results could simplify the process of parametric optimization for architectured materials, offering a less resource-intensive approach to optimization since parameter changes do not necessitate lattice regeneration.

Keywords Homogenization · Architectured materials · Lattice structures · Modelling · Elasticity

---

F. Ongaro · J. Dirrenberger (✉)
PIMM Lab, Arts et Métiers, Cnam, CNRS UMR 8006, 151 Bd de l'Hôpital, 75013 Paris, France
e-mail: justin.dirrenberger@ensam.eu

K. Mathis · F. Masson
CNES-Launchers Directorate, 52 Rue Jacques Hillairet, 75612 Paris Cedex, France
e-mail: kevin.mathis@cnes.fr

F. Masson
e-mail: frederic.masson@cnes.fr

F. Willot et al. (eds.), *Continuum Models and Discrete Systems*,
Springer Proceedings in Mathematics & Statistics 457,
https://doi.org/10.1007/978-3-031-58665-1_8

1 Introduction

1.1 Architectured Materials: Overview

Architectured materials, composed of two or more materials and/or spaces [3], offer a unique blend of mechanical properties such as stiffness, strength, and toughness at a lower density. This combination allows for previously conflicting performances to coexist within a single homogeneous material.

Their distinctive configuration, resulting from their discrete nature, makes architectured materials highly suitable for engineering applications across various industries, including aerospace, automotive, marine, and construction. For instance, the light weight of microtrusses or lattice structures is perfect for core materials in lightweight, high-performance sandwich panels in aerospace components and sports equipment. Additionally, their low compressive strength coupled with high deformation capacity makes them excellent for shock mitigation and energy absorption in impulsive events [12].

However, their application hinges on the development of suitable constitutive models to uncover their effective properties. Two sets of parameters influence their mechanical behavior: one relates to the material composition, and the other to the geometric and topological characteristics of the microstructure. Transitioning from a microscopic discrete description to a coarse continuous model is challenging. To address this, concepts of energy equivalence and micro-to-macro force and displacement relations are typically employed. Common simplifications like linear elasticity, material isotropy, and the assumption that the microstructure behaves according to classical beam theory are often used. These simplifications provide explicit stress-strain relations and aid in understanding the fundamental mechanical aspects of these materials.

1.2 State of the Art

A significant number of researchers have explored the mechanical modeling of architectured materials, leading to a diverse array of techniques that have been extensively discussed in recent literature. Key contributions, offering various methods and assumptions, can be found in the review papers authored by [9, 20, 30]. Additionally, the technical details concerning the analysis of structural lattices and the formulation of their equivalent constitutive equations have been comprehensively reviewed by [24]. This review also includes a comprehensive list of references for further reading and study.

The most recognized and extensively utilized micromechanical model in the field of architectured materials is attributed to the groundbreaking work of [16, 17]. Their research primarily concentrated on the deformation mechanism of a single cell when subjected to various types of external loads. Specifically, Gibson and colleagues derived first-order power-law relations that link the parameters of the microstructure with the macroscopic properties of a broad spectrum of architectured materials. This was achieved by assuming infinitesimal strains and employing standard beam theory to model the edges of the lattice structure.

References [8, 21] proposed an alternative methodology to address the critical transition from microscale to macroscale and to formulate the constitutive model for two-dimensional microstructures undergoing in-plane deformations. Similar to the approach by [16, 17], their method also idealizes the discrete lattice as a series of Euler-Bernoulli beams. The macroscopic description in their approach is derived using an energy-based technique.

In the context of homogenization theory, [18] adopted a unique approach by interpreting the discrete lattice through the lens of finite difference formalism. This method involves applying Taylor's series expansions to the nodal displacements and rotations. This technique is particularly motivated by the multi-scale nature of the problem being addressed, allowing for a more accurate representation of the behavior of the discrete lattice when transitioning to an equivalent continuum model.

The application of crystal theory, as outlined by [6], was instrumental for researchers like [13, 19, 28], in estimating the effective stiffness of various two-dimensional lattice materials. These materials were modeled as pin-jointed, infinite microtruss structures created by tessellating a unit cell into a periodic modular pattern. The approach they employed is based on the Cauchy-Born assumption [5]. This assumption allows for the expression of the lattice's nodal displacements in terms of the macroscopic strain field applied. From this, the homogenized properties of the lattice material can be derived, effectively linking the microscale structure to its macroscopic mechanical behavior.

The mechanical behavior of three-dimensional lattice materials has also been studied using force- and energy equivalence-based techniques. Zhu et al. [31], for instance, formulated the Young's modulus, shear modulus, and Poisson's ratio of the tetrakaidecahedral lattice in terms of its relative density. Furthermore, the mechanical performance of the octet-truss lattice has been explored both theoretically and numerically, as detailed in [7, 10]. These studies not only derived the effective elastic properties of the lattice but also established collapse surfaces for buckling and plastic yielding. The validity of these findings was further reinforced through experimental observations, ensuring a comprehensive understanding of the lattice's mechanical performance.

A broader examination of three-dimensional periodic lattices is presented in [1, 2, 25, 27]. These authors utilized a finite element-based homogenization method to evaluate the macroscopic elastic properties of both open- and closed-cell lattices. Their studies focused on various well-known topologies of the unit cell, such as Body-Centered Cubic (BCC), Face-Centered Cubic (FCC), and octet structures. This approach allowed for a detailed and comprehensive understanding of how different lattice topologies influence the overall elastic properties of the material.

1.3 Scope and Outline of the Present Paper

Though the above-mentioned authors numerically and theoretically analysed the mechanical behavior of architectured materials, up to date closed-form expressions for the effective elastic moduli have been derived in a very limited number of cases. To make a contribution to this incomplete research area and to provide some useful tools for practical applications, this paper suggests a modelling technique to derive explicit relations for the macroscopic elastic properties of orthotropic hexatruss lattice materials which were initially introduced by [11] for the cubic case, although not yet studied in the literature for the orthotropic case. For the sake of conciseness, we focused in this work on the stiffness in the three principal directions, as we could measure them experimentally and were of interest for the industrial application considered in this work.

The work is organised in 4 sections, including this introduction. Initially, by modelling the examined lattice as a sequence of three-dimensional Euler-Bernoulli beams, an energy-based approach coupled with the Cauchy-Born assumption provide, in Sect. 2, closed-form expressions for the macroscopic elastic moduli. The outcome of the experimental campaign to verify the derived analytical equations is described in Sect. 3. A very good agreement generally emerges. Finally, Sect. 4 summarises the main findings. This paper is an extension of the previously available conference paper [23].

2 The Hexatruss Lattice: Theoretical Modelling and Homogenization of the Discrete System

2.1 Problem Statement

This paper focuses on the three-dimensional auxetic microstructure presented by [11], generated by tessellating a hexatruss unit cell (Fig. 1) through the space along three independent periodic vectors, $\mathbf{l}_1$, $\mathbf{l}_2$ and $\mathbf{l}_3$. In the global reference system, defined by the unit orthonormal vectors $\boldsymbol{e}_1$, $\boldsymbol{e}_2$, $\boldsymbol{e}_3$ and by the coordinate system (X, Y, Z), the components of the lattice vectors are

$$\mathbf{l}_1 = \left[L_1 \; 0 \; 0 \right]^T, \quad \mathbf{l}_2 = \left[0 \; L_2 \; 0 \right]^T, \quad \mathbf{l}_3 = \left[0 \; 0 \; L_3 \right]^T, \tag{1}$$

with L_1, L_2, L_3 the overall dimensions of the cell (the lattice size) in the directions $\boldsymbol{e}_1$, $\boldsymbol{e}_2$, $\boldsymbol{e}_3$, respectively.

The resulting system can be conceived as a collection of discrete elements connected at nodes locations periodically arranged. In particular, two classes of nodes can be identified: the internal nodes and the boundary nodes. The first connect elements of only a single cell while the second connect elements of confining cells. Also, due to the periodicity of the lattice configuration, the boundary nodes are corresponding

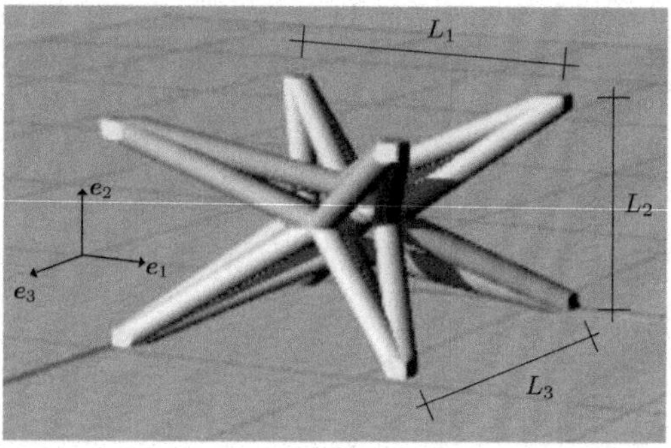

Fig. 1 The hexatruss unit cell

along the lattice vectors and, as outlined in the following sections, such nodes will be subjected to the imposition of appropriate constraints in order to preserve the periodicity of the deformed lattice.

Finally, from a mechanical point of view, the examined microstructure is treated as an interconnected network of three-dimensional Euler-Bernoulli beams having circular cross-section of radius r and made of a linear elastic, homogeneous and isotropic material with Young modulus E_s, Poisson ration ν_s and shear modulus G_s.

2.2 *Continuum Modelling*

2.2.1 Overview

At the macroscopic scale, the effective properties of a lattice material can be derived from the geometric and mechanical parameters of the microstructure. However, analysing large size volumes on a microstructural level to gain an accurate estimation of the local fields is unsuitable and, in some cases, may involve considerable efforts. Thus, the typical approach to the continuum modelling of architectured materials includes the selection of a *Representative Volume Element* (RVE) that is then subjected to the imposition of appropriate boundary conditions to model different loading situations.

By decomposing the displacement field on the material points of the RVE, $\boldsymbol{x} \in V$, into the main part $\mathbf{E} \cdot \boldsymbol{x}$ and the zero-mean fluctuation part $\boldsymbol{u}^*$,

$$\boldsymbol{u} := \mathbf{E} \cdot \boldsymbol{x} + \boldsymbol{u}^*, \qquad \forall \boldsymbol{x} \in V, \tag{2}$$

with

$$\mathbf{E} := \begin{bmatrix} E_{11} & E_{12} & E_{13} \\ E_{12} & E_{22} & E_{23} \\ E_{13} & E_{23} & E_{33} \end{bmatrix} \tag{3}$$

the macroscopic strain tensor, periodic conditions enforce the periodicity of the fluctuation field and the anti-periodicity of the traction field $\boldsymbol{t}$ on the RVE's boundary. Namely,

$$\begin{cases} \boldsymbol{u}^*_+ = \boldsymbol{u}^*_-, & \forall \boldsymbol{x}_+ \in \partial V_+ \text{ and matching } \boldsymbol{x}_- \in \partial V_-, \\ \boldsymbol{\Sigma} \cdot \boldsymbol{n}_+ := \boldsymbol{t}_+ = \boldsymbol{t}_- =: \boldsymbol{\Sigma} \cdot \boldsymbol{n}_-, & \forall \boldsymbol{x}_+ \in \partial V_+ \text{ and matching } \boldsymbol{x}_- \in \partial V_-, \\ \partial V_+ \cup \partial V_- = \partial V, \\ \partial V_+ \cap \partial V_- = \varnothing, \end{cases} \tag{4}$$

where $\boldsymbol{n}_+$ and $\boldsymbol{n}_-$ stand, respectively, for the unit normal to ∂V_+ and ∂V_-,

$$\boldsymbol{\Sigma} := \begin{bmatrix} \Sigma_{11} & \Sigma_{12} & \Sigma_{13} \\ \Sigma_{12} & \Sigma_{22} & \Sigma_{23} \\ \Sigma_{13} & \Sigma_{23} & \Sigma_{33} \end{bmatrix} \tag{5}$$

the macroscopic stress tensor.

Periodic boundary conditions will be selected in Sect. 2.2.3 to investigate the mechanical response of the examined microstructure, whose periodic configuration allows us to identify the RVE with the repetitive unit cell of the tessellation [22].

2.2.2 The Hexatruss Unit Cell as RVE: Geometric Description and Energetics

The hexatruss unit cell is composed by the eight boundary nodes 1, 2, 3, 4, 5, 6, 7, 8 linked to the internal ones 9, 10, 11, 12, 13, 14 by the line elements

$$1-9,\ 2-9,\ 6-9,\ 5-9 \quad \text{(Face Front)}$$

$$4-10,\ 3-10,\ 7-10,\ 8-10 \quad \text{(Face Back)}$$

$$1-11,\ 2-11,\ 3-11,\ 4-11 \quad \text{(Face Bottom)}$$

$$5-12,\ 6-12,\ 7-12,\ 8-12 \quad \text{(Face Top)}$$

$$1-13,\ 4-13,\ 8-13,\ 5-13 \quad \text{(Face Left)}$$

$$2-14,\ 3-14,\ 7-14,\ 6-14 \quad \text{(Face Right)}$$

that, in the global reference system $(\boldsymbol{e}_1, \boldsymbol{e}_2, \boldsymbol{e}_3)$, are described by the vectors

$$\mathbf{b}_{A-B} = \mathbf{p}_B - \mathbf{p}_A, \tag{6}$$

with $\mathbf{p}_A$ and $\mathbf{p}_B$, respectively, the position vectors of the extreme nodes of the element, A and B the index listed in Table 1. For sake of clarity, it is worth noting that, in Eq. (6), $\mathbf{p}_A$ and $\mathbf{p}_B$ are expressed in the global reference system so that, to differentiate the global variables from the local ones (cf. Sect. 1.2), uppercase letters, i.e., indices A and B, are used. For the interested reader, the components of the $\mathbf{b}_{A-B}$ vectors are listed in Appendix A.

As anticipated in Sect. 1.1, from a mechanical point of view, the connecting elements are represented as three-dimensional Euler-Bernoulli beams and this assumption, along with an energy-based approach, lead to closed-form expressions for the effective elastic moduli of the examined lattice material.

Specifically, in terms of energy, for any given deformation the elastic energy representative of the whole discrete structure, W, can be evaluated from that of the beams composing the unit cell of the periodic array. Namely, after writing the equation for the elastic strain energy of the Euler-Bernoulli beam in the global reference system and summing the energetic contribution of each beam, it emerges

$$W = \frac{1}{2}\mathbf{D}^T \cdot \mathbf{K}\mathbf{D}, \tag{7}$$

with

$$\mathbf{D} := \begin{bmatrix} \mathbf{D}_1 \\ \mathbf{D}_2 \\ \cdots \\ \mathbf{D}_{14} \end{bmatrix} = \begin{bmatrix} U_1 \; V_1 \; W_1 \; \Phi_{X1} \; \Phi_{Y1} \; \Phi_{Z1} \; \cdots \; \Phi_{Z14} \end{bmatrix}^T \tag{8}$$

the global displacements vector, collecting the displacements and rotations of all nodes of the unit cell and $\mathbf{K}$ the global stiffness matrix, assembled by adopting standard techniques of the finite element method [14].

2.2.3 Elastic Energy and Effective Moduli of the Equivalent Continuum

The continuum form of W can be derived by adopting the Cauchy-Born relation [27, 29], providing a viable approach to express the displacements of each node of the unit cell as a function of the macroscopic strain field applied, $\mathbf{E}$.

Specifically, if we introduce the vector $\mathbf{D}_R$ collecting the displacements of the generic Rh node within the cell and the corresponding position vector $\mathbf{p}_R$, the Cauchy-Born relation, in its general form, can be expressed as

$$\mathbf{D}_R = \mathbf{D}_0 + \mathbf{E} \cdot (\mathbf{p}_R - \mathbf{p}_0), \tag{9}$$

with $\mathbf{D}_0$ and $\mathbf{p}_0$, in turn, the displacements vector and position vector of an appropriately selected unit cell's independent node. Also, by enforcing the equilibrium conditions on the nodal forces [28], it is possible to calculate the displacements $\mathbf{D}_0$ in terms of $\mathbf{E}$.

Notwithstanding alternative choices of the independent node are possible [27], in this work the boundary node 4 has been selected as independent node, leading to

$$\mathbf{D}_R = \mathbf{D}_4 + \mathbf{E} \cdot (\mathbf{p}_R - \mathbf{p}_4), \qquad R = 1, 2, \ldots, 14. \tag{10}$$

This assumption provides an easy way to impose kinematic conditions on the boundary nodes of the cell,

$$\begin{aligned} \mathbf{D}_1 &= \mathbf{D}_4 + \mathbf{E} \cdot \mathbf{l}_3, & \mathbf{D}_2 &= \mathbf{D}_4 + \mathbf{E} \cdot (\mathbf{l}_1 + \mathbf{l}_3), \\ \mathbf{D}_3 &= \mathbf{D}_4 + \mathbf{E} \cdot \mathbf{l}_1, & \mathbf{D}_5 &= \mathbf{D}_4 + \mathbf{E} \cdot (\mathbf{l}_2 + \mathbf{l}_3), \\ \mathbf{D}_6 &= \mathbf{D}_4 + \mathbf{E} \cdot (\mathbf{l}_1 + \mathbf{l}_2 + \mathbf{l}_3), \\ \mathbf{D}_7 &= \mathbf{D}_4 + \mathbf{E} \cdot (\mathbf{l}_1 + \mathbf{l}_2), & \mathbf{D}_8 &= \mathbf{D}_4 + \mathbf{E} \cdot \mathbf{l}_2, \end{aligned} \tag{11}$$

that coincide, as anticipated in Sect. 2.2.1, with the application of periodic boundary conditions in order to enforce the periodicity of the deformed configuration.

Substituting Eqs. (10) into (7) and dividing the expression that turns out from the calculation by the volume of the unit cell, $V = L_1 L_2 L_3$, give the strain energy density in the continuum approximation, w. It emerges that w is a quadratic form of the infinitesimal strains E_{ij},

$$w = w\left(E_{ij}\right), \quad i, j = 1, 2, 3, \tag{12}$$

and, as in classical continuum mechanics, evaluating the Hessian of the strain energy density with respect to the deformation components enables to derive the entries

$$C_{ijkl} = \frac{\partial^2 w}{\partial E_{ij}\, \partial E_{kl}}, \quad i, j, k, l = 1, 2, 3 \tag{13}$$

of the fourth-order tensor of effective elastic moduli of the lattice, $\boldsymbol{C}$, that relates the macroscopic stress, Σ_{ij}, and strain, E_{ij}, components as follows:

$$\Sigma_{ij} = C_{ijkl} E_{kl}, \quad i, j, k, l = 1, 2, 3. \tag{14}$$

Using Voigt's notation, the effective elastic moduli depending on the geometrical and mechanical parameters of the microstructure can be derived:

$$C_{ij} = C_{ij}\left(E_s,\, \nu_s,\, L_1,\, L_2,\, L_3,\, r,\, d_1,\, d_2,\, d_3\right), \quad i, j = 1, 2, ..., 6, \tag{15}$$

with E_s and ν_s the mechanical characteristics of the constituent material, r the radius of the beams' cross-section, L_k and d_k, $k = 1, 2, 3$.

The effective moduli to be considered correspond to the principal directions of the cell $\boldsymbol{e}_1, \boldsymbol{e}_2, \boldsymbol{e}_3$:

$$C_X := C\left(\boldsymbol{e}_1\right), \qquad C_Y := C\left(\boldsymbol{e}_2\right), \qquad C_Z := C\left(\boldsymbol{e}_3\right), \tag{16}$$

where, for sake of clarity, the subscripts X, Y, Z have been used to denote the effective principal stiffness, $C_{(\cdot)}$, with reference to the axes $\boldsymbol{e}_1, \boldsymbol{e}_2, \boldsymbol{e}_3$.

It emerges that the microstructural parameters strongly affect the effective properties listed in Eq. (16) and also in this case explicit relations, reported in Appendix B, can be obtained:

$$
\begin{aligned}
C_X &= \frac{\partial^2 w}{\partial E_{11}\, \partial E_{11}} = C_X\left(E_s,\, \nu_s,\, L_1,\, L_2,\, L_3,\, r,\, d_1,\, d_2,\, d_3\right), \\
C_Y &= \frac{\partial^2 w}{\partial E_{22}\, \partial E_{22}} = C_Y\left(E_s,\, \nu_s,\, L_1,\, L_2,\, L_3,\, r,\, d_1,\, d_2,\, d_3\right), \qquad (17) \\
C_Z &= \frac{\partial^2 w}{\partial E_{33}\, \partial E_{33}} = C_Z\left(E_s,\, \nu_s,\, L_1,\, L_2,\, L_3,\, r,\, d_1,\, d_2,\, d_3\right).
\end{aligned}
$$

From this point of view, the advantage of the proposed theory is that an explicit relation between the microstructural parameters and the macroscopic moduli can be obtained. In a practical context, this offers three advantages. Firstly, a full understanding of the interplay between the macroscopic properties and the underlying microstructure, suggesting an alternative route to the time-consuming experimental investigations for predicting the macroscopic stiffness. Secondly, an assistance for the designer in the selection of the microstructural parameters, dimension of the unit cell, radius of the beams' cross-section, position of the internal nodes, in order to reach a prescribed value of macroscopic stiffness. Finally, the possibility to formulate a more efficient and less computationally expensive optimization problem in order to derive the microstructural parameters satisfying a given set of stiffness requirements. In particular, the optimization problem that would be established would coincide with the optimization of the analytical expressions in Eq. (17) so that, to evaluate the optimal parameters, the re-generation and corresponding analysis of the lattice are not required at each iteration.

3 Validation of the Theoretical Results by Experimental Confrontation

In this section, the analytical results proposed in Eq. (17) are verified by comparison with those obtained by performing experimental tests.

To further establish the validity of the proposed modelling approach, the theoretical findings in Eq. (17) are compared with the outcome of the experimental tests presented in the following.

Also in this case, for simplicity, we will refer to E_X, E_Y and E_Z as the apparent Young moduli of the lattice along the principal directions of the unit cell, $\boldsymbol{e}_1$, $\boldsymbol{e}_2$ and $\boldsymbol{e}_3$, making the assumption that the effective elastic moduli in principal directions defined in Sect. 2.2.3 are similar in value to the experimentally measured Young moduli, therefore assuming a Poisson ratio close to zero, which is reasonable in the principal direction for an hexatruss lattice [11].

3.1 Description of the Samples

The mechanical tests have been conducted on 3D-printed $3 \times 3 \times 3$ unit cell hexatruss lattices and, to obtain an estimation of the effective stiffness in the directions $\boldsymbol{e}_1$, $\boldsymbol{e}_2$ and $\boldsymbol{e}_3$, three configurations have been investigated (Fig. 2).

- Configuration $C1$: to consider the direction $\boldsymbol{e}_1$ and derive E_X,
- Configuration $C2$: to consider the direction $\boldsymbol{e}_2$ and derive E_Y,
- Configuration $C3$: to consider the direction $\boldsymbol{e}_3$ and derive E_Z.

Aiming at obtaining a good quality of the printed samples, a unit cell with overall dimensions

$$L_1 = 9.6\,\text{mm}, \qquad L_2 = 11.52\,\text{mm}, \qquad L_3 = 7.68\,\text{mm} \tag{18}$$

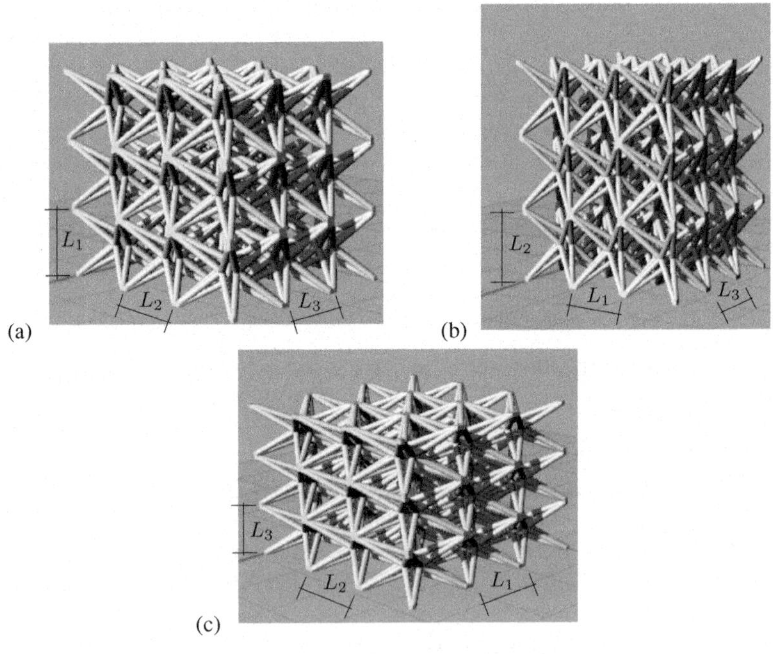

Fig. 2 The three different configurations of the samples involved in the mechanical tests: **a** configuration C1, **b** configuration C2, **c** configuration C3

has been selected, leading to

- Configuration $C1$ (length × width × height): 34.56 mm × 23.04 mm × 28.8 mm,
- Configuration $C2$ (length × width × height): 28.8 mm × 23.04 mm × 34.56 mm,
- Configuration $C3$ (length × width × height): 28.8 mm × 34.56 mm × 23.04 mm.

In terms of radius of the beams' cross-section, r, and position of the internal nodes, parameters d_1, d_2 and d_3, the following values have been adopted:

$$\begin{gathered} r = 0.45\,\text{mm}, \\ d_1 = 2.8\,\text{mm}, \quad d_2 = 2.2\,\text{mm}, \quad d_3 = 3.3\,\text{mm}. \end{gathered} \tag{19}$$

Regarding the constituent material, two different polymers have been used to print the samples: PLA (Polylactic Acid) and PET (Polyethylene Terephthalate) having, on order, Young modulus of

$$E_{PLA} = 1.3\,\text{GPa}, \qquad E_{PET} = 0.8\,\text{GPa}, \tag{20}$$

values obtained by conducting traction tests according to the ASTM D638-14 standard [4]. Also, as suggested in the literature [15, 26], a Poisson ratio of

$$\nu_{PLA} = 0.33, \qquad \nu_{PET} = 0.43\,\text{GPa} \tag{21}$$

has been respectively assumed.

Finally, to account for possible errors associated with the 3D-printing process, three copies of each configuration have been printed and tested.

3.2 *Type of Tests*

To characterise the 3D-printed samples, the following mechanical tests have been performed (Fig. 3).

- uniaxial compression tests at room temperature,
- displacement-controlled tests at a displacement rate of 0.5 mm/min,
- only elastic deformations involved, $\varepsilon_{\max} = 1\,\%$,
- no interfaces between the sample and the test bench,
- no constraints applied to movement (lateral displacements allowed).

and the testing machine *Instron 5881* has been employed.

Fig. 3 Experimental validation: compression tests

3.3 Methodology Adopted

As anticipated in Sect. 3.2, the aim of the tests was to evaluate the effective stiffness of the hexatruss lattice in the three principal directions of the unit cell, $\boldsymbol{e}_1$, $\boldsymbol{e}_2$ and $\boldsymbol{e}_3$, obtained by analysing the force-displacement and stress-strain curves experimentally obtained.

The force-displacement curves, in particular, have been derived by recording, during the test, both the displacement, ΔL_{Ci}, $i = 1, 2, 3$, and resulting load, F_{Ci}, $i = 1, 2, 3$, applied to the sample. By knowing the sample geometry, it was then possible to generate the stress-strain curves, being the engineering strain, ε_{Ci}, and the engineering stress, σ_{Ci}, given by

$$\varepsilon_{Ci} := \frac{\Delta L_{Ci}}{L_{Ci}} \quad \text{and} \quad \sigma_{Ci} := \frac{F_{Ci}}{A_{Ci}}, \qquad i = 1, 2, 3, \tag{22}$$

with L_{Ci} and A_{Ci}, in turn, the initial height of the sample and its initial cross-sectional area normal to the loading direction.

Finally, since we are working in a linear elastic regime, the slope of the stress-strain curves provides the Young moduli E_{C1}, E_{C2}, E_{C3} corresponding to the three configurations tested $C1$, $C2$, $C3$:

$$E_{C1} := \frac{\sigma_{C1}}{\varepsilon_{C1}}, \quad E_{C2} := \frac{\sigma_{C2}}{\varepsilon_{C2}}, \quad E_{C3} := \frac{\sigma_{C3}}{\varepsilon_{C3}}. \tag{23}$$

As stated, for simplicity, they are assumed to coincide with the effective Young moduli of the lattice along its principal directions $\boldsymbol{e}_1$, $\boldsymbol{e}_2$, $\boldsymbol{e}_3$:

$$E_{C1} \equiv E_X, \quad E_{C2} \equiv E_Y, \quad E_{C3} \equiv E_Z. \tag{24}$$

3.4 Results and Discussion

The outcome of the experimental campaign is gathered in Appendix C as Figs. 4 and 5, in terms of force-displacement curves, and in Figs. 6 and 7, in terms of stress-strain curves. The derived Young moduli are reported in Tables 1 and 2, where the comparison between the experimental results and the theoretical predictions in Eq. (17) is also included.

It should be noted that the curves illustrated in Figs. 4, 5, 6 and 7 and the experimental values listed in Tables 1 and 2 represent, for each configuration, the average response of the three samples tested (cf. Sect. 3.2). Also, in Tables 1 and 2, the analytical values are obtained by substituting into Eq. (17) the parameters listed in Eqs. (18), (19) and the material properties, that coincide with the geometrical and mechanical characteristics of the unit cell adopted in the experimental tests.

As it can be seen, it generally emerges a good agreement between the theoretical and experimental results and, as expected, the theoretical values overestimate the experimental ones of averagely 2–3%. From this point of view, the difference is mainly related to the quality of the 3D-printed samples that, differently from the analytical model where the beams are assumed to be perfectly homogeneous and without defects, in some cases present some imperfections, i.e., small parts missing, not perfectly homogeneous or beams not perfectly connected at the nodes.

However, taking into account the very close comparison in Tables 1 and 2, it can be said that the experimental tests confirm the validity of our modelling strategy. This can also be observed in Figs. 6 and 7, where the experimental stress-strain curves are compared with the theoretical ones. The latter, in particular, are obtained from the analytical values of Table 1, Fig. 6, and Table 2, Fig. 7.

4 Conclusions

Architectured materials, due to their discrete configuration, present enhanced mechanical properties such as stiffness, strength, and toughness at a lower density. This unique attribute, combined with the ability to customize their mechanical properties, renders them highly promising for a wide range of engineering applications, including aerospace, automotive, marine, and construction sectors. Unlike traditional homogeneous materials, architectured materials necessitate specialized constitutive models to elucidate the intricate relationship between microstructural parameters and macroscopic properties. Although various techniques and methods have been proposed in recent literature, only a few studies offer closed-form expressions for the effective mechanical properties. These expressions are crucial for designers to select materials that best meet specific requirements.

Addressing this research gap, this paper focuses on the analysis of a three-dimensional orthotropic lattice composed of hexatruss cells. Using linear elasticity and modeling the microstructure as a series of three-dimensional Euler-Bernoulli

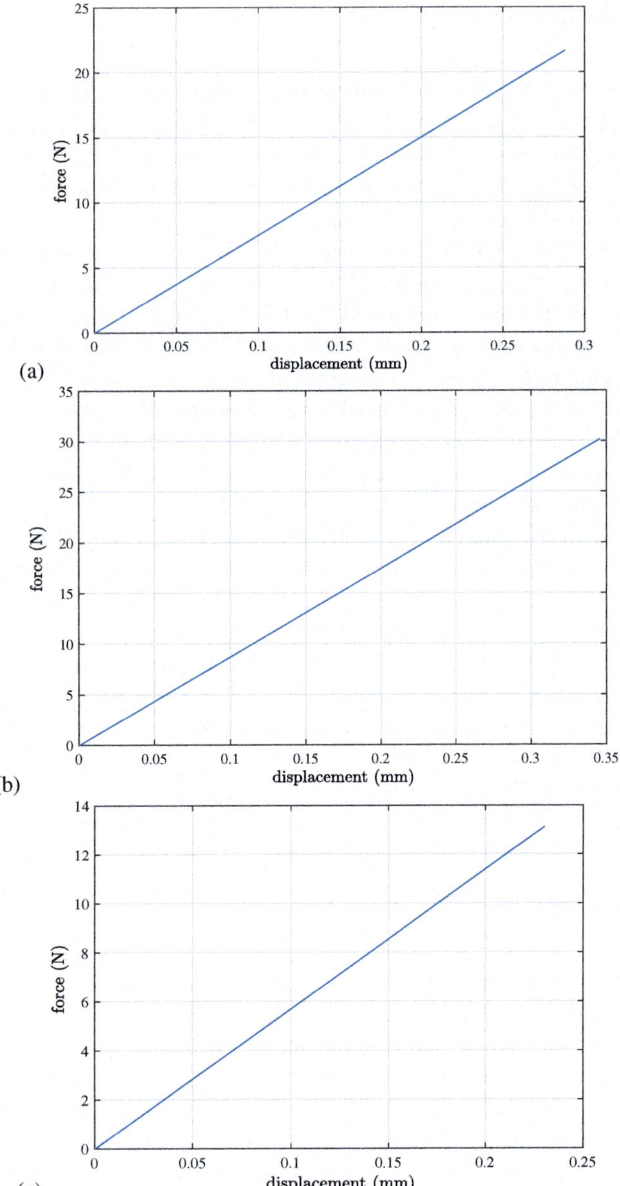

Fig. 4 PLA samples, compression tests: experimental force-displacement curves for **a** configuration C1, **b** configuration C2, **c** configuration C3

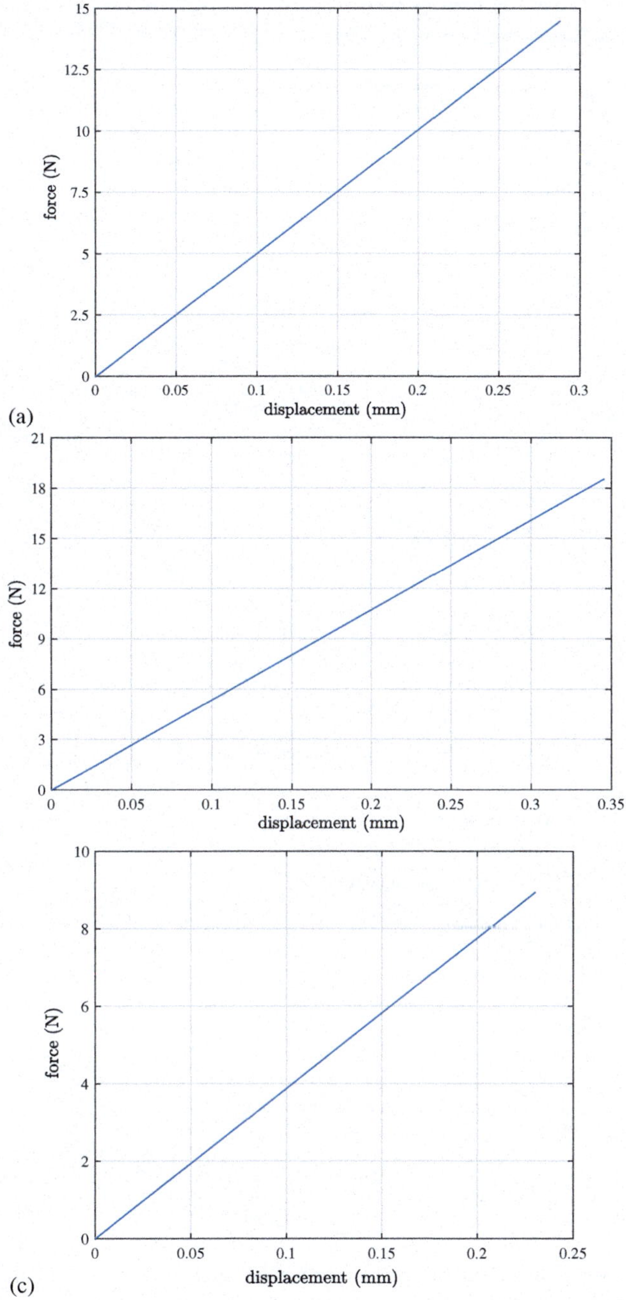

Fig. 5 PET samples, compression tests: experimental force-displacement curves for **a** configuration C1, **b** configuration C2, **c** configuration C3

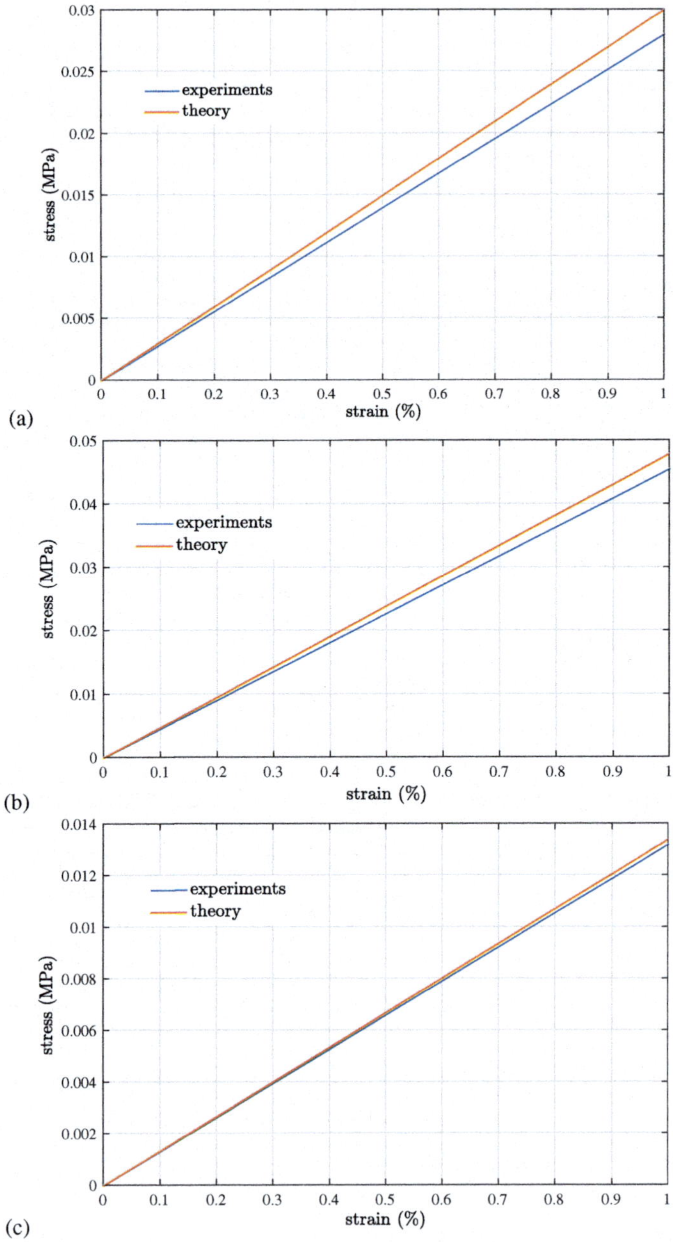

Fig. 6 PLA samples, compression tests: experimental stress-strain curves for **a** configuration C1, **b** configuration C2, **c** configuration C3

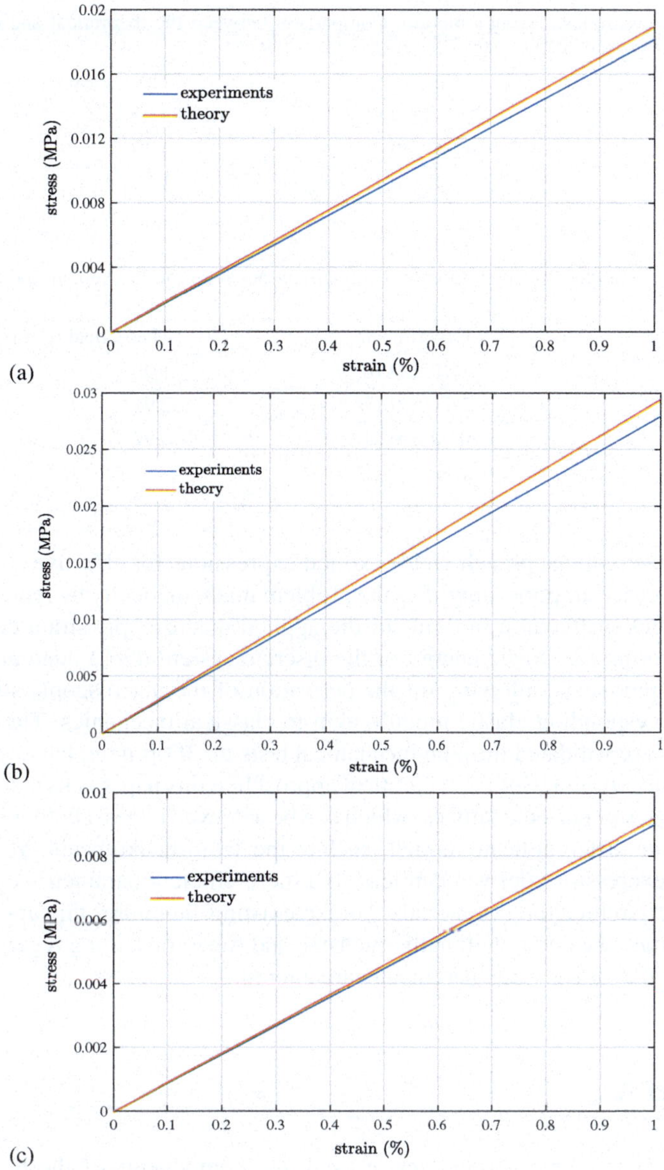

Fig. 7 PET samples, compression tests: experimental stress-strain curves for **a** configuration C1, **b** configuration C2, **c** configuration C3

Table 1 PLA samples, Young's moduli. Comparison between the theoretical and experimental results

	Experiments	Theoretical results
E_X (MPa)	2.72 ± 0.34	3.09
E_Y (MPa)	4.55 ± 0.25	4.79
E_Z (MPa)	1.32 ± 0.11	1.34

Table 2 PET samples, Young's moduli. Comparison between the theoretical and experimental results

	Experiments	Theoretical results
E_X (MPa)	1.81 ± 0.10	1.90
E_Y (MPa)	2.71 ± 0.17	2.95
E_Z (MPa)	0.90 ± 0.04	0.92

beams, an energetic approach yields explicit expressions for effective stiffness. The Cauchy-Born assumption simplifies the problem mathematically by representing the nodal degrees of freedom in terms of the applied macroscopic strain components. This transforms the elastic energy of the discrete system into a quadratic function of these components, allowing for the derivation of the macroscopic stiffness tensor and corresponding elastic moduli, akin to classical mechanics. The theoretical predictions are validated through mechanical tests on 3D-printed lattices.

A notable advantage of the theoretical approach in this paper is its extensibility to different types of periodic lattices, which can be achieved by tessellating various unit cells or using different beam models, such as the Timoshenko beam. Moreover, the analytical expressions derived can lead to a more efficient parametric optimization problem for architectured materials. Unlike existing literature, this approach does not necessitate the computationally intensive and time-consuming regeneration and analysis of the lattice when parameters are altered.

Appendix A

In the global reference system ($\boldsymbol{e}_1$, $\boldsymbol{e}_2$, $\boldsymbol{e}_3$), the components of the $\mathbf{b}_{A-B}$ vectors introduced in Sect. 1.3 are

$$
\mathbf{b}_{1-9} = \begin{bmatrix} L_1/2 \\ L_2/2 \\ -d_1 \end{bmatrix}, \quad \mathbf{b}_{2-9} = \begin{bmatrix} -L_1/2 \\ L_2/2 \\ -d_1 \end{bmatrix}, \quad \mathbf{b}_{6-9} = \begin{bmatrix} -L_1/2 \\ -L_2/2 \\ -d_1 \end{bmatrix}, \quad \mathbf{b}_{5-9} = \begin{bmatrix} L_1/2 \\ -L_2/2 \\ -d_1 \end{bmatrix},
$$

$$
\mathbf{b}_{4-10} = \begin{bmatrix} L_1/2 \\ L_2/2 \\ d_1 \end{bmatrix}, \quad \mathbf{b}_{3-10} = \begin{bmatrix} -L_1/2 \\ L_2/2 \\ d_1 \end{bmatrix}, \quad \mathbf{b}_{7-10} = \begin{bmatrix} -L_1/2 \\ -L_2/2 \\ d_1 \end{bmatrix}, \quad \mathbf{b}_{8-10} = \begin{bmatrix} L_1/2 \\ -L_2/2 \\ d_1 \end{bmatrix},
$$

$$
\mathbf{b}_{1-11} = \begin{bmatrix} L_1/2 \\ d_2 \\ -L_3/2 \end{bmatrix}, \quad \mathbf{b}_{2-11} = \begin{bmatrix} -L_1/2 \\ d_2 \\ -L_3/2 \end{bmatrix}, \quad \mathbf{b}_{3-11} = \begin{bmatrix} -L_1/2 \\ d_2 \\ L_3/2 \end{bmatrix}, \quad \mathbf{b}_{4-11} = \begin{bmatrix} L_1/2 \\ d_2 \\ L_3/2 \end{bmatrix},
$$

$$
\mathbf{b}_{5-12} = \begin{bmatrix} L_1/2 \\ -d_2 \\ -L_3/2 \end{bmatrix}, \quad \mathbf{b}_{6-12} = \begin{bmatrix} -L_1/2 \\ -d_2 \\ -L_3/2 \end{bmatrix}, \quad \mathbf{b}_{7-12} = \begin{bmatrix} -L_1/2 \\ -d_2 \\ L_3/2 \end{bmatrix}, \quad \mathbf{b}_{8-12} = \begin{bmatrix} L_1/2 \\ -d_2 \\ L_3/2 \end{bmatrix},
$$

$$
\mathbf{b}_{1-13} = \begin{bmatrix} d_3 \\ L_2/2 \\ -L_3/2 \end{bmatrix}, \quad \mathbf{b}_{4-13} = \begin{bmatrix} d_3 \\ L_2/2 \\ L_3/2 \end{bmatrix}, \quad \mathbf{b}_{8-13} = \begin{bmatrix} d_3 \\ -L_2/2 \\ L_3/2 \end{bmatrix}, \quad \mathbf{b}_{5-13} = \begin{bmatrix} d_3 \\ -L_2/2 \\ -L_3/2 \end{bmatrix},
$$

$$
\mathbf{b}_{2-14} = \begin{bmatrix} -d_3 \\ L_2/2 \\ -L_3/2 \end{bmatrix}, \quad \mathbf{b}_{3-14} = \begin{bmatrix} -d_3 \\ L_2/2 \\ L_3/2 \end{bmatrix}, \quad \mathbf{b}_{7-14} = \begin{bmatrix} -d_3 \\ -L_2/2 \\ L_3/2 \end{bmatrix}, \quad \mathbf{b}_{6-14} = \begin{bmatrix} -d_3 \\ -L_2/2 \\ -L_3/2 \end{bmatrix}, \tag{25}
$$

with L_1, L_2, L_3, in turn, the dimensions of the unit cell in the directions $\boldsymbol{e}_1$, $\boldsymbol{e}_2$, $\boldsymbol{e}_3$ and d_1, d_2, d_3 the parameters specifying the position of the internal nodes.

Appendix B

The complete expressions of the effective elastic moduli in the directions $\boldsymbol{e}_1$, $\boldsymbol{e}_2$ and $\boldsymbol{e}_3$ have been derived using Matlab, and are respectively,

$$
\begin{aligned}
C_X &= p_{12}\left(\left(AL_1^2\, p_5 + 48I\left(p_6 + 4d_2^2\, p_1^{5/2}\right)\right) p_3^{5/2} + 64\, d_3^6\, A\,(p_1 p_2)^{5/2} + d_3^4\, p_4\,(16\,A + 192I)\right) E_s, \\
C_Y &= p_{12}\left(4d_3^2\,(48I + AL_2^2)\,(p_1 p_2)^{5/2} + 48I p_7 + A\,(p_8 + p_9)\right) E_s, \\
C_Z &= p_{12}\left(192 d_3^2 I\,(p_1 p_2)^{5/2} + A\, p_1\, p_2\, p_3\, p_{10} + 48I p_{11}\right) E_s,
\end{aligned} \tag{26}
$$

where, to simplify the notation, $A := \pi r^2$, $I := \pi r^4/4$, E_s the Young's modulus of the lattice's constituent material and

$$
\begin{aligned}
&p_1 = 4d_1^2 + L_1^2 + L_2^2, \qquad p_2 = 4d_2^2 + L_1^2 + L_3^2, \qquad p_3 = 4d_3^2 + L_2^2 + L_3^2, \\
&p_4 = p_1^{5/2} p_2^{5/2} \left(L_2^2 + L_3^2\right), \qquad p_5 = p_2^{5/2} \left(4d_1^2 + L_2^2\right) + L_1^2 \left(p_1^{5/2} + p_2^{5/2}\right) + p_1^{5/2} \left(L_3^2 + 4d_2^2\right), \\
&p_6 = \left(p_2^{5/2} \left(4d_1^2 + L_2^2\right) + L_3^2 p_1^{5/2}\right), \\
&p_7 = \left(4d_2^2 L_1^2 p_1^{5/2} + \left(4d_1^2 + L_1^2\right) p_2^{5/2}\right) p_3^{5/2} + L_3^2 p_1^{5/2} \left(p_2^{5/2} + 4d_2^2 p_3^{5/2}\right), \\
&p_8 = 16 d_2^4 \left(4d_2^2 + L_1^2 + L_3^2\right) p_1^{5/2} p_3^{5/2} + L_2^4 p_2^{5/2} \left(p_1^{5/2} + p_3^{5/2}\right), \\
&p_9 = L_2^2 p_2^{5/2} \left(L_3^2 p_1^{5/2} + \left(4d_1^2 + L_1^2\right) p_3^{5/2}\right), \qquad p_{10} = 16 d_1^4 p_2^{3/2} p_3^{3/2} + L_3^2 p_1^{3/2} \left(p_2^{3/2} + p_3^{3/2}\right), \\
&p_{11} = \left(4d_2^2 p_1^{5/2} + L_1^2 \left(p_1^{5/2} + 4d_1^2 p_2^{5/2}\right)\right) p_3^{5/2} + L_2^2 p_2^{5/2} \left(p_1^{5/2} + 4d_1^2 p_3^{5/2}\right), \\
&p_{12} = \frac{16\, A \sqrt{p_1} + \sqrt{p_2} + \sqrt{p_3}}{\left(L_2 L_3\right)^2 \left(p_1 p_2 p_3\right)^{5/2}}.
\end{aligned}
\tag{27}
$$

Appendix C

See Figs. 4, 5, 6 and 7.

References

1. Albertini, F., Dirrenberger, J., Molotnikov, A., Sollogoub, C.: Computational investigation of the effective mechanical behavior for 3D pre-buckled auxetic lattices. J. Appl. Mech. **86**, 111003 (2019)
2. Albertini, F., Dirrenberger, J., Sollogoub, C., Maconachie, T., Leary, M., Molotnikov, A.: Experimental and computational analysis of the mechanical properties of composite auxetic lattice structures. Addit. Manuf. **47**, 102351 (2021)
3. Ashby, M.F.: Designing architectured materials. Scripta Materialia **68**(1), 4–7 (2013)
4. ASTM. D638-14: Standard test method for tensile properties of plastics (2014)
5. Born, M., Huang, K.: Dynamical Theory of Crystal Lattices. Oxford University Press (1954)
6. Brillouin, L.: Wave Propagation in Periodic Structures. Dover (1953)
7. Challapalli, A.: Continuum model for effective properties of orthotropic octet-truss lattice through additive manufacturing. Int. J. Comput. Methods Eng. Sci. Mech. 1–11 (2020). https://doi.org/10.1080/15502287.2020.1711829
8. Chen, J.Y., Huang, Y., Ortiz, M.: Fracture analysis of cellular materials: a strain gradient model. J. Mech. Phys. Solids **46**(5), 789–828 (1998)
9. Christensen, R.M.: Mechanics of cellular and other low-density materials. Int. J. Solids Struct. **37**, 93–104 (2000)
10. Deshpande, V.S., Fleck, N.A., Ashby, M.F.: Effective properties of the octet-truss lattice material. J. Mech. Phys. Solids **49**(8), 1747–1769 (2001)
11. Dirrenberger, J., Forest, S., Jeulin, D.: Effective elastic properties of auxetic microstructures: anisotropy and structural applications. Int. J. Mech. Mater. Des. **9**(1), 21–33 (2013). https://doi.org/10.1007/s10999-012-9192-8

12. D'Mello, R.J., Waas, A.M.: In-plane crush response and energy absorption of circular cell honeycomb filled with elastomer. Compos. Struct. **106**, 491–501 (2013)
13. Elsayed, M.S., Pasini, D.: Analysis of the elastostatic specific stiffness of 2d stretching-dominated lattice materials. Mech. Mater. **42**(7), 709–725 (2010)
14. Ferreira, A.J.M.: MATLAB Codes for Finite Element Analysis. Springer, Berlin (2009)
15. Ferreira, R.T.L., Amatte, I.C., Dutra, T.A., Bürger, D.: Experimental characterization and micrography of 3d printed pla and pla reinforced with short carbon fibers. Compos. Part B: Eng. **124**, 88–100 (2017)
16. Gibson, L.J.: Modelling the mechanical behavior of cellular materials. Mater. Sci. Eng. A **110**, 1–36 (1989)
17. Gibson, L.J., Ashby, M.F.: Cellular Solids, 2nd edn. Cambridge University Press (1999)
18. Gonella, S., Ruzzene, M.: Homogenization and equivalent in-plane properties of two-dimensional periodic lattices. Int. J. Solids Struct. **45**(10), 2897–2915 (2008)
19. Hutchinson, R.G., Fleck, N.A.: The structural performance of the periodic truss. J. Mech. Phys. Solids **54**(4), 756–782 (2006)
20. Kraynik, A.M., Neilsen, M.K., Reinelt, D.A., Warren, W.E.: Foam micromechanics. In: Foams and Emulsions: Proceedings of the NATO Advanced Study Institute on Foams, Emulsions and Cellular Materials. Kluwer (1998)
21. Kumar, R.S., McDowell, D.L.: Generalized continuum modeling of 2-d periodic cellular solids. Int. J. Solids Struct. **41**, 7399–7422 (2004)
22. Nemat-Nasser, S., Hori, M.: Micromechanics: Overall Properties of Heterogenous Materials. North-Hollan d Publishing (1993)
23. Ongaro, F., Mathis, K., Masson, F., Dirrenberger, J.: Architectured materials for space applications: a computational tool for the parametric optimization of a three-dimensional lattice subjected to stiffness constraints. In: 8th European Conference for Aeronautics and Aerospace Sciences (EUCASS). Madrid, Spain (2019)
24. Ostoja-Starzewski, M.: Lattice models in micromechanics. Appl. Mech. Rev. **55**(6), 35–60 (2002)
25. Refai, K., Montemurro, M., Brugger, C., Saintier, N.: Determination of the effective elastic properties of titanium lattice structures. Mech. Adv. Mater. Struct. **27**(23), 1966–1982 (2020)
26. Seitz, J.T.: The estimation of mechanical properties of polymers from molecular structure. J. Appl. Polym. Sci. **49**, 1331–1351 (1993)
27. Vigliotti, A., Pasini, D.: Stiffness and strength of tridimensional periodic lattices. Comput. Methods Appl. Mech. Eng. **229**, 27–43 (2012)
28. Vigliotti, A., Pasini, D.: Linear multiscale analysis and finite element validation of stretching and bending dominated lattice materials. Mech. Mater. **46**, 57–68 (2012)
29. Vigliotti, A., Pasini, D.: Structural optimization of lattice materials. In: Proceedings of the ASME 2011 (2012c)
30. Warren, W.E., Kraynik, A.M.: Linear elastic behavior of a low-density kelvin foam with open cells. Mech. Mater. **64**, 787–794 (1997)
31. Zhu, H.X., Knott, J.F., Mills, N.J.: Analysis of the elastic properties of open-cell foams with tetrakaidecahedral cells. J. Mech. Phys. Solids **45**, 319–325 (1997)

Experimental Characterization and Numerical Modeling of the Porosity Formation Mechanisms in Thermoplastic Laminates at High Temperature

David Philippe, Benoit Vieille, and Fabrice Barbe

Abstract Subjecting thermoplastic-based laminates to high thermal energy, such as a flame, leads to the gradual deterioration of the matrix, involving solid-state changes and significant fluctuations in thermomechanical properties. Despite these changes and the presence of substantial temperature gradients, the laminates can still bear mechanical loads, even after the matrix has melted. When temperatures exceed the melting point, the primary mechanism of matrix thermal decomposition involves the formation of voids. While these voids weaken the material from a mechanical perspective, they also act as thermal insulators, protecting the matrix on the side facing away from the heat source. To accurately describe the evolution of thermomechanical properties whereas one face of the laminates is exposed to a flame, it is crucial to understand the kinetics of void formation. Experimental investigations revealed that porosity content and swelling are strongly dependent on the time and temperature of thermal exposure. A mesoscopic Finite Element model was thus developed, representing porosities at a structural level. The formation of voids and the associated swelling were simulated using a probabilistic approach, guiding the progressive transformation of elements into voids based on their thermal state.

Keywords Thermoplastic composites · Thermomechanics · High temperature · Multiscale · Multiphysics

D. Philippe · B. Vieille · F. Barbe (✉)
Groupe de Physique des Matériaux, CNRS, Normandie Université, University of Rouen, INSA Rouen, 76000 Rouen, France
e-mail: fabrice.barbe@insa-rouen.fr

D. Philippe
e-mail: david.philippe@insa-rouen.fr

B. Vieille
e-mail: benoit.vieille@insa-rouen.fr

F. Willot et al. (eds.), *Continuum Models and Discrete Systems*,
Springer Proceedings in Mathematics & Statistics 457,
https://doi.org/10.1007/978-3-031-58665-1_9

1 Introduction

This work discusses the challenges faced by aeronautical structural parts made from polymer matrix composite materials when exposed to high thermal energy conditions. These conditions can range from ambient temperature to the temperature at which the polymer starts to decompose (around 450 °C). The ability of the composite to withstand these conditions depends on the degradation of its constituent properties, particularly the polymer matrix. To understand the mechanics and thermal transfers in such scenarios, a combination of experimental and numerical analyses is required to replicate critical operating conditions like fires in aircraft engines.

In the continuation of previous works [1–3], this study focuses on the behavior of a specific composite material, Quasi-Isotropic Carbon fiber reinforced PolyPhenylene Sulfide (QI C/PPS) laminates, as they are exposed to heat. It discusses how various changes occur in the material, particularly in the polymer matrix, due to temperature fluctuations. Whereas the carbon fiber thermomechanical behavior remains almost temperature independent below 700 °C [4], the thermoplastic polymer matrix undergoes state transitions at different temperatures: glassy at ∼90 °C, melting at ∼270 °C and then thermal decomposition at ∼450 °C, above which pyrolysis takes place [5–7]. As a consequence, in the context of a directional thermal aggression such as a flame, high thermal gradients will take place, involving the different solid-state transformations of the matrix; it leads in particular to the formation of cavities which although they weaken the mechanical properties act as a thermal insulator for the polymer matrix away from the flame. Accurate descriptions and models of the material's thermomechanical behavior in this context need to consider time and temperature dependencies, as well as the heterogeneous nature of the composite's meso-structure, including the local porosity content.

To reach these objectives, this study first proposes an experimental characterization of the porosity formation in order to identify and quantify the various involved mechanisms: nucleation and growth of porosities in the polymer matrix zones, fiber bundle/matrix interface debonding, intra-fiber bundle porosity formation, and the accompanying mechanism of cavity pressurization (due to pyrolysis gas) leading to swelling on the laminates macroscale. The results are then used as inputs for the development of a numerical model reproducing the temperature dependent phenomenon of porosity formation on the local scale of the laminates. The kinetics of porosity formation during the thermal decomposition of the matrix has been widely covered experimentally using mass loss measurements [7–11]. Analytical laws have been found to numerically reproduce this kinetics [1, 12–14]. However, these approaches only consider the porosity content as a variable of the numerical model, while an explicit representation of the porosity distribution is required to accurately take into account their influence on the laminates behavior [1, 15, 16]. The proposed numerical approach consists in explicitely representing the biphase meso-structure of the laminates (woven fiber bundles and matrix) and in explicitely introducing the nucleation and growth of porosities from a temperature and time dependent kinetics law identified from the experiments. Various possibilities to be examined in future studies are presented to account for the experimental results regarding the fiber bundle/matrix interface debonding and the intra-fiber bundle porosities.

2 Porosity Formation Mechanisms

During the exposure to a temperature above that of polymer matrix decomposition, porosities gradually form within the material, with a macroscale kinetics law significantly depending on time and temperature [3]. To have a view on the corresponding porosity network on the local scale of the laminates, Fig. 1a presents a 3D representation, as obtained from X-Ray tomography analysis, of the porosities distribution after 5 min of isothermal treatment at 530 °C. A 2D representation is provided in Fig. 1b: a 2D plane at half thickness of the laminates (i.e. a plane parallel to the plate of the laminates) has been extracted and segmentation was set such as to show only the polymer and the porosities. The X-Ray tomography and optical microscopy analyses of samples isothermally decomposed for different temperatures and times (see Ref. [3]) have enabled to draw up a classification of the porosities according to four main categories:

- 1: Small circular porosities which have just nucleated.
- 2: Intermediate circular porosities which have grown/coalesced.
- 3: Large macro-porosites which originate (at least partly) from the interface debonding.
- 4: Intra-fiber bundle porosities whose shape are mainly longitudinal along the fiber direction.

It is to be noted that, until they reach a certain size, the porosities located in the matrix-rich areas appear to have a spherical shape, as expected from gas bubbles forming within melted polymer. However, the growth and coalescence processes are not entirely understood, as well as their interaction with the delamination. Is there a minimum/maximum size for a nucleated porosity? Does it occur instantaneously and

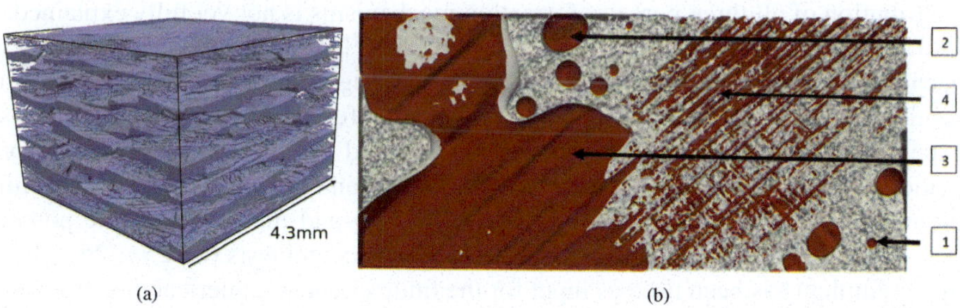

Fig. 1 X-Ray tomography observations of a laminates sample after 5 min exposure at 530 °C: **a** extracted porosities in 3D of a $4.3 \times 4.3 \times 2.21$ mm^3 domain, **b** porosities (in red) and matrix (in grey) in a 2D slice at half thickness. Numbers 1 to 4 refer to the proposed classification of porosities

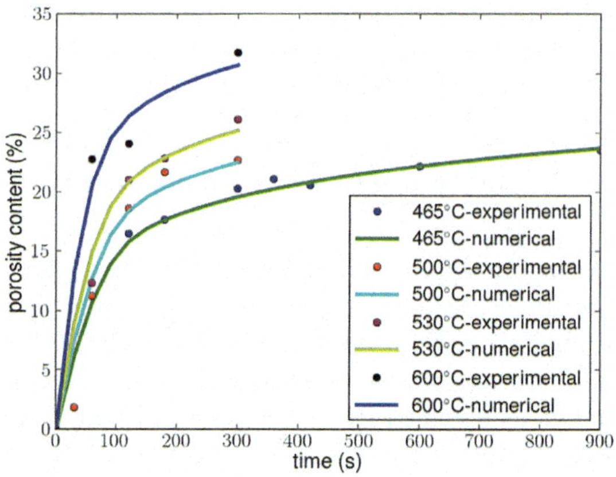

Fig. 2 Comparison of the experimental/numerical evolution over time of the porosity content η_p (volume fraction of porosities) for different temperatures of exposure. The numerical results correspond to the model accounting only for the nucleation mechanism, extracted from [3]

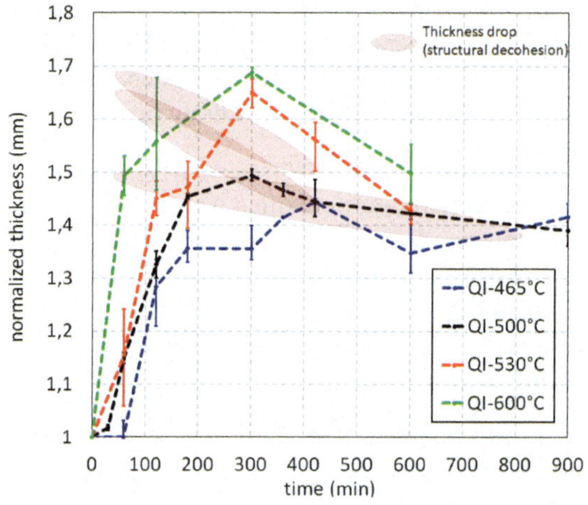

Fig. 3 Evolution over time of the laminates thickness for different temperatures of exposure, extracted from [3]

is followed by the growth? How does the external surface of the porosities exactly grow, or does it consist in a coalescence of many smaller pockets? Hence, the exact contribution of all three porosity formation mechanisms is not yet fully explained.

The evolution of the porosity content in C/PPS QI laminates with temperature has been established in previous works [3] (Fig. 2), along with the thickness expansion resulting from the porosity formation (Fig. 3). The following sections describe the experimental study and numerical modeling carried out on the four porosity formation mechanisms aforementioned i.e. nucleation and growth of porosities within the matrix, fiber bundle/matrix interface debonding and intra-fiber bundle porosities. The porosity content resulting from the experimental tests (on 25 × 25 × 2.21 mm^3 samples) has been used as input for the finite element numerical modeling at a mesoscopic scale (Fig. 4).

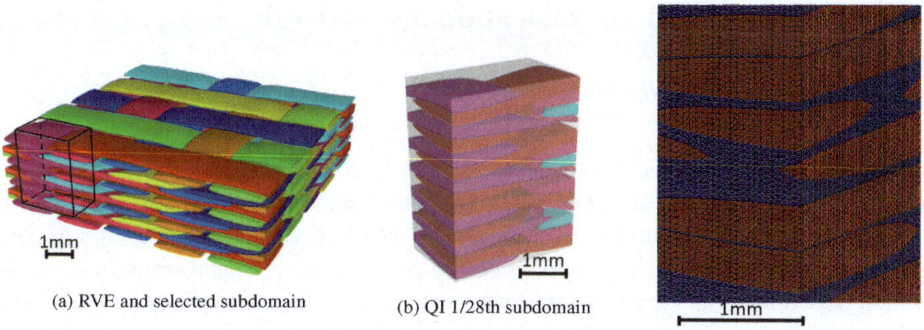

(a) RVE and selected subdomain

(b) QI 1/28th subdomain

(c) Zoom on the subdomain mesh

Fig. 4 3D mesoscopic numerical model of the laminates. Colors are used to differentiate the fiber bundles [3]

3 Modeling Based on Nucleation only

The nucleation modeling is extensively described in [3]. The transformation of matrix elements into porosity ones relies on a probability depending on time and temperature. Let us note η_p the porosity content in the laminates (i.e. the volume fraction of porosities), Δt the time increment of a finite element calculation and $p\,(t, T, \Delta t)$ the probability of a matrix element to transform into a porosity within the time Δt. The kinetics law of the porosity formation reads:

$$\begin{aligned} &\dot{\eta}_p\,(t, T) = G_N\,(t, T)\left(1 - \eta_p\,(t, T)\right) \qquad \forall t \geq 0,\ \forall T \geq T_{\mathrm{d}} \\ &\text{where} \quad G_N\,(t, T) = \frac{p\,(t, T, \Delta t)}{\Delta t} \end{aligned} \tag{1}$$

$G_N\,(t, T)$ is the rate function enabling to control the kinetics of porosity formation in the specific case where porosities only form by nucleation. As to enable the identification of $G_N\,(t, T)$ from macroscale measurements of $\eta_p(t, T)$ in a large spectrum of temperatures and times, a phenomenological form of $G_N\,(t, T)$ was chosen, as expressed in Eq. 2, calling upon parameters having separate effects on $\eta_p(t, T)$.

$$G_N\,(t, T) = \frac{a_1 + a_2 \left(\dfrac{T - T_d}{T^{ref} - T_d}\right)^b}{t + \tau} \tag{2}$$

The five parameters of $G_N\,(t, T)$ have been identified by an optimization process where, at each iteration, the integration of Eq. 1 from the Euler's method is compared to the considered reference experimental $\eta_p(t, T)$. Figure 2 shows that the numerical model accurately reproduces the experimental porosity formation kinetics for various temperatures of exposure. A limitation is observed for the highest temperatures due to experimental uncertainties as explained in [3].

4 Modeling Based on Nucleation and Growth

4.1 Principles of the Modeling

As X-Ray tomography observations have shown, cavities can be present in the form of small or large voids. These latter case are obtained from the mechanism of growth. To introduce this mechanism in the modeling, it has been chosen to assign to each element of polymer matrix a supplementary probability of transforming into cavity which depends on its location with respect to existing cavities (or nuclei) following this principle: given a distribution of nuclei and the relative field of porosity density, the larger the porosity density in the surrounding of the considered element, the larger the probability to transform into a cavity.

Let $p_{growth}\,(t, T, \Delta t)$ be this probability at time t, temperature T and during the time increment Δt. The total probability of an element to transform into a cavity is then defined by Eq. 3.

$$p\,(t, T, \Delta t) = p_{\text{nucleation}}\,(t, T, \Delta t) + p_{\text{growth}}\,(t, T, \Delta t) \tag{3}$$

p_{growth} at a given matrix element is determined from the density of porosities in its direct neighborhood. Let a domain occupied by matrix element be divided into cubic *boxes* where the local porosity densities are computed. As represented in Fig. 5, each box contains of the order of few matrix elements and three scales are considered to determine the probability of transformation of an element:

- Scale of an element: matrix/porosity finite element of the model.
- Scale of a *box*: it contains a fixed number of matrix elements defining its size. The example provided in this section considers 5 elements in each box. A difference is made between the box of interest containing the matrix element and the surrounding boxes.
- Scale of a *local domain*: ensemble of boxes surrounding a given matrix element.

The boxes constituting a *local domain* are classified according to their distance from the central box: the 1st order is the box of interest, the 2nd order ones are the 26 boxes in the direct neighborhood, the 3rd order ones are the next 98, etc. The effect of the neighborhood of a given element on its transformation probability is maximum for the elements located inside the box of interest and decreases as the order of classification increases.

Let us note n the number defining the classification order of neighbours and d_i the porosity density in the boxes i forming the local domain of a given matrix element. The probability of transformation of an element inside the box of interest is determined from a weighted average of the porosity densities in the local domain according to Eq. 4, where ω_i stand for the weight constants for $i \in \left[(2k-3)^3+1, (2k-1)^3\right]$, $k \in \{2, ..., n\}$.

$$p_{\text{growth}} = \sum_{i=1}^{(2n-1)^3} \omega_i d_i \tag{4}$$

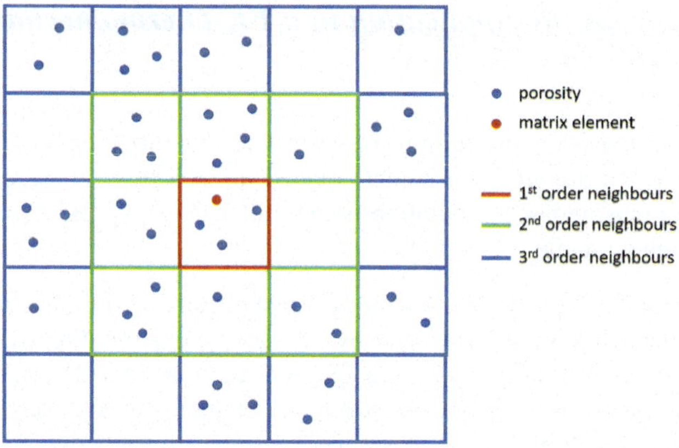

Fig. 5 2D representation of the *local domain* around a given *box* of interest and the associated classification of the neighborhood from 1st to 3rd order

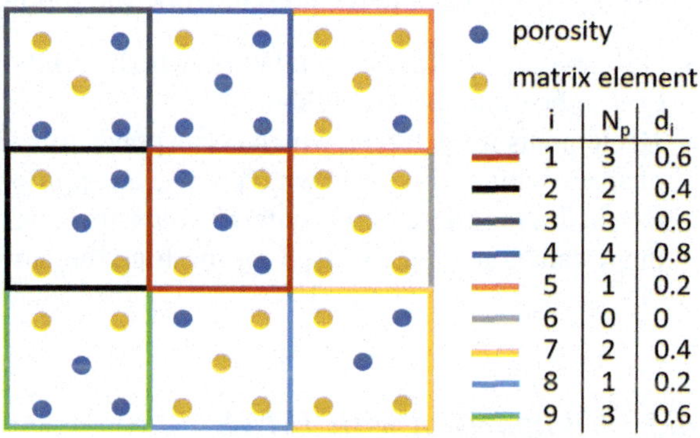

Fig. 6 2D illustration of the porosity distribution within the boxes of a 2nd order local domain considering a box size of 5 elements. Calculation of the porosity density of the boxes forming the local domain

The weight constants $(\omega_i)_{i\in[3^{k-1}+1,3^k],k\in\{1,\ldots,n\}}$ then constitute the set of parameters governing the kinetics of growth. The form they have been given in order to introduce the dependence of the growth rate to temperature and time, and the identification that could be made from the experimental results of porosity content evolution through time for the different temperatures of exposure are presented in Sect. 4.4.

In order to illustrate how this model works, let us consider the case where a box contains five elements, being either matrix or porosity, and $n = 2$, as depicted in 2D in Fig. 6, along with the corresponding densities of the constituent boxes of the local domain.

Let us assume further that $\omega_1 = 0.2$ and $\omega_{2,\ldots,9} = 0.1$. Equation 4 then gives a probability of transformation in the central box of interest $p_{\text{growth}} = 0.38$.

4.2 Strategy for the Application in a FE Thermomechanical Computation

The question first rests on the compromise between the maximum order of influence considered and the amount of elements placed in each box. These considerations must rely on a reasonable computation time while preserving the chemico-physical aspect of the phenomenon:

- On the one hand, an ideal system would consist in considering each box containing only one matrix element influenced by the other elements directly connected to it. It is however not numerically feasible. On the other hand, an extremely cost effective discretization considering too many elements in each box would not be representative enough.
- Taking into account too many orders would result in (i) large calculation times and (ii) loss of accuracy as elements would interact from a large distance.

As a first approach in the present work, the following configuration was chosen:

- Only 1st and 2nd order neighbors, which results in a local domain made of the central box and its 26 neighbors,
- 10 elements in each box, which corresponds to 2–3 elements along each direction of space within a box, and which is also an appropriate number of almost identical tetrahedra for filling a cube without voids; this consideration makes sense as regard the tetrahedral elements used for meshing the laminates.

4.3 Comparison with JMAK

The JMAK model (Johnson-Mehl-Avrami-Kolmogorov) provides a reference kinetics law of the macroscale porosity content since it relies also on the mechanisms of nucleation and growth of a product phase within a parent one. From the instantaneous initial nucleation of a certain nuclei volume fraction $z\,(t = 0)$, and given an isotropic and time-independent growth rate, the rate of transformation to the product phase can be integrated in the form of Eq. 5, where $z\,(t)$ (assimilated to $\eta_p\,(t)$) stands for the volume fraction of the product phase:

$$z\,(t) = 1 - \exp\left(-K t^n\right) \tag{5}$$

K is a constant depending on $z\,(t = 0)$ and the interface growth velocity.
n is an integer no larger than 4, usually set at $D + 1$, D being the dimension of the geometrical model. In the present case, $n = 4$ since our model is three-dimensional.

Let us set as an example $z\,(t = 0) = 0.1\%$ in the numerical model and $K = 1.3E{-9}$. Let us suppose that the nucleation only occurs at the initial time in the proposed finite element growth model so that only the growth governs the porosity

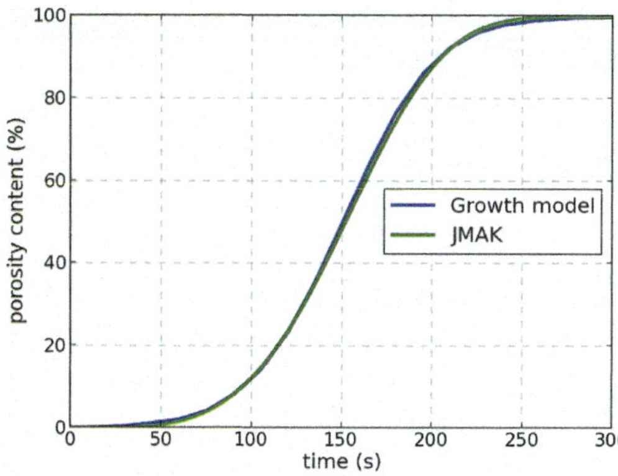

Fig. 7 Comparison of the porosity content evolution over time as obtained by the JMAK model and the nucleation-and-growth model ($\eta_p\ (t = 0) = 0.1\%$)

content evolution. A cubic domain of matrix containing 1,000,000 elements was used. It was found that the weights of the present approach (as described in Eq. 6) can be adjusted as to fit the JMAK-predicted evolution (Fig. 7). This shows that the proposed numerical growth model, although established on a local scale, has the capacity to reproduce JMAK-like macroscale kinetics.

4.4 Growth Model Identification

In the modeling of porosity formation from nucleation and growth, $G_N\ (t, T)$ must be lowered with respect to its identification in the modeling based only on nucleation. In the case of exposure where porosities form large pockets (high temperature and rather long time of exposure), the porosity formation can be assumed to be governed by the growth mechanism. To address this situation, as a first approach to demonstrate the capabilities of the growth model, the nucleation probability was arbitrarily divided by 2.5 so that 40% of the porosity content corresponds to nuclei and 60% originate from the growth.

As mentioned in Sect. 4.1, it is necessary to account for the second order neighbors for the welfare of the modeling. However their influence (or weight) can be reduced to a minimum. To that end, it was chosen arbitrarily to set the weight of the central box 80 times higher than the surrounding ones. Doing that, the combined weight of the 26 surrounding boxes is around 3 times lower and represents 25% of the total weight.

As for the determination of G_N, the weights ω_i must follow a similar trend: a high value at the initial times to initiate the growth process, before decreasing with time. The proposed definition of ω_i reads:

$$\omega_i\ (t, T) = \frac{1}{\left(\tau_{1,i} + \tau_{2,i}\left(\frac{T - T_d}{T_{\mathrm{ref}} - T_d}\right)\right)t^{c}} \qquad \text{with}\ t > 0 \qquad (6)$$

Table 1 Parameters used to define $\omega_i\ (t, T)$ as given by Eq. 6

Parameter	Unit	Value		Main role
		$i = 1$	$i = \{2 \ldots 9\}$	
τ_1	Time (s)	$1.37e{-}4$	$1.09e{-}2$	Affects ω_i for $T = T_{ref}$
τ_2	Time (s)	$5.77e{-}6$	$4.62e{-}4$	Dependence of ω_i to T
c	No unit	2.2	2.2	Non-linear dependence of ω_i to t
T_{ref}	Temperature (°C)	530	530	Reference temperature

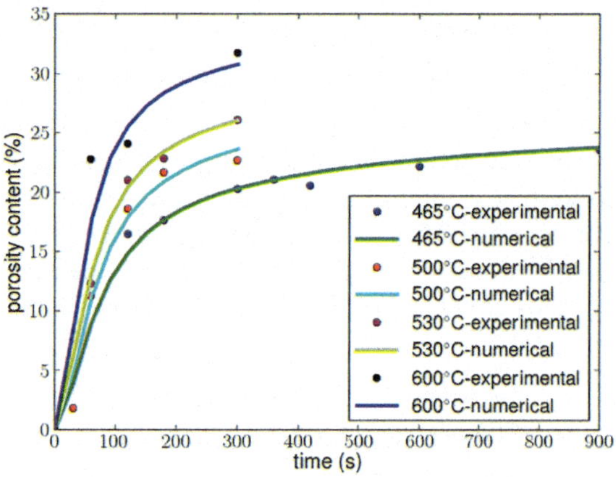

Fig. 8 Comparison of the experimental/numerical evolutions over time of η_p for different temperatures of exposure considering the model based on nucleation and growth

As the growth occurs after the nucleation and the first iteration is at the end of the first time step of the finite element calculation, $t > 0$. The various parameters and their roles are presented in Table 1.

The parameters from the definition $\omega_i\ (t, T)$ were identified in order to fit the experimental data, and the obtained porosity content evolution is depicted on Fig. 8. It appears that the optimized parameters accurately replicate the experimental results similarly to what has been obtained with the model based only on nucleation, along with the same uncertainties for the highest temperature level.

4.5 Nucleation Based Versus Nucleation-and-Growth Based Modeling

As shown in Sects. 3 and 4.4, could it be by nucleation only or by nucleation and growth, both mechanisms enable to reproduce correctly the evolution of the porosity content through time for the different tested temperatures of exposure. But these models intrisically lead to different morphologies of cavities, as can be seen on Fig. 9. It presents the distribution of finite elements of porosities in a cubic mesh

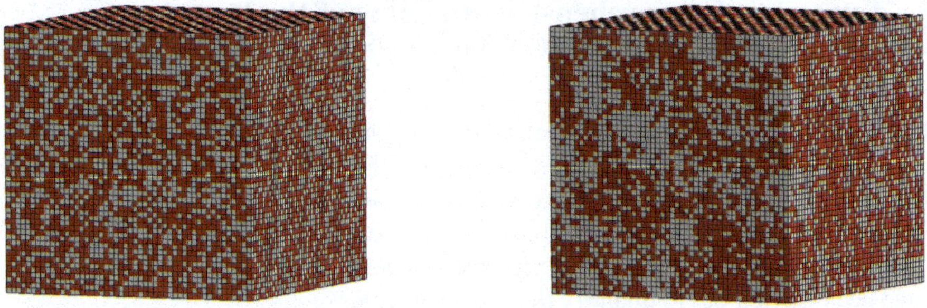

(a) (b)

Fig. 9 Porosity distribution (in red) for a 54% porosity content in a domain of matrix elements (100%): **a** nucleation only, **b** with an initial nucleation of 25% porosities then growth

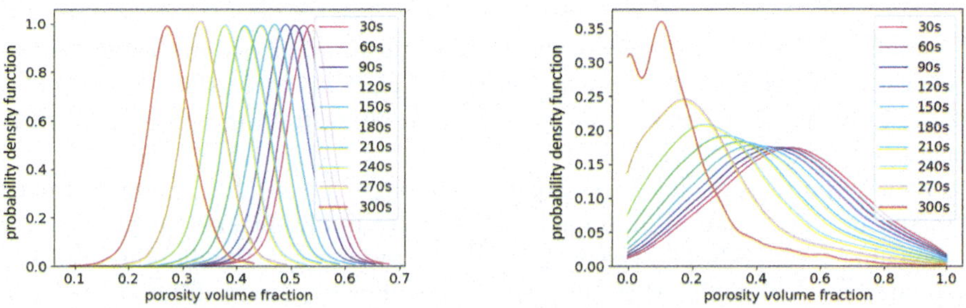

(a) (b)

Fig. 10 Evolution through time of the distribution of the porosity density in each box over time. Porosities formed through **a** nucleation, **b** nucleation and growth

domain constituted of matrix only, where 54% of porosities have been generated from each model for a given isothermal exposure. Whereas porosities are uniformally distributed with the nucleation model, the nucleation-and-growth model leads to large clusters of porosities embedded within large matrix areas as well.

Another way to highlight the difference in the spatial distribution of porosities from a model to the other is provided by the statistical distributions of the local porosity density (or porosity volume fraction). Figure 10 presents these distributions at different instants of an isothermal exposure. Whereas the nucleation-based model produces -evidently- normal distributions translating to high values of the porosity volume fraction as time of exposure increases, keeping a constant standard deviation, the nucleation-and-growth model leads to distribution which evolve through time in terms of shape, average value and standard deviation. The most important change in the distribution is observed at the earliest steps of the porosity formation.

5 Towards a Comprehensive Modeling of the Porosity Formation

The identification of the numerical models accounting for nucleation-only or for nucleation-and-growth have been established with respect to the experimental measurements of the porositiy content evolutions. No conclusive model has yet been developed to reproduce the fiber bundle/matrix interface debonding and the intra-fiber bundle porosity. This section presents the preliminary experimental tests which have been carried out to quantify these phenomena.

5.1 Fiber Bundle/Matrix Interface Debonding

Generally, under thermal aggression, the mechanism of debonding at the interface between a fiber bundle and the matrix first takes place at the interface between plies because this is where most of the porosities resulting from the processing of the material are present [17, 18].

Figure 3 presents the thickness expansion through time for different temperatures of exposure. For some of the samples characterized, the thickness expansion was originating from both the pressurization of the pyrolysis gas within the porosities and interface debonding. These two phenomena thus need to be differentiated as to how they influence the swelling of the laminates.

In order to quantify the role of each mechanism, laminates samples have been exposed to temperatures in the range from 250 to 450 °C. It was indeed observed that interface debonding can occur for temperatures lower than that of the thermal decomposition (as can be seen on Fig. 11 at 350 °C).

The same protocol as the one used in [3] has been followed using a furnace and three samples for each time/temperature conditions. The interface decohesion can then be isolated and the trend of its evolution over time and temperature determined. The obtained trend was then assumed to stay similar throughout the decomposition process. As a result, the thickness expansion originating from the interface debonding was obtained for the whole range of temperatures [250; 600] °C, and the one

Fig. 11 Optical microscope observation of the porosities resulting from the debonding of C/PPS laminates after 5min of thermal exposure at 350 °C

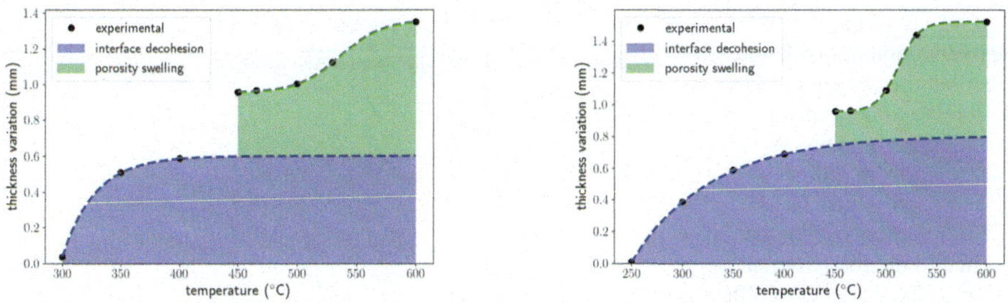

Fig. 12 Evolution of the thickness expansion over time, differentiating the pressurization and debonding mechanisms after **a** 3 min, **b** 5 min of exposure

resulting from the pressurization was directly deduced. The experimental results of this method is presented on Fig. 12 for various times of exposure. The debonding-related expansion was found to have a converging tendency, while a clear increase was observed at the onset of the thermal decomposition. Although no numerical modeling of the debonding has been performed, cohesive elements could be used in future works to reproduce this mechanism.

5.2 *Intra-fiber Bundle Porosity*

In order to accurately model the thermomechanical behavior of a laminates in future works, it is necessary to account for the porosity formation within the fiber bundles. Indeed, it will reduce the mechanical loading capacity through (i) matrix being replaced by voids and (ii) lowering of the fiber cohesion which is maintained by the matrix.

The porosity content within the fiber bundles has been determined from 2D optical microscope observations extracted from the core of the samples. A grey scale threshold has been used. The results presented in Fig. 13 show the progressive increase of the porosity content over time and temperature. One can note in particular that the total amount of porosities within a fiber bundle reaches 20–25% of the total volume of matrix in the fiber bundle.

To include this mechanism in the modeling remains an open question. From the point of view of the thermal problem, it might be sufficient to approximate the fiber bundle conductivity from a mixture rule or from an homogenization scheme. From the mechanical point of view, in Ref. [19] it is proposed to directly relate the mechanical properties of the fiber bundle to the porosity content, as in the classical theory of the damage of materials. The relevance of such an approach is however questionable: yet it might lead to a rather correct approximation concerning the axial stiffness of a fiber bundle, its accuracy concerning the shear modulus of the fiber bundle is called into question since the porosities are distributed according to a network aligned in the

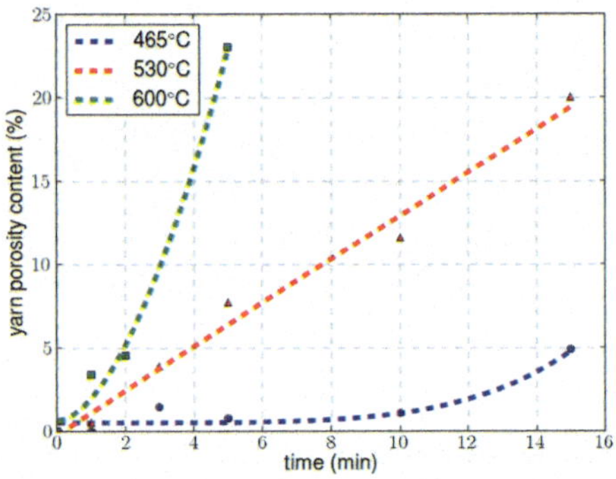

Fig. 13 Evolution of the porosity content inside the fiber bundles over time for different temperatures in C/PPS QI laminates, as measured from optical microscopy scans. The measured porosity content is related to the total amount of matrix within a fiber bundle

direction of the fibers. The development of a solution to account for this mechanism in the modeling will be the subject of future works.

6 Conclusion

In the context of one-sided exposure to high thermal energy sources like flames, the tensile strength of polymer matrix composite laminates decreases significantly, but it remains substantial for a considerable duration, preventing catastrophic failure. A key factor in this degradation is the pyrolysis of the matrix. This article takes part to a series of modeling developments for the progressive thermal decomposition of the polymer matrix within a composite laminate, focusing on: (1) Describing thermal gradients accurately, (2) Modeling the kinetics of porosity formation during matrix decomposition to account for the rapid evolution of the polymer state, (3) Explicitly representing the presence of voids that act as thermal insulators, (4) Developing a model allowing the thermomechanical coupling to be simulated. Its first originality lies in the extension of a previous model (in [3]), accounting only for the nucleation of cavities, to the mechanism of growth of cavities. Along with these new developments, based on X-Ray tomography and optical microscopy analyses, the principal mechanisms driving the formation of porosities and the swelling of the laminate have been differentiated: nucleation and growth inside the matrix rich regions, debonding at the interface between a fiber bundle and the matrix, intra-fiber bundle porosity formation. Possible methodologies to include the mechanisms of debonding and intra-bundle porosity formation are discussed, paving the ground for future developments.

References

1. Carpier, Y., Vieille, B., Coppalle, A., Barbe, F.: Polym. Compos. **41**(9), 3552 (2020). https://doi.org/10.1002/pc.25641
2. Carpier, Y., Alia, A., Vieille, B., Barbe, F.: Polym. Degrad. Stab. **186**, 109525 (2021). https://doi.org/10.1016/j.polymdegradstab.2021.109525
3. Philippe, D., Vieille, B., Barbe, F.: Compos. Part B: Eng. **266**, 111026 (2023). https://doi.org/10.1016/j.compositesb.2023.111026, https://www.sciencedirect.com/science/article/pii/S1359836823005292
4. Feih, S., Mouritz, A.: Compos. A Appl. Sci. Manuf. **43**(5), 765 (2012). https://doi.org/10.1016/j.compositesa.2011.06.016. Office of Naval Research (ONR): Composites in Fire
5. Bibinger, J., Eibl, S., Gudladt, H.J.: Appl. Compos. Mater. **29**(5), 1817 (2022)
6. Vetter, T.M., Eibl, S., Gudladt, H.J.: Appl. Compos. Mater. **28**(2), 427 (2021)
7. Ma, C.C.M., Hsia, H.C., Liu, W.L., Hu, J.T.: J. Thermoplast. Compos. Mater. **1**(1), 39 (1988). https://doi.org/10.1177/089270578800100104
8. Moldoveanu, S.C.: Analytical Pyrolysis of Synthetic Organic Polymers, Techniques and Instrumentation in Analytical Chemistry, vol. 25, pp. 31–107. Elsevier (2005). https://doi.org/10.1016/S0167-9244(05)80003-4
9. Yao, F., Zheng, J., Qi, M., Wang, W., Qi, Z.: Thermochim. Acta **183**, 91 (1991). https://doi.org/10.1016/0040-6031(91)80448-R
10. Patel, P., Hull, T.R., Lyon, R.E., Stoliarov, S.I., Walters, R.N., Crowley, S., Safronava, N.: Polym. Degrad. Stab. **96**(1), 12 (2011). https://doi.org/10.1016/j.polymdegradstab.2010.11.009
11. Levchik, G.F., Si, K., Levchik, S.V., Camino, G., Wilkie, C.A.: Polym. Degrad. Stab. **65**(3), 395 (1999). https://doi.org/10.1016/S0141-3910(99)00028-2
12. Day, M., Budgell, D.: Thermochim. Acta **203**, 465 (1992). https://doi.org/10.1016/0040-6031(92)85217-J
13. Starink, M.: Thermochim. Acta **404**(1), 163 (2003). https://doi.org/10.1016/S0040-6031(03)00144-8
14. Endres, W., Lechner, M.D., Steinberger, R.: Macromol. Mater. Eng. **288**(6), 525–530 (2003). https://doi.org/10.1002/mame.200390050. Cited by: 18
15. Mouritz, A., Gibson, A.G.: Fire Properties of Polymer Composite Materials, Solid Mechanics and Its Applications, vol. 143. Springer (2006). https://doi.org/10.1007/978-1-4020-5356-6
16. Carpier, Y., Vieille, B., Barbe, F., Coppalle, A.: Compos. A Appl. Sci. Manuf. **162**, 107165 (2022). https://doi.org/10.1016/j.compositesa.2022.107165
17. Amedewovo, L., Levy, A., de Parscau du Plessix, B., Aubril, J., Arrive, A., Orgéas, L., Le Corre, S.: Compos. Part A Appl. Sci. Manuf. **167**, 107412 (2023). https://doi.org/10.1016/j.compositesa.2022.107412, https://www.sciencedirect.com/science/article/pii/S1359835X22005930
18. Sihn, S., Pitz, J., Vernon, J.P.: Compos. Struct. **322**, 117422 (2023). https://doi.org/10.1016/j.compstruct.2023.117422, https://www.sciencedirect.com/science/article/pii/S0263822323007687
19. Mackenzie, J.: Proc. Phys. Soc. Sect. B **63**(1), 2 (1950)

An Adaptive Version of the Eyre-Milton Solution Scheme for FFT-Based Homogenization of Composites

Karam Sab and Jérémy Bleyer

Abstract A new solution scheme for homogenization of composites is studied in this paper. This scheme introduced in Sab et al. [10] is obtained by adapting Eyre-Milton's one [1]. It can efficiently handle the case of infinitely double-contrasted composites containing both pores and infinitely rigid inclusions. While a very simple micro structure was previously considered by the authors, a much more realistic micro structure is studied here, and more insights on the role of the reference medium introduced by the Lippmann-Schwinger equation are given.

Keywords Computational homogenization · FFT-based method · Iterative scheme · Linear elasticity · Composite materials

1 Introduction

In this paper, a new solution scheme for homogenization of composites is studied. This scheme, introduced in Sab et al. [10], is obtained by adapting Eyre-Milton's one [1]. It can efficiently handle the case of infinitely double-contrasted composites containing both pores and infinitely rigid inclusions. Indeed, all the existing so-called "FFT-based numerical homogenization techniques", first introduced by Moulinec and Suquet [8, 9], are unable to handle such an extreme case. The reader is referred to the recent review [13] for a more detailed description of this numerical technique which is based on the Lippmann-Schwinger equation. This equation makes use of a reference homogeneous medium and it is solved iteratively by various iterative solvers [1, 4, 6, 7, 11, 12, 14]. While a very simple micro structure was considered in Sab et al. [10], we will study in this paper a much more realistic micro structure

K. Sab (✉) · J. Bleyer
Laboratoire Navier, Univ Gustave Eiffel, Ecole des Ponts, CNRS, Navier, 77454 Marne-la-Vallée, France
e-mail: karam.sab@enpc.fr

J. Bleyer
e-mail: jeremy.bleyer@enpc.fr

F. Willot et al. (eds.), *Continuum Models and Discrete Systems*,
Springer Proceedings in Mathematics & Statistics 457,
https://doi.org/10.1007/978-3-031-58665-1_10

and we will give more insights on the role of the reference medium introduced by the Lippmann-Schwinger equation.

2 The Homogenization Problem

We consider homogenization of a periodic linearly elastic composite material containing rigid inclusions and pores. The unit-cell in the d-dimensional real space $\mathbb{R}^d$ ($d = 2, 3$) is denoted by $\Omega = [0, L_1] \times \cdots \times [0, L_d]$ where $L_i > 0$ are the side lengths of Ω. A field X is Ω-periodic if $X(y_1, \ldots, y_d) = X(y_1 + L_1, \ldots, y_d) = \cdots = X(y_1, \ldots, y_d + L_d)$ at any point $\boldsymbol{y} = (y_1, \ldots, y_d) \in \mathbb{R}^d$. It is assumed that the stiffness fourth-order tensor field $\boldsymbol{C}$ is Ω-periodic and exhibits both minor ($C_{ijkl} = C_{ijlk} = C_{jikl}$) and major ($C_{ijkl} = C_{klij}$) symmetries. More precisely, let Ω^{r}, Ω^{p} and Ω^{m} denote the domains of Ω occupied by, respectively, the rigid inclusions where $\boldsymbol{C}$ is infinite, the pores where $\boldsymbol{C}$ is null, and the heterogeneous matrix where $\boldsymbol{C}$ is finite and not null.

Let $\boldsymbol{L}^2_{\mathrm{sym}}(\Omega)$ denote the space of symmetric second-order tensor fields $\boldsymbol{y} \mapsto \boldsymbol{e}(\boldsymbol{y})$ ($e_{ij}(\boldsymbol{y}) = e_{ji}(\boldsymbol{y}), i, j = 1, \ldots, d$), which are Ω-periodic and square integrable over Ω. $\boldsymbol{S}$ is the sub-space of the $\boldsymbol{e} \in \boldsymbol{L}^2_{\mathrm{sym}}(\Omega)$ which are divergence-free in the sense of distributions over $\mathbb{R}^d$, $e_{ij,j} = 0$, and $\boldsymbol{D}$ is the sub-space of the $\boldsymbol{e} \in \boldsymbol{L}^2_{\mathrm{sym}}(\Omega)$ which are the symmetric part of the gradient of an Ω-periodic displacement vector field $\boldsymbol{u} = (u_i)$, $e_{ij} = \frac{1}{2}(u_{i,j} + u_{j,i})$.

Let $\boldsymbol{E}$ denote a macroscopic strain tensor (symmetric, second-order tensor). Then, the homogenization problem becomes: find $(\boldsymbol{e}_E, \boldsymbol{\sigma}_E) \in \boldsymbol{D} \times \boldsymbol{S}$ such that:

$$\boldsymbol{E} + \boldsymbol{e}_E = 0 \text{ in } \Omega^{\mathrm{r}}, \quad \boldsymbol{\sigma}_E = 0 \text{ in } \Omega^{\mathrm{p}} \quad \text{and} \quad \boldsymbol{\sigma}_E = \boldsymbol{C} : (\boldsymbol{E} + \boldsymbol{e}_E) \text{ in } \Omega^{\mathrm{m}}. \tag{1}$$

where ":" is the double contraction operator.

Under a uniform ellipticity condition on $\boldsymbol{C}$ in Ω^{m}, and some regularity conditions on the geometry of the domains Ω^{r} and Ω^{p}, which are discussed in Sab et al. [10], the above problem has a unique solution up to a couple $(\boldsymbol{e}, \boldsymbol{\sigma}) \in \boldsymbol{D} \times \boldsymbol{S}$ with $(\boldsymbol{e}, \boldsymbol{\sigma}) = (0, 0)$ in Ω^{m}.

Finally, the homogenized stiffness tensor $\boldsymbol{C}_{\mathrm{hom}}$ is uniquely defined using the following equality which holds true for any $\boldsymbol{E}$:

$$\boldsymbol{E} : \boldsymbol{C}_{\mathrm{hom}} : \boldsymbol{E} = \langle \boldsymbol{\sigma}_E : \boldsymbol{e}_E \rangle_{\Omega^{\mathrm{m}}} \tag{2}$$

where $\langle \bullet \rangle_\Omega$ is the volume average operator over Ω.

3 The Lippmann–Schwinger Equation

In the remainder of this paper, $\boldsymbol{C}_0$ denotes an arbitrary reference material, which is a uniform, positive-definite ($\boldsymbol{e} : \boldsymbol{C}_0 : \boldsymbol{e}$ is strictly positive unless $\boldsymbol{e} = 0$), fourth-order tensor with both minor and major symmetries. Then, the $\boldsymbol{C}_0$-norm is defined over $\boldsymbol{L}^2_{\text{sym}}(\Omega)$ as follows:

$$\|\boldsymbol{e}\|_{C_0} = \sqrt{\langle \boldsymbol{e} : \boldsymbol{C}_0 : \boldsymbol{e} \rangle_\Omega} \tag{3}$$

Owing to the symmetry and positive definiteness of $\boldsymbol{C}_0$, the $\boldsymbol{C}_0$-norm $\|\bullet\|_{C_0}$ is equivalent to the natural norm $\|\boldsymbol{e}\|_{\boldsymbol{L}^2_{\text{sym}}} = \sqrt{\langle \boldsymbol{e} : \boldsymbol{e} \rangle_\Omega}$.

For any $\boldsymbol{\tau} \in \boldsymbol{L}^2_{\text{sym}}(\Omega)$, the following problem

$$\text{Find } \boldsymbol{e}_\tau \in \boldsymbol{D} \text{ such that } \boldsymbol{C}_0 : \boldsymbol{e}_\tau + \boldsymbol{\tau} \in \boldsymbol{S} \tag{4}$$

has a unique solution $\boldsymbol{e}_\tau$ that depends linearly on $\boldsymbol{\tau}$. The Green operator $\boldsymbol{\Gamma}^0$ associated to $\boldsymbol{C}_0$ is defined as the linear operator that maps τ onto $-\boldsymbol{e}_\tau$ [2, 3, 16]:

$$\boldsymbol{e}_\tau = -\boldsymbol{\Gamma}^0 * \boldsymbol{\tau} \tag{5}$$

The Green operator $\boldsymbol{\Gamma}^0$ has an analytical expression in Fourier space, and hence the convolution $\boldsymbol{\Gamma}^0 * \boldsymbol{\tau}$ can be easily and efficiently computed when FFT algorithms are used. The $\boldsymbol{\Gamma}^0$ operator thus defined enjoys a number of properties [4], among them:

$$\boldsymbol{\Gamma}^0 * \boldsymbol{\tau} = 0 \quad \Longleftrightarrow \quad \boldsymbol{\tau} \in \boldsymbol{S} \tag{6}$$

It is useful to introduce the following decomposition of any tensor field in $\boldsymbol{L}^2_{\text{sym}}(\Omega)$. For all $\boldsymbol{e} \in \boldsymbol{L}^2_{\text{sym}}(\Omega)$, we define:

$$\boldsymbol{e}^D = \boldsymbol{\Gamma}^0 * (\boldsymbol{C}_0 : \boldsymbol{e}) \quad \text{and} \quad \boldsymbol{e}^S = \boldsymbol{e} - \boldsymbol{e}^D \tag{7}$$

with the following properties:

$$\boldsymbol{e}^D \in \boldsymbol{D}, \quad \boldsymbol{C}_0 : \boldsymbol{e}^S \in \boldsymbol{S} \quad \text{and} \quad \|\boldsymbol{e}\|^2_{C_0} = \|\boldsymbol{e}^D\|^2_{C_0} + \|\boldsymbol{e}^S\|^2_{C_0}. \tag{8}$$

The above properties can be established using the well-known decomposition of $\boldsymbol{L}^2_{\text{sym}}(\Omega)$, endowed with this natural norm and scalar product, into the two orthogonal sub-spaces $\boldsymbol{S}$ and $\boldsymbol{D}$ (see e.g. Milton [5]):

$$\boldsymbol{L}^2_{\text{sym}}(\Omega) = \boldsymbol{S} \overset{\perp}{\oplus} \boldsymbol{D} \tag{9}$$

It should be emphasized that multiplying $\boldsymbol{C}_0$ by a strictly positive real does not change the decomposition (7).

Finally, we have in hand all the tools we need to introduce the residual $\boldsymbol{X}$ of any couple of strain-stress fields, $(\varepsilon, \boldsymbol{\sigma}) \in \boldsymbol{L}^2_{\text{sym}}(\Omega) \times \boldsymbol{L}^2_{\text{sym}}(\Omega)$ which comply to the constitutive equations, i.e. such that:

$$\varepsilon = 0 \text{ in } \Omega^{\mathrm{r}}, \quad \boldsymbol{\sigma} = 0 \text{ in } \Omega^{\mathrm{p}} \quad \text{and} \quad \boldsymbol{\sigma} = \boldsymbol{C} : \varepsilon \text{ in } \Omega^{\mathrm{m}}, \tag{10}$$

then, its corresponding residual field is:

$$\boldsymbol{X}(\boldsymbol{y}) = \boldsymbol{E} - \varepsilon(\boldsymbol{y}) - \boldsymbol{\Gamma}^0(\boldsymbol{y}) * [(\boldsymbol{\sigma}(\boldsymbol{y}) - \boldsymbol{C}_0 : \varepsilon(\boldsymbol{y})], \tag{11}$$

and the Lippmann–Schwinger equation reads:

$$\boldsymbol{X} = 0. \tag{12}$$

Actually, $(\varepsilon, \boldsymbol{\sigma})$ is a solution of the homogenization problem if, and only if, it solves the Lippmann–Schwinger equation. Indeed, by definition, we have $\varepsilon^D = \boldsymbol{\Gamma}^0 * [\boldsymbol{C}_0 : \varepsilon]$, and $\boldsymbol{X}$ can be decomposed as:

$$\boldsymbol{X} = \boldsymbol{X}^S + \boldsymbol{X}^D, \quad \boldsymbol{X}^S = \boldsymbol{E} - \varepsilon^S = (\boldsymbol{E} - \varepsilon)^S, \quad \boldsymbol{X}^D = -\boldsymbol{\Gamma}^0 * \boldsymbol{\sigma} = -\left(\boldsymbol{C}_0^{-1} : \boldsymbol{\sigma}\right)^D. \tag{13}$$

The conclusion is obtained by noticing that $\boldsymbol{X} = 0$ is equivalent to $\boldsymbol{X}^D = 0$ and $\boldsymbol{X}^S = 0$, and that $\boldsymbol{X}^D = 0$ is equivalent to $\boldsymbol{\sigma} \in \boldsymbol{S}$ and $\boldsymbol{X}^S = 0$ is equivalent to $\varepsilon = \boldsymbol{E} + \varepsilon^D$.

4 The New Scheme

The new scheme solves the Lippmann–Schwinger equation in an iterative way. It is started with a chosen couple of strain and stress fields in the unit-cell $(\varepsilon^0, \boldsymbol{\sigma}^0)$ verifying (10), and its corresponding residual $\boldsymbol{X}^0$ is computed with formula (11). For $n \geq 0$, if $\boldsymbol{X}^n = 0$ then the couple $(\varepsilon^n, \boldsymbol{\sigma}^n)$ is a solution of the homogenization problem and the iteration is stopped. Otherwise, compute $\boldsymbol{X}^{n+1}$, ε^{n+1} and $\boldsymbol{\sigma}^{n+1}$ as follows:

$$\boldsymbol{Z}^n(\boldsymbol{y}) = \boldsymbol{\alpha}(\boldsymbol{y}) : \boldsymbol{X}^n(\boldsymbol{y}) + \boldsymbol{\Gamma}^0(\boldsymbol{y}) * [(\boldsymbol{C}^D(\boldsymbol{y}) - \boldsymbol{C}^S(\boldsymbol{y})) : \boldsymbol{X}^n(\boldsymbol{y})] \tag{14}$$

$$\lambda^n = \frac{\langle \boldsymbol{X}^n : \boldsymbol{C}_0 : \boldsymbol{Z}^n \rangle_\Omega}{\langle \boldsymbol{Z}^n : \boldsymbol{C}_0 : \boldsymbol{Z}^n \rangle_\Omega} \tag{15}$$

$$\boldsymbol{X}^{n+1} = \boldsymbol{X}^n - \lambda^n \boldsymbol{Z}^n = \boldsymbol{X}^0(\boldsymbol{y}) - \sum_{k=0}^{n} \lambda^k \boldsymbol{Z}^k(\boldsymbol{y}), \tag{16}$$

$$\boldsymbol{\varepsilon}^{n+1} = \boldsymbol{\varepsilon}^n + \lambda^n \boldsymbol{\alpha} : \boldsymbol{X}^n = \boldsymbol{\varepsilon}^0(\boldsymbol{y}) + \boldsymbol{\alpha}(\boldsymbol{y}) : \sum_{k=0}^{n} \lambda^k \boldsymbol{X}^k(\boldsymbol{y}), \tag{17}$$

$$\boldsymbol{\sigma}^{n+1} = \boldsymbol{\sigma}^n + \lambda^n \boldsymbol{C}^D : \boldsymbol{X}^n = \boldsymbol{\sigma}^0(\boldsymbol{y}) + \boldsymbol{C}^D(\boldsymbol{y}) : \sum_{k=0}^{n} \lambda^k \boldsymbol{X}^k(\boldsymbol{y}). \tag{18}$$

where the fourth-order tensor fields $\boldsymbol{\alpha}$, $\boldsymbol{C}^S$ and $\boldsymbol{C}^D$ are given by:

$$\begin{gathered} \forall \boldsymbol{y} \in \Omega^{\mathrm{r}}, \quad \boldsymbol{\alpha}(\boldsymbol{y}) = 0, \quad \boldsymbol{C}^D = 2\boldsymbol{C}_0, \quad \boldsymbol{C}^S = 0, \\ \forall \boldsymbol{y} \in \Omega^{\mathrm{p}}, \quad \boldsymbol{\alpha}(\boldsymbol{y}) = 2\boldsymbol{I}, \quad \boldsymbol{C}^D = 0, \quad \boldsymbol{C}^S = 2\boldsymbol{C}_0, \\ \forall \boldsymbol{y} \in \Omega^{\mathrm{m}}, \quad \boldsymbol{\alpha}(\boldsymbol{y}) = 2(\boldsymbol{C}(\boldsymbol{y}) + \boldsymbol{C}_0)^{-1} : \boldsymbol{C}_0, \quad \boldsymbol{C}^D = \boldsymbol{C} : \boldsymbol{\alpha}, \quad \boldsymbol{C}^S = \boldsymbol{C}_0 : \boldsymbol{\alpha}, \end{gathered} \tag{19}$$

where $\boldsymbol{I}$ is the fourth-order identity tensor operating on symmetric second-order tensors.

By construction, the couple $(\boldsymbol{\varepsilon}^n, \boldsymbol{\sigma}^n)$ complies to the constitutive law (10). So it is logical to stop the iterations when the $\boldsymbol{C}_0$-norm of its corresponding residual $\boldsymbol{X}^n$ is small enough.

It is proved in Sab et al. [10] that, under the same already mentioned uniform ellipticity conditions and geometric regularity that ensure the existence of a solution to the homogenization problem, the convergence of the above scheme is guaranteed for any choice of the reference material $\boldsymbol{C}_0$ and any starting couple $(\boldsymbol{\varepsilon}^0, \boldsymbol{\sigma}^0)$ verifying (10),

Notice that $\boldsymbol{\alpha}$, $\boldsymbol{C}^S$ and $\boldsymbol{C}^D$ in the pores (respectively, rigid inclusions) are the limit values, as $\boldsymbol{C}$ goes to zero (respectively, infinity), of $\boldsymbol{\alpha}$, $\boldsymbol{C}^S$ and $\boldsymbol{C}^D$ in the heterogeneous matrix.

In the absence of pores and rigid inclusions, $\Omega = \Omega^{\mathrm{m}}$, the above scheme will coincide with the well-known so-called basic scheme first proposed by Moulinec and Suquet [8, 9] if $\boldsymbol{\alpha}$ is set to $\boldsymbol{I}$ everywhere in the unit cell, and hence $\boldsymbol{C}^S = \boldsymbol{C}_0$ and $\boldsymbol{C}^D = \boldsymbol{C}$, and λ^n is set to one for all n. Moreover, it will coincide with the Eyre–Milton scheme [1] if $\boldsymbol{\alpha}$ is still defined by (19), and hence $\boldsymbol{C}^S = \boldsymbol{C}_0 : \boldsymbol{\alpha}$ and $\boldsymbol{C}^D = \boldsymbol{C} : \boldsymbol{\alpha}$, and λ^n is set to one for all n. Actually, λ^n is selected in the above scheme so as to minimize the next residual $\boldsymbol{X}^{n+1}$. That is why we call the resulting iterative scheme the *Adaptive Eyre–Milton* scheme (AEM).

5 The Modified Lippmann-Schwinger Equation

The simplest initialization in the AEM scheme is to set to zero both initial strain and initial stress:

$$\boldsymbol{\varepsilon}^0(\boldsymbol{y}) = 0, \quad \boldsymbol{\sigma}^0(\boldsymbol{y}) = 0, \quad \boldsymbol{X}^0(\boldsymbol{y}) = \boldsymbol{E}. \tag{20}$$

With this zero-initialization, the AEM scheme converges to a solution of the homogenization problem which writes:

$$\boldsymbol{\varepsilon}_E = \boldsymbol{\alpha}(\boldsymbol{y}) : \boldsymbol{\varepsilon}_E^*(\boldsymbol{y}), \quad \boldsymbol{\sigma}_E = \boldsymbol{C}^D(\boldsymbol{y}) : \boldsymbol{\varepsilon}_E^*(\boldsymbol{y}), \tag{21}$$

where $\boldsymbol{\varepsilon}_E^*$ is given by:

$$\boldsymbol{\varepsilon}_E^*(\boldsymbol{y}) = \sum_{k \geq 0} \lambda^k \boldsymbol{X}^k(\boldsymbol{y}). \tag{22}$$

The solution (21) is such that its residual (11) is null. Hence, inserting (21) in (11) and taking into account the equations $\boldsymbol{C}^S = \boldsymbol{C}_0 : \boldsymbol{\alpha}$ and $\boldsymbol{C}^S + \boldsymbol{C}^D = 2\boldsymbol{C}_0$, it is found that $\boldsymbol{\varepsilon}_E^*$ is a solution to the following modified Lippmann-Schwinger equation:

$$\boldsymbol{\alpha} : \boldsymbol{\varepsilon}^* + \left(2(\boldsymbol{I} - \boldsymbol{\alpha}) : \boldsymbol{\varepsilon}^*\right)^D = \boldsymbol{E}. \tag{23}$$

Actually, $\boldsymbol{\varepsilon}_E^*$ is the unique solution of the above equation under the same regularity conditions which ensures the existence of a solution to the initial homogenization problem, as proved in Sab et al. [10].

Equation (21) appears as a suitable change of variables involving the use of the bounded tensor field $\boldsymbol{\alpha}$ used in the above introduced modified Lippmann-Schwinger equation.

Using simple algebra leads to the following expression of $\boldsymbol{\varepsilon}_E^*$:

$$\boldsymbol{\varepsilon}_E^* = \frac{1}{2}\left((\boldsymbol{C}^0)^{-1} : \boldsymbol{\sigma}_E + \boldsymbol{\varepsilon}_E\right) \tag{24}$$

and hence,

$$\langle \boldsymbol{\varepsilon}_E^* \rangle_\Omega = \frac{1}{2}\left((\boldsymbol{C}^0)^{-1} : \boldsymbol{C}_{\text{hom}} + \boldsymbol{I}\right) : \boldsymbol{E}. \tag{25}$$

Noteworthy, we will have $\langle \boldsymbol{\varepsilon}_E^* \rangle_\Omega = \boldsymbol{E}$ if the reference medium $\boldsymbol{C}^0$ coincides with the homogenized one $\boldsymbol{C}_{\text{hom}}$. That is why it is recommended to choose the reference medium in the same order of magnitude as the homogenized one.

If the reference medium is much stiffer than the homogenized one, it is seen from (24) that $\boldsymbol{\varepsilon}_E^*$ is (almost) equal to $\boldsymbol{\varepsilon}_E/2$ everywhere. In this case, the AEM scheme amounts to set $\boldsymbol{\alpha} = 0$ in the rigid inclusions and $\boldsymbol{\alpha} = 2\boldsymbol{I}$ everywhere else.

On the other hand, if the reference medium is much softer than the homogenized one, than $\boldsymbol{\alpha}$ will be very close to zero in the heterogeneous matrix, and equal to zero in the rigid inclusions. This means that the same change of variables is adopted

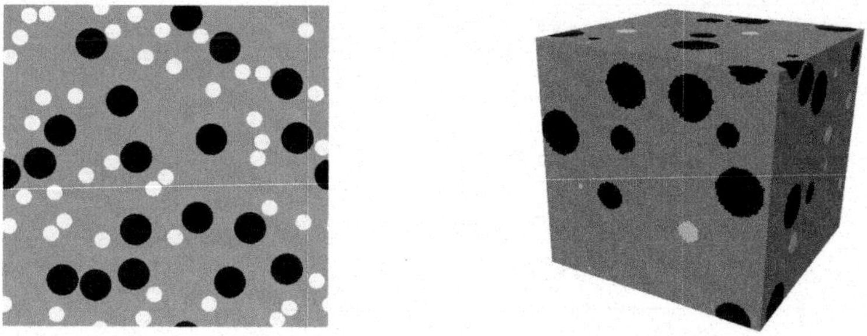

(a) 2D case (1024 × 1024) (b) 3D case (128 × 128 × 128)

Fig. 1 Random microstructures consisting of pores (white) and rigid inclusions (black)

in both the rigid inclusions phase and the heterogeneous matrix, and it wouldn't be surprising if the scheme's performance were poor.

6 Illustrative Applications

In this section, we consider a 2D microstructure consisting of random circular pores of radius 0.025 (phase 2) and larger rigid inclusions of radius 0.05 (phase 3) embedded in a matrix (phase 1), see Fig. 1a. The matrix has a Poisson ratio $\nu_1 = 0.2$ whereas the pores and the inclusions are modeled with a medium with zero and infinite stiffness respectively. The RVE is discretized with a 1024×1024 grid and the loading consists of a uniaxial unit strain $\boldsymbol{E} = \boldsymbol{e}_x \otimes \boldsymbol{e}_x$. For the Green operator discretization scheme, we use the finite differences scheme proposed in Willot [15] which results in improved accuracy of the local fields over more classical discretization strategies. Finally, we choose the matrix properties for the reference medium.

These estimates are based on the orthogonal decomposition of the residual (13), which leads to the following expression:

$$\|\boldsymbol{X}\|^2_{\boldsymbol{C}_0} = \|\, (\boldsymbol{E} - \boldsymbol{\varepsilon})^S \,\|^2_{\boldsymbol{C}_0} + \|\, \left(\boldsymbol{C}_0^{-1} : \boldsymbol{\sigma}\right)^D \,\|^2_{\boldsymbol{C}_0}. \tag{26}$$

There are two contributions to the residual: the error in compatibility, $\|\, (\boldsymbol{E} - \boldsymbol{\varepsilon})^S \,\|_{\boldsymbol{C}_0}$, and the error in equilibrium, $\|\, \left(\boldsymbol{C}_0^{-1} : \boldsymbol{\sigma}\right)^D \,\|_{\boldsymbol{C}_0}$. When these errors are not zero, their relative importance depends on the choice of the reference medium $\boldsymbol{C}_0$. Since in the AEM scheme the strain is not compatible, nor the stress is in equilibrium, one has to ensure that both the projection of $\boldsymbol{E} - \boldsymbol{\varepsilon}$ on S and the projection of $\boldsymbol{C}_0^{-1} : \boldsymbol{\sigma}$ on D are small enough. Given a certain tolerance `tol`, this can be achieved when:

$$\Delta = \max(\Delta^S; \Delta^D) \leq \texttt{tol} \tag{27}$$

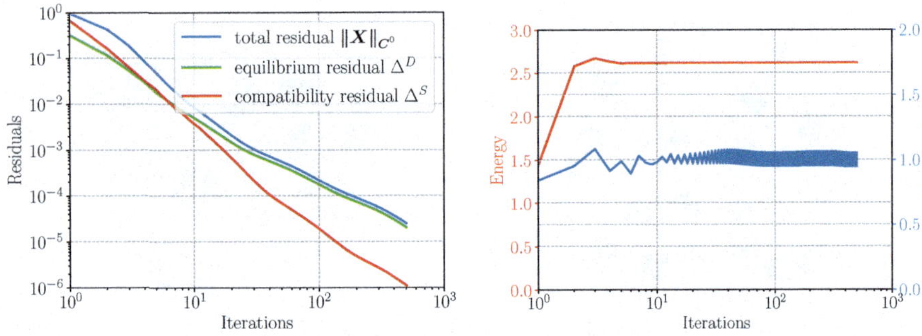

(a) Residuals convergence (b) Energy and step size

Fig. 2 Convergence of the AEM algorithm for a 2D random microstructure

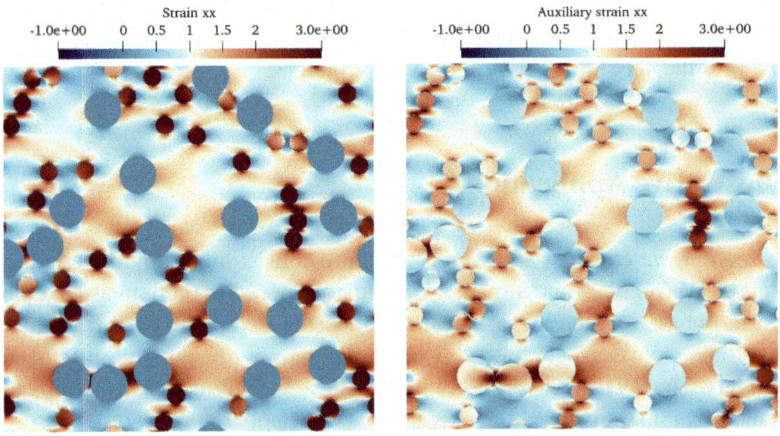

Fig. 3 Horizontal strain ε_{xx} (left) and auxiliary strain ε^*_{xx} at convergence ($E_{xx} = 1$)

where we define the normalized compatibility and equilibrium residuals:

$$\Delta^S = \frac{\| (\boldsymbol{E} - \boldsymbol{\varepsilon})^S \|_{\boldsymbol{C}_0}}{\|\boldsymbol{E} - \boldsymbol{\varepsilon}\|_{\boldsymbol{C}_0}} \tag{28}$$

$$\Delta^D = \frac{\| \left(\boldsymbol{C}_0^{-1} : \boldsymbol{\sigma}\right)^D \|_{\boldsymbol{C}_0}}{\|\boldsymbol{C}_0^{-1} : \boldsymbol{\sigma}\|_{\boldsymbol{C}_0}} \tag{29}$$

Both quantities can be computed from the intermediate computation steps of the AEM iteration as described in Sab et al. [10].

Figure 2a reports the convergence of the total absolute residual $\|\boldsymbol{X}^n\|$ and both normalized compatibility and equilibrium residuals Δ^S, Δ^D as functions of the iteration number. We observe in Fig. 2a that the convergence of the AEM algorithm is very good, even in this case of infinite contrast with true pores and stiff inclusions. This is further confirmed by Fig. 2b which reports the convergence of the energy as well as the evolution of the adaptive step size λ. Notably, the energy converges extremely rapidly within only a few tens of iterations.

To gain a deeper insight into why the AEM algorithm is efficient, Fig. 3 illustrates the evolution of the horizontal strain field ε_{xx} as well as the corresponding auxiliary strain field ε^*_{xx} after 500 iterations. It can be observed that the auxiliary strain field $\boldsymbol{\varepsilon}^*$ exhibits a similar strain level as that of the matrix inside the pores and the inclusions. As a result, it clearly acts as a rescaled strain level by removing the degeneracy induced by the infinite or zero stiffness of inclusions and pores.

Finally, we consider now the 3D microstructure of Fig. 1b, still consisting of small pores and larger rigid inclusions. We assess the performance of the AEM algorithm with respect to several classical schemes such as the basic Moulinec-Suquet (MS) scheme, the standard Eyre-Milton (EM) scheme and the Conjugate-Gradient (CG) scheme. For the CG scheme, we solve the following symmetric linear system:

$$\Delta \boldsymbol{C}^{-1} : \boldsymbol{\tau} + \boldsymbol{\Gamma}^0 * \boldsymbol{\tau} = \boldsymbol{E} \tag{30}$$

for the polarization $\boldsymbol{\tau} = \Delta \boldsymbol{C} : \boldsymbol{\varepsilon}$. All schemes are implemented in the same framework using the Janus library (https://github.com/sbrisard/janus). In each case, we compute the average energy and the total residual norm $\|\boldsymbol{X}\|_{C_0}$. Since only AEM is able to handle the exact infinite-contrast case, we use a stiffness of $10^6\mathbb{C}_1$ (resp. $10^{-6}\mathbb{C}_1$) for the rigid (resp. porous) phase for MS, EM and CG to mimic the situation of an infinite contrast. The reference medium is taken as the matrix properties for EM (corresponding to the geometric average between soft and stiff phases) whereas we use an arithmetic average for MS and CG, as commonly suggested.

Figure 4 reports the evolution of the energy and the various residuals for all tested algorithms. We observe that AEM exhibits a much smaller total residual than all other methods. Besides, it is the only method that is able to reach a stabilized value of the energy in less than 500 iterations since MS and EM yield extremely bad estimates of the energy whereas CG shows a very slowly decreasing evolution of the energy.

Finally, the influence of the reference medium is assessed by varying the value of the shear modulus G_0. Figure 5 reports the convergence of the energy and the equilibrium and compatibility residuals when choosing a reference medium much softer than the matrix ($G_0 = 0.1G_1$ and $G_0 = 0.01G_1$), much stiffer ($G_0 = 10G_1$ and $G_0 = 100G_1$) or of similar stiffness ($G_0 = G_1$ and $G_0 = 2G_1$). For reference, the homogenized shear modulus is around $G_{\text{hom}} = 1.3G_0$.

As anticipated in the previous section, we observe that choosing a reference medium which is much softer strongly deteriorates the convergence speed, in particular the compatibility residual stalls during the first iterations. Conversely, choosing a much stiffer reference medium also impacts the reduction of the equilibrium residual in the early stages but it has a much less important impact on the energy convergence than choosing a soft reference material. Overall, we find that the algorithm behaviour is quite robust with respect to the choice of the reference medium provided that it is not too soft with respect to the dominant phase.

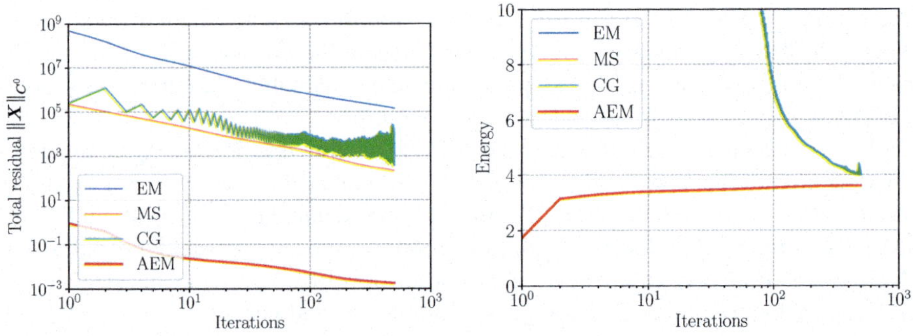

(a) Total residual (b) Energy

Fig. 4 Convergence of the different methods on the random 3D microstructure

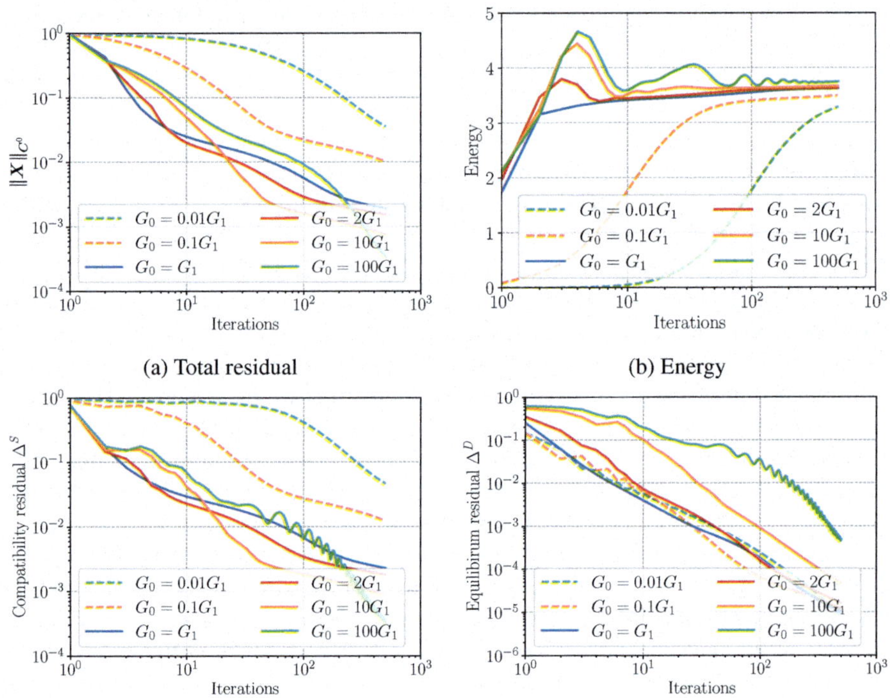

(a) Total residual (b) Energy

(c) Compatibility residual (d) Equilibrium residual

Fig. 5 Convergence of the AEM algorithm when varying the reference medium

Acknowledgements The authors would like to thank Etienne Combes for his preliminary works on the method and Sébastien Brisard for his help in implementing the method in the Janus package.

References

1. Eyre, D.J., Milton, G.W.: A fast numerical scheme for computing the response of composites using grid refinement. The European Physical Journal - Applied Physics **6**(01), 41–47 (1999). https://doi.org/10.1051/epjap:1999150
2. Korringa, J.: Theory of elastic constants of heterogeneous media. Journal of Mathematical Physics **14**(4), 509–513 (1973). https://doi.org/10.1063/1.1666346
3. Kröner, E.: On the Physics and Mathematics of Self-Stresses. In: Zeman, J.L., Ziegler, F. (eds.) Topics in Applied Continuum Mechanics, pp. 22–38. Springer Verlag Wien, Vienna (1974)
4. Michel, J.C., Moulinec, H., Suquet, P.: A computational scheme for linear and non-linear composites with arbitrary phase contrast. International Journal for Numerical Methods in Engineering **52**(1–2), 139–160 (2001). https://doi.org/10.1002/nme.275
5. Milton GW (2002) The Theory of Composites. No. 6 in Cambridge Monographs on Applied and Computational Mathematics, Cambridge University Press, Cambridge; New York
6. Monchiet, V., Bonnet, G.: A polarization-based FFT iterative scheme for computing the effective properties of elastic composites with arbitrary contrast. International Journal for Numerical Methods in Engineering **89**(11), 1419–1436 (2012). https://doi.org/10.1002/nme.3295
7. Moulinec, H., Silva, F.: Comparison of three accelerated FFT-based schemes for computing the mechanical response of composite materials. International Journal for Numerical Methods in Engineering **97**(13), 960–985 (2014). https://doi.org/10.1002/nme.4614
8. Moulinec, H., Suquet, P.: A fast numerical method for computing the linear and nonlinear mechanical properties of composites. Comptes rendus de l'Académie des sciences Série II, Mécanique, physique, chimie, astronomie **318**(11), 1417–1423 (1994)
9. Moulinec, H., Suquet, P.: A numerical method for computing the overall response of nonlinear composites with complex microstructure. Computer Methods in Applied Mechanics and Engineering **157**(1–2), 69–94 (1998). https://doi.org/10.1016/S0045-7825(97)00218-1
10. Sab, K., Bleyer, J., Brisard, S., Dolbeau, M.: An FFT-based adaptive polarization method for infinitely contrasted media with guaranteed convergence. Comput. Methods Appl. Mech. Eng. **427**, 117012 (2024). https://doi.org/10.1016/j.cma.2024.117012
11. Schneider, M.: An FFT-based fast gradient method for elastic and inelastic unit cell homogenization problems. Computer Methods in Applied Mechanics and Engineering **315**, 846–866 (2017). https://doi.org/10.1016/j.cma.2016.11.004
12. Schneider, M.: On the Barzilai-Borwein basic scheme in FFT-based computational homogenization. International Journal for Numerical Methods in Engineering **118**(8), 482–494 (2019). https://doi.org/10.1002/nme.6023
13. Schneider, M.: A review of nonlinear FFT-based computational homogenization methods. Acta Mechanica **232**(6), 2051–2100 (2021). https://doi.org/10.1007/s00707-021-02962-1
14. Schneider, M., Wicht, D., Böhlke, T.: On polarization-based schemes for the FFT-based computational homogenization of inelastic materials. Computational Mechanics **64**(4), 1073–1095 (2019). https://doi.org/10.1007/s00466-019-01694-3
15. Willot, F.: Fourier-based schemes for computing the mechanical response of composites with accurate local fields. Comptes Rendus Mécanique **343**(3), 232–245 (2015)
16. Zeller, R., Dederichs, P.H.: Elastic Constants of Polycrystals. Physica Status Solidi (B) **55**(2), 831–842 (1973). https://doi.org/10.1002/pssb.2220550241

A Stabilised Micropolar Theory Derived from a Periodic Beam Lattice

Harm Askes, Mariateresa Lombardo, and Duc C. D. Nguyen

Abstract Starting from the finite element equations of a periodic beam lattice, a continuum theory is derived via Taylor series expansions. The continualised equations of motion are of the micropolar type and thus contain stiffness and inertia contributions in terms of translational as well as rotational degrees of freedom. Scrutiny of the underlying energy functionals and of the model's response to excitation via harmonic waves confirms that the continualised model is unstable. In order to stabilise the model, Padé approximation is applied to the rotational equation of motion. As a result of this process, micro-inertia terms appear in the model that are expressed as the spatial gradients of the standard rotational inertia terms. It is demonstrated by means of an analysis of dispersive waves that the Padé approximation not only stabilises the model but also improves the accuracy with which the continuum model approximates the original discrete model.

Keywords Beam lattice · Continualisation · Micropolar theory

1 Introduction

The question of how to relate a discrete material model to a continuum material model is long-standing and goes at least back to Cauchy who used Taylor series to translate the equations of motion of a crystal into continuum equations [11]. The use of Taylor series has particular advantages for regular geometries such as periodic arrays of particles [12, 22] or lattice structures consisting of beams [5, 10]. With the growing popularity of additive manufacturing methods, there is a renewed interest in designing

H. Askes (✉)
Faculty of Engineering Technology, University of Twente, Enschede, Netherlands
e-mail: h.askes@utwente.nl

M. Lombardo
Hilti AG, Kaufering, Germany

D. C. D. Nguyen
Department of Civil and Structural Engineering, University of Sheffield, Sheffield, UK

F. Willot et al. (eds.), *Continuum Models and Discrete Systems*,
Springer Proceedings in Mathematics & Statistics 457,
https://doi.org/10.1007/978-3-031-58665-1_11

and optimising periodic lattice structures, in which modelling and simulation play an important role. Thus, it remains relevant to investigate suitable modelling strategies for periodic lattices, and in order to be able to carry out simulations of realistic specimens, both discrete and continuum modelling strategies must be pursued—including the way the two modelling approaches may be related to one another.

In this contribution, we will focus on periodic lattice structures modelled with elementary engineering beam theory, thus including translational as well as rotational degrees of freedom. These structures have been studied for many decades, mostly leading to continuum theories of the so-called micropolar type [1, 5, 9, 10, 14, 16, 23, 24]. Nevertheless, certain questions remain on the stability of the resulting continuum equations [9, 18]. It turns out that the source of instability is the sign of a higher-order spatial derivative term in the rotational equation of motion. In other areas of generalised continua, Padé approximations or other methods of asymptotic equivalence have been applied to replace unstable higher-order gradients with stable higher-order gradients, thereby typically substituting strain gradients with inertia gradients [3, 19, 20, 25, 28], and this approach will be explored here in the context of micropolar theories derived from a beam lattice.

Below, we will summarise the derivation of a micropolar continuum model from a periodic beam lattice and we will demonstrate the instability of the obtained model. A particular straightforward application of Padé approximations will be shown to be effective in eliminating the unstable terms and replacing them by (asymptotically equivalent) micro-inertia terms. The effectiveness of the new model, not only in demonstrating its stability but also its improved accuracy in capturing the behaviour of the underlying discrete model, will be shown by means of an analysis of dispersive waves.

2 Discrete Model

The starting point of the analysis is the definition of a discrete model for the beam lattice. The Euler-Bernoulli thin beam assumptions are followed, which leads to an element stiffness matrix $\mathbf{K}_{\text{el}}$ as

$$\mathbf{K}_{\text{el}} = \begin{bmatrix} \frac{EA}{L} & 0 & 0 & -\frac{EA}{L} & 0 & 0 \\ 0 & \frac{12EI}{L^3} & \frac{6EI}{L^2} & 0 & -\frac{12EI}{L^3} & \frac{6EI}{L^2} \\ 0 & \frac{6EI}{L^2} & \frac{4EI}{L} & 0 & -\frac{6EI}{L^2} & \frac{2EI}{L} \\ -\frac{EA}{L} & 0 & 0 & \frac{EA}{L} & 0 & 0 \\ 0 & -\frac{12EI}{L^3} & -\frac{6EI}{L^2} & 0 & \frac{12EI}{L^3} & -\frac{6EI}{L^2} \\ 0 & \frac{6EI}{L^2} & \frac{2EI}{L} & 0 & -\frac{6EI}{L^2} & \frac{4EI}{L} \end{bmatrix} \tag{1}$$

where E is the Young's modulus of the material, L indicates the length of each individual strut, and A and I are the cross-sectional area and the cross-sectional second moment of area, respectively. We assume square cross sections with dimension h, so that $A = h^2$ and $I = \frac{1}{12}h^4$. In addition, Rayleigh's correction of Euler-Bernoulli beam theory is used, implying the inclusion of rotational inertia. Assuming a lumped mass distribution leads to an element mass matrix $\mathbf{M}_{\text{el}}$ given by

$$\mathbf{M}_{\text{el}} = \frac{1}{2}\rho L \cdot \text{diag}\,[A, A, I, A, A, I] \tag{2}$$

with ρ the mass density. Assuming of square grid of beams with uniform properties, the equations of motion for a generic node with coordinates (m, n) read

$$2\rho AL\ddot{u}_{m,n} = \frac{EA}{L}\left(u_{m-1,n} - 2u_{m,n} + u_{m+1,n}\right) + \frac{12EI}{L^3}\left(u_{m,n-1} - 2u_{m,n} + u_{m,n+1}\right) - \frac{6EI}{L^2}\left(\theta_{m,n-1} - \theta_{m,n+1}\right) \tag{3a}$$

$$2\rho AL\ddot{v}_{m,n} = \frac{EA}{L}\left(v_{m,n-1} - 2v_{m,n} + v_{m,n+1}\right) + \frac{12EI}{L^3}\left(v_{m-1,n} - 2v_{m,n} + v_{m+1,n}\right) + \frac{6EI}{L^2}\left(\theta_{m-1,n} - \theta_{m+1,n}\right) \tag{3b}$$

$$2\rho IL\ddot{\theta}_{m,n} = \frac{6EI}{L^2}\left(u_{m,n-1} - u_{m,n+1} - v_{m-1,n} + v_{m+1,n}\right) - \frac{2EI}{L}\left(\theta_{m-1,n} + \theta_{m+1,n} + \theta_{m,n-1} + \theta_{m,n+1} + 8\theta_{m,n}\right) \tag{3c}$$

where u and v are the displacement components and θ is the micro-rotation.

3 Continualisation and Truncation: A Micropolar Theory

An equivalent continuum model can be obtained from Eqs. (3) by replacing the discrete displacements with equivalent continuum fields, that is $u_{m,n} = u(x, y)$ and $u_{m\pm1,n\pm1} = u(x \pm L, y \pm L)$. Next, Taylor series expansions are used according to

$$u(x \pm L, y) = u(x, y) \pm L\frac{\partial u(x, y)}{\partial x} + \frac{1}{2}L^2\frac{\partial^2 u(x, y)}{\partial x^2} \pm \frac{1}{6}L^3\frac{\partial^3 u(x, y)}{\partial x^3} + \ldots \tag{4a}$$

$$u(x, y \pm L) = u(x, y) \pm L\frac{\partial u(x, y)}{\partial y} + \frac{1}{2}L^2\frac{\partial^2 u(x, y)}{\partial y^2} \pm \frac{1}{6}L^3\frac{\partial^3 u(x, y)}{\partial y^3} + \dots \tag{4b}$$

Similar expansions for v and θ then lead to three equations of motion given by

$$\rho^{\mathrm{M}}\ddot{u} = E^{\mathrm{M}}\left(\mathcal{D}_x^{\mathrm{e}}u_{,xx} + \lambda^2\mathcal{D}_y^{\mathrm{e}}u_{,yy} + \lambda^2\mathcal{D}_y^{\mathrm{o}}\theta_{,y}\right) \tag{5a}$$

$$\rho^{\mathrm{M}}\ddot{v} = E^{\mathrm{M}}\left(\mathcal{D}_y^{\mathrm{e}}v_{,yy} + \lambda^2\mathcal{D}_x^{\mathrm{e}}v_{,xx} - \lambda^2\mathcal{D}_x^{\mathrm{o}}\theta_{,x}\right) \tag{5b}$$

$$\frac{1}{12}\rho^{\mathrm{M}}\lambda^2L^2\ddot{\theta} = E^{\mathrm{M}}\lambda^2\left(\mathcal{D}_x^{\mathrm{o}}v_{,x} - \mathcal{D}_y^{\mathrm{o}}u_{,y} - 2\theta - \frac{1}{6}L^2\left(\mathcal{D}_x^{\mathrm{e}}\theta_{,xx} + \mathcal{D}_y^{\mathrm{e}}\theta_{,yy}\right)\right) \tag{5c}$$

where $\lambda = h/L$ is a slenderness parameter and a variable following a comma denotes a partial derivative with respect to this variable. The macroscopic mass density ρ^{M} and Young's modulus E^{M} are obtained by taking volume averages and are given by $\rho^{\mathrm{M}} = 2\rho\lambda$ and $E^{\mathrm{M}} = E\lambda$. Furthermore, the following short-hand notations are introduced:

$$\mathcal{D}_x^{\mathrm{e}} = 1 + \frac{1}{12}L^2\frac{\partial^2}{\partial x^2} + \frac{1}{360}L^4\frac{\partial^4}{\partial x^4} + \dots \tag{6a}$$

$$\mathcal{D}_x^{\mathrm{o}} = 1 + \frac{1}{6}L^2\frac{\partial^2}{\partial x^2} + \frac{1}{120}L^4\frac{\partial^4}{\partial x^4} + \dots \tag{6b}$$

with similar expressions for $\mathcal{D}_y^{\mathrm{e}}$ and $\mathcal{D}_y^{\mathrm{o}}$.

When deriving a model from energy functionals, it is important not to truncate the Taylor series too early in the process, as this may lead to inconsistencies and incomplete expressions—see for instance [10] who corrected the earlier work of [5] based on the requirement that continualisation of the energy functionals should lead to the same results as continualisation of the equations of motion. In order to decide how to truncate multiple equations consistently, an objective approach is to derive a dispersion relation so that all terms of equal order of magnitude (expressed, say, in terms of powers of the product of strut length L and wave number) may be grouped together. A systematic study and the derivation of a consistent hierarchy of models will be presented elsewhere, but here we will simply use identical truncations of all $\mathcal{D}$ in all three equations of motion. Thus, we obtain

$$\rho^{\mathrm{M}}\ddot{u} = E^{\mathrm{M}}\left(u_{,xx} + \lambda^2 u_{,yy} + \lambda^2\theta_{,y}\right) \tag{7a}$$

$$\rho^{\mathrm{M}}\ddot{v} = E^{\mathrm{M}}\left(v_{,yy} + \lambda^2 v_{,xx} - \lambda^2\theta_{,x}\right) \tag{7b}$$

$$\frac{1}{12}\rho^{\mathrm{M}}\lambda^2L^2\ddot{\theta} = E^{\mathrm{M}}\lambda^2\left(v_{,x} - u_{,y} - 2\theta - \frac{1}{6}L^2\left(\theta_{,xx} + \theta_{,yy}\right)\right) \tag{7c}$$

The structure of these equations follow those of [10], extended with inertia, and differ from those in [5] in the sign of the $\nabla^2\theta$ term in Eq. (7c). The structure of Eqs. (7) is also in accordance with so-called *micropolar theory* except, again, for the sign of the

$\nabla^2\theta$ term [15]. This sign discrepancy has significant implications for the stability of the model, as will be shown below.

4 Stability Considerations

Stability of the model can be verified by inspection of the energy functionals, by which the stability of each individual term can be verified, or by an analysis of dispersive waves, by which it can be verified whether certain unstable terms are outweighed by certain other stable terms.

4.1 Positive Definiteness of Energy Functionals

The energy functionals associated with Eqs. (7) are given by

$$\mathcal{T} = \frac{1}{2}\rho^{\mathrm{M}}\left(\dot{u}^2 + \dot{v}^2 + \frac{1}{12}\lambda^2 L^2 \dot{\theta}^2\right) \tag{8a}$$

for the kinetic energy density $\mathcal{T}$ and

$$\mathcal{U} = \frac{1}{2}E^{\mathrm{M}}\left(u_{,x}^2 + v_{,y}^2 + \lambda^2\left(u_{,y} + \theta\right)^2 + \lambda^2\left(v_{,x} - \theta\right)^2 - \frac{1}{6}\lambda^2 L^2\left(\theta_{,x}^2 + \theta_{,y}^2\right)\right) \tag{8b}$$

for the potential energy density $\mathcal{U}$ [18]. Since all terms are expressed as squares, the negative sign preceding the contribution proportional to $\frac{1}{6}\lambda^2 L^2$ indicates a loss of positive definiteness and a loss of stability [9, 18].

Kumar and McDowell argue that a non-positive definite energy *density* could be acceptable as long as its volume integral is positive definite [18, p. 7414]. Such would be the case if destabilising terms can be proven to be outweighed by other contributions that are stabilising. To verify whether this is the case here, we will carry out a dispersion analysis next.

4.2 Dispersion Analysis

A coupled set of harmonic waves is used as trial functions for Eqs. (7), i.e.

$$\begin{bmatrix} u(x,y,t) \\ v(x,y,t) \\ \theta(x,y,t) \end{bmatrix} = \begin{bmatrix} U \\ V \\ \Theta \end{bmatrix} \cdot \exp\left(\mathrm{i}\left(\omega t - k_x x - k_y y\right)\right) \tag{9}$$

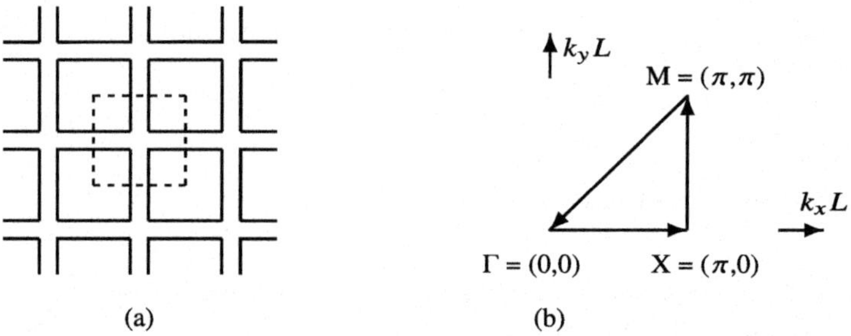

Fig. 1 Periodic lattice with unit cell (**a**) and its dimensionless irreducible First Brillouin zone (**b**)

where U, V and Θ are unknown amplitudes, ω is the angular frequency, k_x and k_y are the components of the wave number k such that $k^2 = k_x^2 + k_y^2$, and i is the imaginary unit. Assuming an infinite domain and substituting Eq. (9) into Eqs. (7) yields

$$\left[-\left(\frac{\omega L}{c_e}\right)^2 \mathcal{M} + \mathcal{K}\right]\begin{bmatrix} U \\ V \\ \mathrm{i}L\Theta \end{bmatrix} = 0 \tag{10}$$

where $c_e = \sqrt{E^{\mathrm{M}}/\rho^{\mathrm{M}}}$ is the wave speed. Furthermore, the inertia and stiffness contributions are collected in

$$\mathcal{M} = \mathrm{diag}\left[1,\, 1,\, \frac{1}{12}\lambda^2\right] \tag{11a}$$

and

$$\mathcal{K} = \begin{bmatrix} k_x^2 L^2 + \lambda^2 k_y^2 L^2 & 0 & -\lambda^2 k_y L \\ 0 & k_y^2 L^2 + \lambda^2 k_x^2 L^2 & \lambda^2 k_x L \\ -\lambda^2 k_y L & \lambda^2 k_x L & 2\lambda^2\left(1 - \frac{1}{12}k^2 L^2\right) \end{bmatrix} \tag{11b}$$

Setting the determinant of the system matrix in Eq. (10) to zero yields a cubic equation in terms of ω^2.

Figure 1a shows a unit cell within the periodic beam lattice. Using the centre of the unit cell as the origin of the coordinate system, the irreducible First Brillouin Zone can be defined via the path ΓXMΓ given in Fig. 1b. These combinations of the wave number components k_x and k_y are used to solve the cubic equation in ω^2. The results are a set of three frequency-wave number curves associated with a compression wave, a shear wave and a rotational wave. Figure 2depicts these curves for the First Brillouin Zone. For comparison, the dispersion curves of the discrete model of Eqs. (3) are also plotted.

An important observation from Fig. 2 is that for certain parts of the First Brillouin Zone, only two solutions for ω could be found; this concerns the entire path XM but also parts of paths ΓX and MΓ. To prove this point, path ΓX will be studied in detail.

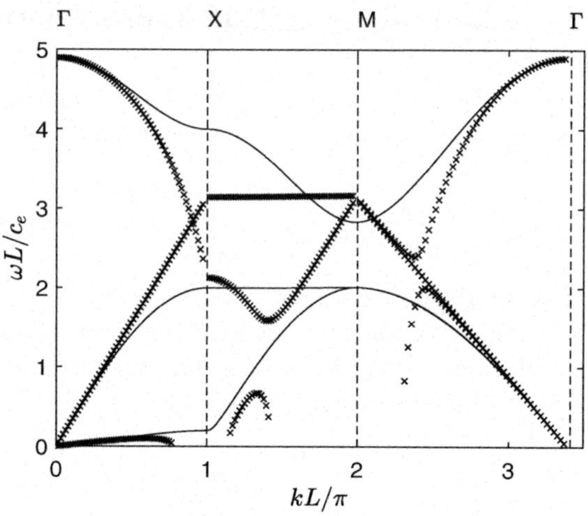

Fig. 2 Dispersion curves in the First Brillouin zone for the discrete model (solid) and the continualised model using Taylor series (×)

Substituting $k_y = 0$ and $k_x = k$ into Eq. (10) results in

$$(k^2L^2 - \overline{\omega}^2)\left((\lambda^2k^2L^2 - \overline{\omega}^2)\left(2\lambda^2 - \frac{1}{6}\lambda^2k^2L^2 - \frac{1}{12}\lambda^2\overline{\omega}^2\right) - \lambda^4k^2L^2\right) = 0 \tag{12}$$

where $\overline{\omega} \equiv \omega L/c_e$ is the dimensionless angular frequency. The first factor leads to $\overline{\omega} = \pm kL$ which is related to the compression wave. Setting the remaining factor equal to zero leads to a quadratic equation in terms of $\overline{\omega}^2$ as

$$a\overline{\omega}^4 + b\overline{\omega}^2 + c = 0 \quad \text{where} \quad \begin{cases} a = \frac{1}{12} \\ b = -2 + \frac{1}{12}k^2L^2\left(2 - \lambda^2\right) \\ c = \lambda^2k^2L^2\left(1 - \frac{1}{6}k^2L^2\right) \end{cases} \tag{13}$$

It can be verified that $b < 0$ and $c < 0$ for $6 < k^2L^2 < 24/(2 - \lambda^2)$. Under these conditions, it follows that $\sqrt{b^2 - 4ac} > -b$ and that there will be one negative root for $\overline{\omega}^2$. When substituted back into Eq. (9), the resulting imaginary values for $\overline{\omega}$ will lead to an unbounded growth, indicating a loss of stability.

The conclusion therefore is that the effects of destabilising terms (as identified from inspection of the energy functionals) are not guaranteed to be neutralised by stabilising terms (as demonstrated via the dispersion analysis). It is thus pertinent to explore stabilisation of the model of Eqs. (7).

5 Stabilisation via Padé Approximations

Padé approximations provide a systematic framework to write a particular function as a rational function, with some approximation error. For instance, one may write

$$1 + \epsilon \approx \frac{1}{1-\epsilon} \quad \text{for} \quad |\epsilon| \ll 1 \tag{14}$$

As explored by Andrianov and coworkers [2–4], this principle may also be applied to partial differential operators. Considering Eqs. (7c) and (8b), it is clear that the problematic term is the second-order spatial derivative of the micro-rotation θ. In the spirit of Eq. (14), we therefore write

$$1 + \frac{1}{12}L^2\nabla^2 \approx \frac{1}{1 - \dfrac{1}{12}L^2\nabla^2} \tag{15}$$

where $\nabla^2 = \frac{\partial^2}{\partial x^2} + \frac{\partial^2}{\partial y^2}$. Equation (15) can also be cast as

$$\left(1 - \frac{1}{12}L^2\nabla^2\right)\left(1 + \frac{1}{12}L^2\nabla^2\right) = 1 - \frac{1}{144}L^4\nabla^4 \approx 1 \tag{16}$$

where fourth-order derivatives have been ignored—note that this is the same truncation made in going from Eqs. (5) to (7).

Equation (16) suggests that the unstable term in Eq. (7c) can be eliminated by multiplying this expression with $\left(1 - \frac{1}{12}L^2\nabla^2\right)$. Thus,

$$\begin{aligned} 0 &= \left(1 - \frac{1}{12}L^2\nabla^2\right)\left\{\frac{\lambda^2 L^2}{12c_e^2}\ddot{\theta} - \lambda^2\left(v_{,x} - u_{,y} - \left(1 + \frac{1}{12}L^2\nabla^2\right)2\theta\right)\right\} \\ &\approx \left(1 - \frac{1}{12}L^2\nabla^2\right)\frac{\lambda^2 L^2}{12c_e^2}\ddot{\theta} - \lambda^2\left(v_{,x} - u_{,y} - 2\theta\right) \end{aligned} \tag{17}$$

where, in line with the truncations made in Sect. 3, third-order derivatives of u and v and fourth-order derivatives of θ have been ignored. Moreover, second-order deriviates of θ have been eliminated in accordance with Eq. (16).

With Eq. (17) replacing Eq. (7c), it is possible to re-evaluate the energy density expressions of Eqs. (8a) and (8b). The kinetic energy density $\mathcal{T}$ is extended with an extra term as

$$\mathcal{T} = \frac{1}{2}\rho^{\mathrm{M}}\left\{\dot{u}^2 + \dot{v}^2 + \frac{1}{12}\lambda^2 L^2\left(\dot{\theta}^2 + \frac{1}{12}L^2\left(\dot{\theta}_{,x}^2 + \dot{\theta}_{,y}^2\right)\right)\right\} \tag{18a}$$

whereas the potential energy density $\mathcal{U}$ is reduced to

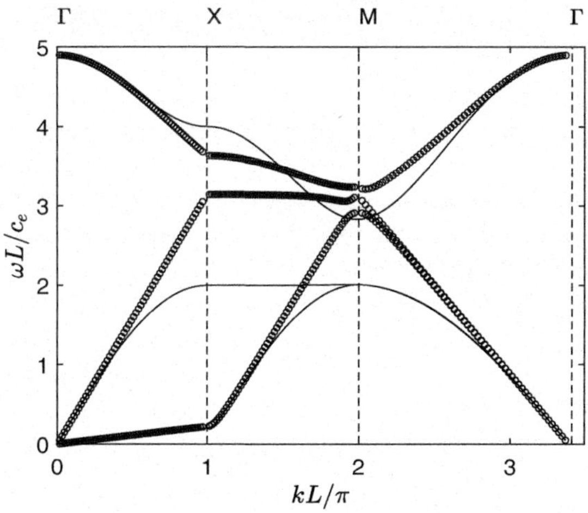

Fig. 3 Dispersion curves in the First Brillouin Zone for the discrete model (solid) and the continualised model using Padé approximations (o)

$$\mathcal{U} = \frac{1}{2} E^{\mathrm{M}} \left(u_{,x}^2 + v_{,y}^2 + \lambda^2 \left(u_{,y} + \theta \right)^2 + \lambda^2 \left(v_{,x} - \theta \right)^2 \right) \tag{18b}$$

It is easily seen that all terms in Eqs. (18a) and (18b) are expressed as squares preceded by a positive sign. Thus, the model should be stable. To verify this, we revisit Eq. (10) where the system matrix is now given via

$$\mathcal{M} = \mathrm{diag} \left[1, 1, \frac{1}{12} \lambda^2 \left(1 + \frac{1}{12} k^2 L^2 \right) \right] \tag{19a}$$

and

$$\mathcal{K} = \begin{bmatrix} k_x^2 L^2 + \lambda^2 k_y^2 L^2 & 0 & -\lambda^2 k_y L \\ 0 & k_y^2 L^2 + \lambda^2 k_x^2 L^2 & \lambda^2 k_x L \\ -\lambda^2 k_y L & \lambda^2 k_x L & 2\lambda^2 \end{bmatrix} \tag{19b}$$

Figure 3 shows the results of resolving the cubic equation that results from setting the determinant of the system matrix in Eq. (10) equal to zero. It can be verified that three real roots for ω are obtained for the entire path ΓXMΓ across the First Brillouin Zone. Thus, the stabilisation through Padé approximations appears to have had the desired effect. In addition, when comparing Figs. 2 and 3, also the accuracy with which the discrete model is approximated has improved through the use of Padé approximations, in particular that of the upper branch. This branch is associated with the rotational mode, and this improvement in accuracy reflects the fact that the Padé approximation was used on the rotational equation of motion.

6 Summary, Discussion and Outlook

A continuum theory of the micropolar type has been derived from a discrete model of a periodic lattice of beam elements. The use of Rayleigh beam theory for the individual beam elements means that the continualised equations of motion include rotational inertia. Inspection of the energy functional and a dispersion analysis have confirmed that the model is unstable. However, stabilisation is possible by applying a Padé approximation. The resulting continuum equations read

$$\rho^{\mathrm{M}}\ddot{u} = E^{\mathrm{M}}\left(u_{,xx} + \lambda^2 u_{,yy} + \lambda^2\theta_{,y}\right) \tag{20a}$$

$$\rho^{\mathrm{M}}\ddot{v} = E^{\mathrm{M}}\left(v_{,yy} + \lambda^2 v_{,xx} - \lambda^2\theta_{,x}\right) \tag{20b}$$

$$\left(1 - \frac{1}{12}L^2\nabla^2\right)\frac{1}{12}\rho^{\mathrm{M}}\lambda^2 L^2\ddot{\theta} = E^{\mathrm{M}}\lambda^2\left(v_{,x} - u_{,y} - 2\theta\right) \tag{20c}$$

Only the last of these has been modified; the first two are identical to those given in Eqs. (7). Within the confines of this book chapter, it is impossible to do justice to all the intricate details of the discrete and continuous modelling aspects of beam lattices, hence we briefly discuss a number of aspects that may warrant further reflection and consideration.

6.1 *Micro-inertia*

The additional inertia term in Eq. (20c) was introduced mathematically via an approximation based on asymptotic equivalence. This may cast doubts on its physical relevance, but note that such terms have been introduced in a variety of earlier works, typically aimed at capturing the microstructural effects of inertia [8, 21, 29]. Such micro-inertia terms can be proven to introduce long-range interactions between non-adjacent material points [6] and thus are firmly rooted in nonlocal mechanics.

6.2 *Classification of the Model*

Whereas the original continualised equations of motion as given in expressions (7) are clearly of the micropolar type as defined by Eringen [15], applying the same terminology to the model of Eqs. (20) is less straightforward: in the derivation of his original micropolar theory, Eringen emphasized the inclusion of spatial gradients of the micro-rotation in the rotational equation of motion. Such terms are lacking in Eq. (20c), but could potentially be restored by multiplying Eq. (7c) with a factor $\left(1 - \alpha L^2\nabla^2\right)$ taking $\alpha > \frac{1}{12}$; the latter approach is not pursued in the present contribution, but the interested reader may want to refer to [25].

A static reduction of Eq. (20c) would imply that the micro-rotation θ is equal to the macro-rotation $\frac{1}{2}\left(v_{,x} - u_{,y}\right)$. Such a continuum theory is known as *couple stress theory* [15] and would allow the elimination of the rotation as an independent degree of freedom [15, 16, 19]. However, the presence of rotational inertia precludes the identification of micro and macro-rotation; thus, couple stress theory does not seem to be the appropriate classification for the model of Eqs. (20). The presence of rotational inertia but the lack of second spatial derivatives of the rotation means the model could be categorised as a *reduced Cosserat* theory [17] extended with micro-inertia.

6.3 Truncation

In the derivation of Eq. (20c), certain truncations have been made that are consistent with the ones used to arrive at Eqs. (7). However, the question remains what the most rational basis is for series truncation. As already mentioned before, it is desirable that continualisation of the equations of motion and continualisation of the energy functionals lead to the same results [10]. This is typically achieved by requiring stationarity of the energy functionals prior to truncation [7]—see also [26] and follow-up discussion [13, 27].

In this paper, the focus has been on deriving a micropolar continuum theory and its subsequent stabilisation. However, inspection of Eq. (12) reveals that the inclusion of the spatial gradients of θ, considered an essential feature of micropolar theory, leads to a mismatch in powers of k^2L^2. Thus, it could be argued that the micropolar theory is not an asymptotically consistent approximation of the underlying beam lattice. The lowest possible truncation that is asymptotically consistent would be of the *reduced Cosserat* type [17]. Such a model may then serve as the basis for the systematic formulation of a hierarchy of models in which higher-order spatial gradients of the displacements and rotation appear.

References

1. Alavi, S., Ganghoffer, J., Sadighi, M., Nasimsobhan, M., Akbarzadeh, A.: Continualization method of lattice materials and analysis of size effects revisited based on Cosserat models. Int. J. Solids Struct. **111**(894), 254–255 (2022)
2. Andrianov, I., Awrejcewicz, J.: Continuous models for 2D discrete media valid for higher-frequency domain. Comput. Struct. **86**, 140–144 (2008)
3. Andrianov, I., Awrejcewicz, J., Barantsev, R.: Asymptotic approaches in mechanics: new parameters and procedures. ASME Appl. Mech. Rev. **56**, 87–110 (2003)
4. Andrianov, I., Bolshakov, V., Danishevskyy, V., Weichert, D.: Higher order asymptotic homogenization and wave propagation in periodic composite materials. Proc. Roy. Soc. A **464**, 1181–1201 (2008)
5. Askar, A., Cakmak, A.: A structural model of a micropolar continuum. Int. J. Eng. Sci. **6**, 583–589 (1968)

6. Askes, H., Gitman, I.: A computational mechanics perspective on long-range interactions in gradient elasticity with microinertia. J. Mech. Behav. Mater. **23**, 37–40 (2014)
7. Askes, H., Metrikine, A.: Higher-order continua derived from discrete media: continualisation aspects and boundary conditions. Int. J. Solids Struct. **42**, 187–202 (2005)
8. Askes, H., Nguyen, D., Tyas, A.: Increasing the critical time step: micro-inertia, inertia penalties and mass scaling. Comput. Mech. **47**, 657–667 (2011)
9. Bacigalupo, A., Gambarotta, L.: A dynamic high-frequency consistent continualization of beam-lattice materials. Compos. Struct. **272**, 114,146 (2021)
10. Bažant, Z., Christensen, M.: Analogy between micropolar continuum and grid frameworks under initial stress. Int. J. Solids Struct. **8**, 327–346 (1972)
11. Cauchy, A.: Mémoire sur les vibrations d'un double système de molécules et de l'éther continu dans un corps cristallisé. In: Oeuvres complètes, 1re Série — Tome II, pp. 338–350. Gauthier-Villars (reprint 1908), Paris (1850)
12. Chang, C., Gao, J.: Second-gradient constitutive theory for granular material with random packing structure. Int. J. Solids Struct. **32**, 2279–2293 (1995)
13. Domenico, D.D., Askes, H., Aifantis, E.: Discussion of "Derivation of Mindlin's first and second strain gradient elastic theory via simple lattice and continuum models" by Polyzos and Fotiadis. Int. J. Solids Struct. **191–192**, 646–651 (2020)
14. Dos Reis, F., Ganghoffer, J.: Construction of micropolar continua from the asymptotic homogenization of beam lattices. Comput. Struct. **112–113**, 354–363 (2012)
15. Eringen, A.: Linear theory of micropolar elasticity. J. Math. Mech. **15**, 909–923 (1966)
16. Gonella, S., Ruzzene, M.: Homogenization and equivalent in-plane properties of two-dimensional periodic lattices. Int. J. Solids Struct. **45**, 2897–2915 (2008)
17. Grekova, E.: Plane waves in the linear elastic reduced Cosserat medium with a finite axially symmetric coupling between volumetric and rotational strains. Math. Mech. Solids **21**, 73–93 (2016)
18. Kumar, R., McDowell, D.: Generalized continuum modeling of 2-D periodic cellular solids. Int. J. Solids Struct. **41**, 7399–7422 (2004)
19. Lombardo, M., Askes, H.: Higher-order gradient continuum modelling of periodic lattice materials. Comput. Mater. Sci. **52**, 204–208 (2012)
20. Metrikine, A., Askes, H.: One-dimensional dynamically consistent gradient elasticity models derived from a discrete microstructure. Part 1: generic formulation. Eur. J. Mech. A/Solids **21**, 555–572 (2002)
21. Milton, G., Willis, J.: On modifications of Newton's second law and linear continuum elastodynamics. Proc. R. Soc. A **463** (2007)
22. Mühlhaus, H.B., Oka, F.: Dispersion and wave propagation in discrete and continuous models for granular materials. Int. J. Solids Struct. **33**, 2841–2858 (1996)
23. Noor, A., Nemeth, M.: Micropolar beam models for lattice grids with rigid joints. Comput. Methods Appl. Mech. Eng. **21**, 249–263 (1980)
24. Ostoja-Starzewski, M.: Lattice models in micromechanics. ASME Appl. Mech. Rev. **55**, 35–60 (2002)
25. Pichugin, A., Askes, H., Tyas, A.: Asymptotic equivalence of homogenisation procedures and fine-tuning of continuum theories. J. Sound Vibr. **313**, 858–874 (2008)
26. Polyzos, D., Fotiadis, D.: Derivation of Mindlin's first and second strain gradient elastic theory via simple lattice and continuum models. Int. J. Solids Struct. **49**, 470–480 (2012)
27. Polyzos, D., Fotiadis, D.: Reply to "Discussion of "Derivation of Mindlin's first and second strain gradient elastic theory via simple lattice and continuum models" by Polyzos and Fotiadis". Int. J. Solids Struct. **191–192**, 652–654 (2020)
28. Rubin, M., Rosenau, P., Gottlieb, O.: Continuum model of dispersion caused by an inherent material characteristic length. J. Appl. Phys. **77**, 4054–4063 (1995)
29. Vardoulakis, I., Aifantis, E.: On the role of microstructure in the behavior of soils: effects of higher order gradients and internal inertia. Mech. Mater. **18**, 151–158 (1994)

Interaction Between Gravity and Gyroscopic Forces: Transient Control of Gyropendulums and Inverted Gyropendulums

A. Kandiah, I. S. Jones, N. V. Movchan, and A. B. Movchan

Abstract In this paper we introduce a mathematical model designed to analyse the interaction between gyroscopic forces and the gravitational field and which has been essential in studying the dynamic response of several classes of chiral metamaterials. For an elementary cell, which includes either a normal or inverted gyroscopic pendulum, we provide the full classification of its transient oscillations linked to the sufficiently high rate of spin. The characteristics of physical chirality in the presence of gravity cover a range of physical phenomena at different scales. The theoretical work is accompanied by the illustrative examples.

Keywords Chiral structures · Pendulums · Gyroscopic spinners · Gravity induced dynamics · Transient oscillations

1 Introduction

In recent years, extensive research has been done on mechanically chiral metamaterials for the control of elastic and acoustic waves in solids [1–18]. Chiral systems exhibit a property that introduces an inherent asymmetry appearing in various physical and biological contexts such as molecular structures and optical phenomena. The notion of chirality in physics and mechanics has been established in the classical literature, in such books as [19–21]. In mechanics, chirality is brought about by the presence of gyroscopic spinners that result in a coupling of the components linked to the motion of the system. In such physical chiral systems, chirality introduces a complexity into the structure by affecting spin dynamics, stability, and responses to external influences [18, 20, 22–31]. Gyroscopic spinners have been used as building blocks in the design of chiral metamaterials. The theoretical background for modelling the dynamic response of gyroscopic spinners connected to elastic multi-structures is included in [13–18]. In addition to a rotational action through the intro-

A. Kandiah (✉) · I. S. Jones · N. V. Movchan · A. B. Movchan
Department of Mathematical Sciences, University of Liverpool, Liverpool L69 7ZL, UK
e-mail: sgakandi@liverpool.ac.uk

F. Willot et al. (eds.), *Continuum Models and Discrete Systems*,
Springer Proceedings in Mathematics & Statistics 457,
https://doi.org/10.1007/978-3-031-58665-1_12

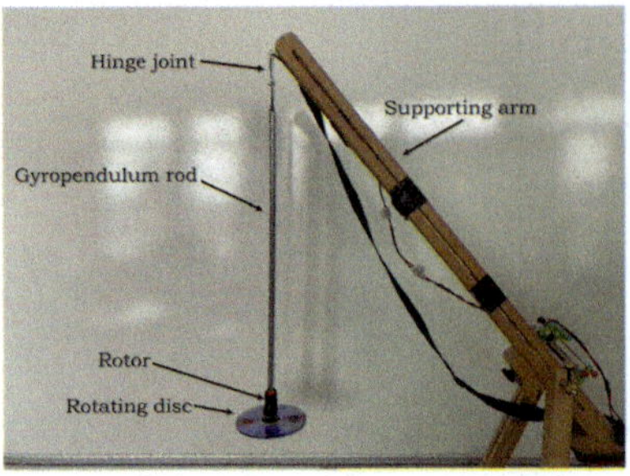

(a)

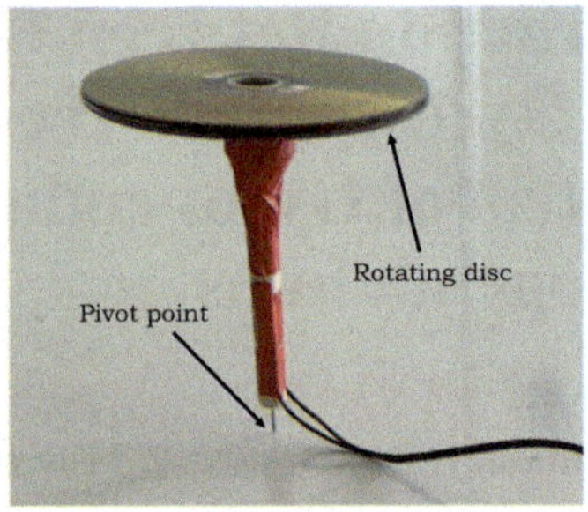

(b)

Fig. 1 Experimental setup of the (**a**) gyropendulum and (**b**) inverted gyropendulum consisting of a rotating disc attached to a rod, which is hinged at its end

duction of an additional force, gravity, a new class of structures is obtained, which we refer to as chiral gravitational multi-structures (see [18]). The transient processes for such systems are of particular interest, due to formal connections with a range of natural phenomena (see, for example [32–35]). The simplest case of such structures, which combines the effect of the gyroscopic action and gravity, is the gyroscopic pendulum (or gyropendulum for brevity) that has been studied in [18]. In this paper, we present a new comparative analysis of the gyropendulum and inverted gyropendulum. The main difference is the orientation of gravity relative to the gyropendulum itself, as illustrated in Fig. 1.

The inverted gyropendulum and the effect of gravity on the dynamics of its motion are analysed. A gyropendulum can be used to model electromagnetic and Coriolis phenomena [19, 32, 36–40]. In the absence of a gyroscopic action, the gyropendulum becomes the familiar simple pendulum. An inverted pendulum subjected to vertical oscillations has been studied in [41, 42], where it was shown that the structure can be stable. The latter is known as Kapitza's pendulum [41], and is used to describe various periodic processes in physics. In the present paper, we show that stable states of the inverted gyropendulum exist in the absence of a vertically oscillating pivot point, but due to the presence of a gyroscopic action introduced through a spinner. The gyropendulum, implemented in the experiment, is shown in Fig. 1a, together with the experimental[1] and theoretical measurements (Fig. 2b) of its smooth trajectory which takes into account the effects of both gyroscopic action and of gravity.

The paper is structured as follows. Section 2 presents the mathematical formulation and analytical solutions of the gyropendulum and inverted gyropendulum.

[1] In Fig. 2, the parameter values used in the experimental setup for the length of the pendulum arm is 0.58 m, and the radius, gyricity and mass of the disc are 0.06 m, -130 rad/s and $m = 0.032$ kg, respectively. In addition, the initial displacement prescribed along the x-direction is 0.0403 m, and the initial displacement in the y-direction is set to 0.0757 m, while the initial velocities in the x- and y-directions are -0.5867 m/s and 0.7067 m/s, respectively. There is a good agreement between the analytical and experimental results in the given time period of 3.8 s, which further confirms the validity of both the analytical approach and proposed linearisation.

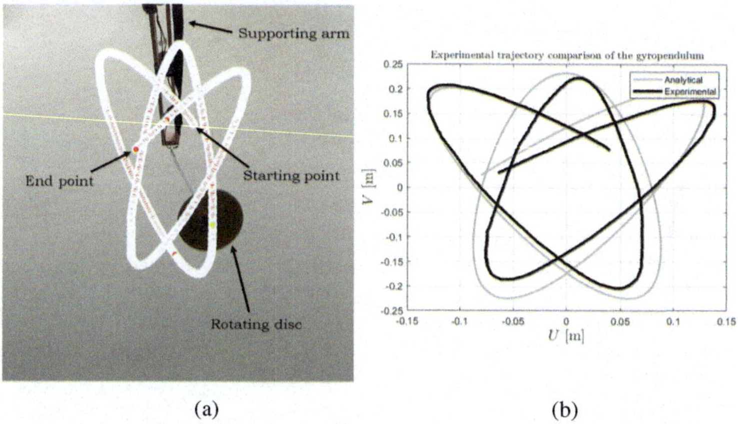

Fig. 2 Observation of the experimental results and analytical approximation for smooth trajectories of the gyropendulum: (**a**) experimental tracing of the trajectory; (**b**) comparison with the analytical approximation

The full classifications of the gyropendulums' motion are given using a novel geometrical methodology, with illustrative examples, in Sect. 3. Section 4 demonstrates the motion of the gyropendulum characterised by approximate polygonal shapes. Finally, the connections with the angular momentum method associated with the transient motion of the inverted gyropendulum are discussed in Sect. 5, followed by the concluding remarks.

2 Transient Motions of Gyropendulums and Inverted Gyropendulums

We focus on analysing the bounded trajectories of the gyropendulum and inverted gyropendulum. The gyropendulum consists of a rigid, massless rod suspended from a pivot at $z = 0$, so that it can swing freely under gravity. At the lower end, the rod is connected to a thin uniform disc which acts as a gyroscopic spinner [18]. The general motion of gyroscopic spinners is usually characterised by the angles of nutation, precession and spin. The analysis here is limited to the regime where the nutation angle θ and its derivatives are considered to be small, resulting in a linearised model.

The time-dependent non-dimensional transverse displacements in the x and y directions at the end of the gyropendulum (see [18]), shown in Fig. 3a, are denoted by $\tilde{U}(\tilde{t})$ and $\tilde{V}(\tilde{t})$, respectively. The governing equations for the motion of the gyropendulum and inverted gyropendulum in non-dimensional form are, respectively, given by

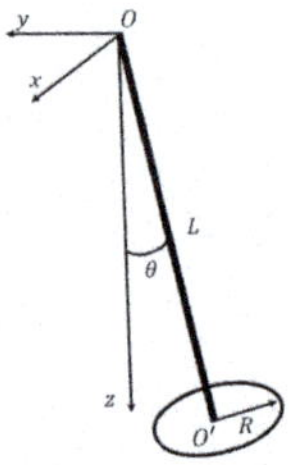

(a) The gyropendulum consisting of a uniform disc attached to the end of a rigid, massless rod.

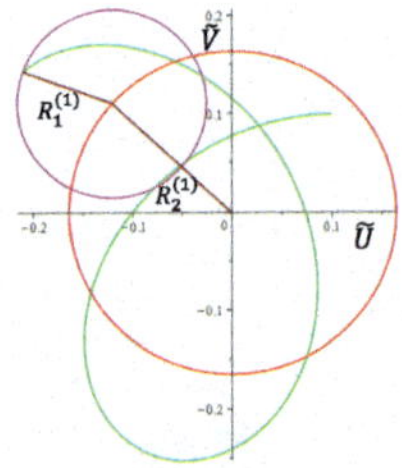

(b) Trajectory formed by two circular motions with $\tilde{U}_0 = 0.1$, $\tilde{V}_0 = 0.1$, $\dot{\tilde{U}}_0 = -0.3$, $\dot{\tilde{V}}_0 = 0$ and $\Gamma = 1$, leading to $R_1^{(1)} = 0.09516$ and $R_2^{(1)} = 0.16415$.

Fig. 3 The gyropendulum and its trajectory

$$\frac{d^2}{d\tilde{t}^2}\begin{pmatrix}\tilde{U}\\ \tilde{V}\end{pmatrix} + \Gamma\begin{pmatrix}0 & 1\\ -1 & 0\end{pmatrix}\frac{d}{d\tilde{t}}\begin{pmatrix}\tilde{U}\\ \tilde{V}\end{pmatrix} + \gamma\begin{pmatrix}\tilde{U}\\ \tilde{V}\end{pmatrix} = \begin{pmatrix}0\\ 0\end{pmatrix}, \tag{2.1}$$

where $\tilde{t}$ is the dimensionless time.[2] The quantity $\Gamma = \Omega I_1/\sqrt{mgLI_0}$ is a measure of the gyroscopic effect with Ω being related to the spin rate. Here, m, g and L denote the mass of the spinner, acceleration due to gravity and length of the pendulum arm, respectively. In addition, I_0 and I_1 represent the moments of inertia along the principal transverse and vertical axes of the spinner combined with the rod, respectively.

The system (2.1) is supplied with the initial conditions

$$\tilde{U}(0) = \tilde{U}_0, \quad \tilde{V}(0) = \tilde{V}_0, \quad \frac{d\tilde{U}}{d\tilde{t}}(0) = \dot{\tilde{U}}_0, \quad \frac{d\tilde{V}}{d\tilde{t}}(0) = \dot{\tilde{V}}_0, \tag{2.2}$$

where $\tilde{U}_0$, $\tilde{V}_0$, $\dot{\tilde{U}}_0$ and $\dot{\tilde{V}}_0$ are given values of the normalised initial displacements and initial velocities.

The quantity $\gamma = \pm 1$ characterises the direction of the gravitational force exerted on the spinner; when $\gamma = 1$ the system (2.1) describes the gyropendulum, while when $\gamma = -1$ the system represents the equations of motion of the inverted gyropendulum. In the present paper, we focus on analysing the bounded motion of such structures. It is noted that the trajectories of the gyropendulum (with $\gamma = 1$) are bounded for all values of the parameter Γ (see, for example, [18]). On the other hand, bounded trajectories[3] of the inverted gyropendulum (with $\gamma = -1$) occur when $|\Gamma| > 2$.

---

[2] The dimensional equations of motion for the transverse displacement components of the gyropendulum and inverted gyropendulum are, in their respective order, given by $\begin{pmatrix}\ddot{U}(t)\\ \ddot{V}(t)\end{pmatrix} + \frac{I_1}{I_0}\Omega\begin{pmatrix}0 & 1\\ -1 & 0\end{pmatrix}\begin{pmatrix}\dot{U}(t)\\ \dot{V}(t)\end{pmatrix} \pm \frac{mgL}{I_0}\begin{pmatrix}U(t)\\ V(t)\end{pmatrix} = \begin{pmatrix}0\\ 0\end{pmatrix}$, and the dimensionless time $\tilde{t} = t\sqrt{mgL/I_0}$ is introduced to formulate the system (2.1), where $\tilde{U}(\tilde{t}) = L^{-1}U(\tilde{t}/\sqrt{mgL/I_0})$ and $\tilde{V}(\tilde{t}) = L^{-1}V(\tilde{t}/\sqrt{mgL/I_0})$.

[3] For $|\Gamma| < 2$, the quantities $\tilde{\lambda}_i$, $i = 1, 2$, are complex (see (2.12)), and the transverse displacements of the inverted gyropendulum consists of exponentially increasing and localised solutions in time

The solution of (2.1) subject to the initial conditions (2.2) can be written in the following form

$$\begin{pmatrix} \tilde{U} \\ \tilde{V} \end{pmatrix} = R_2^{(\gamma)} \begin{pmatrix} \cos\left(\tilde{\nu}_2^{(\gamma)}\tilde{t} + \tau_2^{(\gamma)}\right) \\ \sin\left(\tilde{\nu}_2^{(\gamma)}\tilde{t} + \tau_2^{(\gamma)}\right) \end{pmatrix} + R_1^{(\gamma)} \begin{pmatrix} \cos\left(\tilde{\nu}_1^{(\gamma)}\tilde{t} + \tau_1^{(\gamma)}\right) \\ -\gamma\sin\left(\tilde{\nu}_1^{(\gamma)}\tilde{t} + \tau_1^{(\gamma)}\right) \end{pmatrix}, \tag{2.3}$$

where

$$R_2^{(1)} = \sqrt{\tilde{A}_1^2 + \tilde{A}_2^2}, \quad \tau_2^{(1)} = \begin{cases} \operatorname{sgn}(\tilde{A}_2)\arctan\left(\frac{|\tilde{A}_2|}{|\tilde{A}_1|}\right), & \tilde{A}_1 > 0, \\ \frac{\pi}{2}\operatorname{sgn}(\tilde{A}_2), & \tilde{A}_1 = 0, \\ \operatorname{sgn}(\tilde{A}_2)(\pi - \arctan\left(\frac{|\tilde{A}_2|}{|\tilde{A}_1|}\right)), & \tilde{A}_1 < 0, \end{cases} \tag{2.4}$$

$$R_1^{(1)} = \sqrt{\tilde{A}_3^2 + \tilde{A}_4^2}, \quad \tau_1^{(1)} = \begin{cases} -\operatorname{sgn}(\tilde{A}_4)\arctan\left(\frac{|\tilde{A}_4|}{|\tilde{A}_3|}\right), & \tilde{A}_3 > 0, \\ -\frac{\pi}{2}\operatorname{sgn}(\tilde{A}_4), & \tilde{A}_3 = 0, \\ -\operatorname{sgn}(\tilde{A}_4)(\pi - \arctan\left(\frac{|\tilde{A}_4|}{|\tilde{A}_3|}\right)), & \tilde{A}_3 < 0, \end{cases} \tag{2.5}$$

in which the quantities $\tilde{A}_1$, $\tilde{A}_2$, $\tilde{A}_3$ and $\tilde{A}_4$ are given by

$$\tilde{A}_1 = \frac{\dot{\tilde{V}}_0 + \tilde{\omega}_1\tilde{U}_0}{\tilde{\omega}_1 + \tilde{\omega}_2}, \quad \tilde{A}_2 = \frac{-\dot{\tilde{U}}_0 + \tilde{\omega}_1\tilde{V}_0}{\tilde{\omega}_1 + \tilde{\omega}_2}, \quad \tilde{A}_3 = \frac{-\dot{\tilde{V}}_0 + \tilde{\omega}_2\tilde{U}_0}{\tilde{\omega}_1 + \tilde{\omega}_2}, \quad \tilde{A}_4 = \frac{\dot{\tilde{U}}_0 + \tilde{\omega}_2\tilde{V}_0}{\tilde{\omega}_1 + \tilde{\omega}_2}, \tag{2.6}$$

and

$$R_2^{(-1)} = \sqrt{\tilde{B}_1^2 + \tilde{B}_2^2}, \quad \tau_2^{(-1)} = \begin{cases} -\operatorname{sgn}(\tilde{B}_2)\arctan\left(\frac{|\tilde{B}_2|}{|\tilde{B}_1|}\right), & \tilde{B}_1 > 0, \\ -\frac{\pi}{2}\operatorname{sgn}(\tilde{B}_2), & \tilde{B}_1 = 0, \\ -\operatorname{sgn}(\tilde{B}_2)(\pi - \arctan\left(\frac{|\tilde{B}_2|}{|\tilde{B}_1|}\right)), & \tilde{B}_1 < 0, \end{cases} \tag{2.7}$$

$$R_1^{(-1)} = \sqrt{\tilde{B}_3^2 + \tilde{B}_4^2}, \quad \tau_1^{(-1)} = \begin{cases} \operatorname{sgn}(\tilde{B}_4)\arctan\left(\frac{|\tilde{B}_4|}{|\tilde{B}_3|}\right), & \tilde{B}_3 > 0, \\ \frac{\pi}{2}\operatorname{sgn}(\tilde{B}_4), & \tilde{B}_3 = 0, \\ \operatorname{sgn}(\tilde{B}_4)(\pi - \arctan\left(\frac{|\tilde{B}_4|}{|\tilde{B}_3|}\right)), & \tilde{B}_3 < 0. \end{cases} \tag{2.8}$$

The quantities $\tilde{B}_1$, $\tilde{B}_2$, $\tilde{B}_3$ and $\tilde{B}_4$, with the additional condition that $|\Gamma| > 2$, are as follows:

---

multiplied by oscillatory terms. In contrast, for $|\Gamma| = 2$, there is only one fundamental frequency, which results in solutions that grow linearly in time. When $|\Gamma| > 2$, the displacements of the inverted gyropendulum contain pure oscillatory terms only.

$$\tilde{B}_1 = \frac{\dot{\tilde{V}}_0 - \tilde{U}_0\tilde{\lambda}_1}{\tilde{\lambda}_2 - \tilde{\lambda}_1}, \quad \tilde{B}_2 = \frac{\dot{\tilde{U}}_0 + \tilde{V}_0\tilde{\lambda}_1}{\tilde{\lambda}_2 - \tilde{\lambda}_1}, \quad \tilde{B}_3 = \frac{-\dot{\tilde{V}}_0 + \tilde{U}_0\tilde{\lambda}_2}{\tilde{\lambda}_2 - \tilde{\lambda}_1}, \quad \tilde{B}_4 = \frac{\dot{\tilde{U}}_0 + \tilde{V}_0\tilde{\lambda}_2}{\tilde{\lambda}_2 - \tilde{\lambda}_1}. \tag{2.9}$$

We denote

$$\tilde{\nu}_i^{(\gamma)} = \begin{cases} \tilde{\omega}_i, & \gamma = 1, \\ \tilde{\lambda}_i, & \gamma = -1, \end{cases} \tag{2.10}$$

for $i = 1, 2$, where

$$\tilde{\omega}_1 = \frac{1}{2}\Big(-\Gamma + \sqrt{\Gamma^2 + 4}\Big), \; \tilde{\omega}_2 = \frac{1}{2}\Big(\Gamma + \sqrt{\Gamma^2 + 4}\Big), \tag{2.11}$$

and

$$\tilde{\lambda}_1 = \frac{1}{2}\Big(\Gamma - \sqrt{\Gamma^2 - 4}\Big), \; \tilde{\lambda}_2 = \frac{1}{2}\Big(\Gamma + \sqrt{\Gamma^2 - 4}\Big). \tag{2.12}$$

The representation (2.3) implies that for both the gyropendulum and inverted gyropendulum, the motion can be described by tracing two rigid segments of length $R_2^{(\gamma)}$ and $R_1^{(\gamma)}$. In particular, the motion of the gyropendulum structure is confined to the annulus bounded by the circles of radii $|R_2^{(\gamma)} - R_1^{(\gamma)}|$ and $|R_2^{(\gamma)} + R_1^{(\gamma)}|$. A typical trajectory of the gyropendulum may be seen in Fig. 3b where the circle with radius $R_2^{(1)}$ is fixed with centre at the origin and the circle with radius $R_1^{(1)}$ has a centre moving along the circumference of the first circle. Such a geometrical interpretation of the solution may be used to classify the possible shapes of trajectories for given initial conditions and value of Γ.

Figure 4 shows the illustrative comparison between the experimental measurements and the analytical approximation[4] of the inverted gyropendulum. The trajectory traced in the experimental observation is shown in Fig. 4a, and the plots of both the experimental data and the analytical approximation[5] are shown in Fig. 4b. There is a good agreement between the experimental and analytical results presented in this figure.

3 Geometrical Method of Classification of Trajectories for the Gyropendulum and Inverted Gyropendulum

In this section, we provide the classification of the trajectories of the gyropendulum and inverted gyropendulum into three distinct classes using a geometrical approach based on the analysis of the solutions presented in Sect. 2.

[4] The parameter values used in the analytical solution, given by (2.3) with $\gamma = -1$, are: $\tilde{\lambda}_1 = -6$, $\tilde{\lambda}_2 = -4$, $R_1^{(-1)} = 0.0019$, $R_2^{(-1)} = 0.00525$, $\tau_1^{(-1)} = \pi/7$ and $\tau_2^{(-1)} = 0$.

[5] This path corresponds to the Class 3 motions of the inverted gyropendulum, according to the classification given in Sect. 3.1.2.

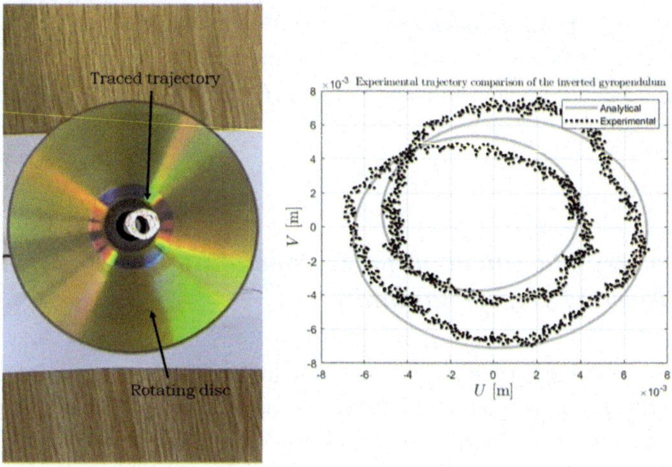

(a) (b)

Fig. 4 Experimental observation and analytical prediction for smooth trajectories of the inverted gyropendulum for the values $R_1^{(-1)}/R_2^{(-1)} = 0.362$ and $\tilde{\lambda}_2/\tilde{\lambda}_1 = 0.667$ with $\Gamma < -2$: (**a**) experimental tracing of the motion; (**b**) comparison with the analytical approximation. The time period of the motion is 3.5 s

The trajectories of the gyropendulum and inverted gyropendulum may have different shapes depending on the chosen initial conditions and the specified value of Γ. Before providing examples of the various classes of trajectories, we make the following observations:

- The trajectories will be closed if the ratio $\tilde{\nu}_2^{(\gamma)}/\tilde{\nu}_1^{(\gamma)}$ is rational, resulting in periodic paths of the structure.
- In the absence of gyroscopic action ($\Gamma = 0$), the motion of the gyropendulum follows elliptical trajectories associated with the linearised motion of a simple pendulum. On the other hand, if $\Gamma = 0$, the inverted gyropendulum becomes an inverted pendulum and is unstable. In our case, for the inverted gyropendulum we consider stable regimes where $|\Gamma| > 2$.
- When $R_2^{(\gamma)} = R_1^{(\gamma)}$, the trajectories pass through the origin.
- Cusps appear in the gyropendulum and inverted gyropendulum trajectories at the points where the transverse velocity components vanish. The radial component vanishes when the line segments are inline. The tangential component of the velocity vanishes when $R_2^{(\gamma)}\tilde{\nu}_2^{(\gamma)} = R_1^{(\gamma)}\tilde{\nu}_1^{(\gamma)}$.
- Changing the initial conditions and/or the value of Γ will change the quantities $R_1^{(\gamma)}$, $R_2^{(\gamma)}$, $\tilde{\nu}_1^{(\gamma)}$, $\tilde{\nu}_2^{(\gamma)}$, $\tau_1^{(\gamma)}$ and $\tau_2^{(\gamma)}$ (see (2.3)). However, if such a change is made that alters only the phases, then the form of the trajectory remains the same as before the change, but the shape is rotated about the origin. Additionally, modifying the values $R_1^{(\gamma)}$ and $R_2^{(\gamma)}$ such that the ratio $R_1^{(\gamma)}/R_2^{(\gamma)}$ is maintained as a constant, for fixed values of Γ, leads to no change in the shape of the trajectory but simply

a spatial scaling. Thus, without loss of generality, we do not consider the effect of changing the phases $\tau_1^{(\gamma)}$ and $\tau_2^{(\gamma)}$ on the motions of the gyropendulum and inverted gyropendulum, and consider the ratio $R_1^{(\gamma)}/R_2^{(\gamma)}$ and Γ when classifying the trajectories.

3.1 Trajectory Classification for Gyropendulums and Inverted Gyropendulums

Based on the considerations above, it is possible to identify three classes of trajectories for both the gyropendulum and inverted gyropendulum by considering various values of the ratio $R_1^{(\gamma)}/R_2^{(\gamma)}$ and Γ. These classes are defined below and the regimes in which they occur are summarised in Tables 1 and 2.

Class 1: trajectories with outward-pointing (gyropendulum) and inward-pointing (inverted gyropendulum) self-intersecting loops. These trajectories do not pass through the origin.

Class 2: occurs at a single value of $R_1^{(\gamma)}/R_2^{(\gamma)}$ yielding two distinct trajectory types:

Class 2a: trajectories with outward-pointing (gyropendulum) and inward-pointing (inverted gyropendulum) cusps, which do not pass through the origin.

Class 2b: trajectories with outward-pointing (gyropendulum) and inward-pointing (inverted gyropendulum) loops passing through the origin, and there are no cusps present.

Class 3: trajectories consisting of smooth curves which do not pass through the origin, with no cusps and no self-intersecting loops. For this class, the difference between gyropendulums and inverted gyropendulums is discussed below.

3.1.1 The Gyropendulum ($\gamma = 1$)

The solution of the gyropendulum, which satisfies the initial conditions (2.2), is given by (2.3) with $\gamma = 1$, which represents a summation of two circular motions with one circle being fixed and the other with its centre following the circumference of the first. In view of the form of (2.3) with $\gamma = 1$ and that both $\tilde{\omega}_1$ and $\tilde{\omega}_2$ are always positive, then the radius vector for the circle of radius $R_2^{(1)}$ is always rotating anticlockwise, while the radius vector of the circle of radius $R_1^{(1)}$ is always rotating clockwise.

In view of this two circle representation and the comments above made on the basis of the two circle geometry, a full classification from the geometry may be deduced based on values of $R_1^{(1)}/R_2^{(1)}$ and Γ as shown in Table 1.

Some representative examples of the gyropendulum trajectories across the range of $R_1^{(1)}/R_2^{(1)}$ are shown in Fig. 5a for the case where $\Gamma > 0$. In addition, the corresponding two circles are shown in each case.

3.1.2 The Inverted Gyropendulum ($\gamma = -1$) in the Region of Stability $|\Gamma| > 2$

Bounded solutions for the gyropendulum exist for all values of Γ, whereas, with regard to (2.12), bounded solutions of the inverted gyropendulum do not exist for all values of the parameter Γ. In particular, when $|\Gamma| > 2$, the trajectory of the inverted gyropendulum, within the linearisation assumptions, remains bounded for any initial conditions prescribed within a specific vicinity of the equilibrium point, and the structure's motion remains within that neighbourhood at all times.

The representation (2.3) for $\gamma = -1$ describes the linearised trajectories of the inverted gyropendulum, which satisfies the initial conditions (2.2). Again, the two circle representation may be used in a similar way to that above, to give a full classification of the trajectories of the inverted gyropendulum based on values of $R_1^{(-1)}/R_2^{(-1)}$ and with $|\Gamma| > 2$ as shown in Table 2.

Illustrative trajectories of the inverted gyropendulum are shown in Fig. 5b for a representative case where $\Gamma > 2$.

The difference between the trajectories of the gyropendulum and inverted gyropendulum is that, for the former, the radius vectors of the circles are moving in opposite directions while for the latter, both radius vectors move in either an anticlockwise

Table 1 Classes of trajectories for the gyropendulum for various values of $R_1^{(1)}/R_2^{(1)}$ and Γ

$\Gamma > 0$	Class	$\Gamma < 0$	Class
$0 < R_1^{(1)}/R_2^{(1)} < 1$	3	$0 < R_1^{(1)}/R_2^{(1)} < \tilde{\omega}_2/\tilde{\omega}_1$	3
$R_1^{(1)}/R_2^{(1)} = 1$	2b	$R_1^{(1)}/R_2^{(1)} = \tilde{\omega}_2/\tilde{\omega}_1$	2a
$1 < R_1^{(1)}/R_2^{(1)} < \tilde{\omega}_2/\tilde{\omega}_1$	1	$\tilde{\omega}_2/\tilde{\omega}_1 < R_1^{(1)}/R_2^{(1)} < 1$	1
$R_1^{(1)}/R_2^{(1)} = \tilde{\omega}_2/\tilde{\omega}_1$	2a	$R_1^{(1)}/R_2^{(1)} = 1$	2b
$R_1^{(1)}/R_2^{(1)} > \tilde{\omega}_2/\tilde{\omega}_1$	3	$R_1^{(1)}/R_2^{(1)} > 1$	3

Table 2 Classes of trajectories for the inverted gyropendulum for various values of $R_1^{(-1)}/R_2^{(-1)}$ and Γ

$\Gamma > 2$	Class	$\Gamma < -2$	Class
$0 < R_1^{(-1)}/R_2^{(-1)} < 1$	3	$0 < R_1^{(-1)}/R_2^{(-1)} < \tilde{\lambda}_2/\tilde{\lambda}_1$	3
$R_1^{(-1)}/R_2^{(-1)} = 1$	2b	$R_1^{(-1)}/R_2^{(-1)} = \tilde{\lambda}_2/\tilde{\lambda}_1$	2a
$1 < R_1^{(-1)}/R_2^{(-1)} < \tilde{\lambda}_2/\tilde{\lambda}_1$	1	$\tilde{\lambda}_2/\tilde{\lambda}_1 < R_1^{(-1)}/R_2^{(-1)} < 1$	1
$R_1^{(-1)}/R_2^{(-1)} = \tilde{\lambda}_2/\tilde{\lambda}_1$	2a	$R_1^{(-1)}/R_2^{(-1)} = 1$	2b
$R_1^{(-1)}/R_2^{(-1)} > \tilde{\lambda}_2/\tilde{\lambda}_1$	3	$R_1^{(-1)}/R_2^{(-1)} > 1$	3

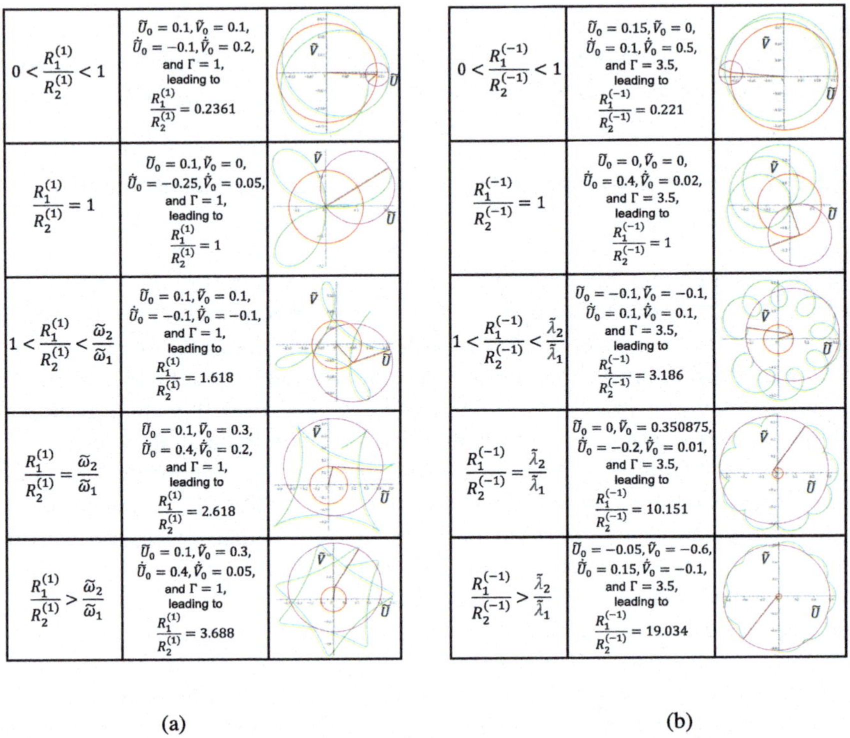

Fig. 5 Examples of trajectories of different classes for (**a**) the gyropendulum for $\Gamma = 1$, and (**b**) the inverted gyropendulum for $\Gamma = 3.5$

(for $\Gamma > 2$) or clockwise (for $\Gamma < -2$) direction. Thus, in this geometric model, the difference between a gyropendulum and inverted gyropendulum is the direction of rotations of the circles.

4 Examples of Trajectories Resembling Regular Polygonal Shapes

Interesting examples, corresponding to smooth curves approximating regular polygonal shapes (class 3), are discussed here.

For the gyropendulum, it is noted that a periodic motion can be initiated so that the trajectory approximates a regular polygon. It is quite extraordinary that such an example is consistent with the observation of the combined action of gravity and gyroscopic forces, discussed in [34, 43, 44], in connection with the hexagonal patterns at the North Pole of Saturn.

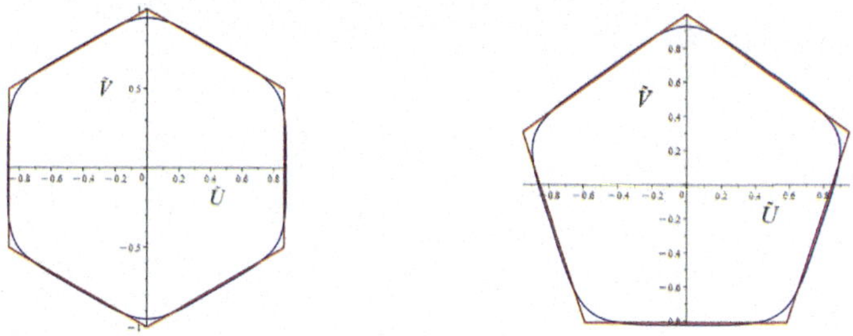

(a) (b)

Fig. 6 The approximations to the prescribed hexagonal and pentagonal orbits. The calculated values of $\varGamma$ and the initial conditions to generate these orbits are (**a**) $\varGamma = 4/\sqrt{5}$, $\tilde{U}_0 = 0$, $\tilde{V}_0 = -0.94837$, $\dot{\tilde{U}}_0 = -0.32624$, $\dot{\tilde{V}}_0 = 0$, and (**b**) $\varGamma = 1.5$, $\tilde{U}_0 = -0.54654$, $\tilde{V}_0 = -0.75225$, $\dot{\tilde{U}}_0 = -0.26550$, $\dot{\tilde{V}}_0 = 0.19290$

For the gyropendulum, the field (2.3) may be linked to a complex Fourier series representation with two frequency components and without the frequency-independent term. It is convenient to write this in complex notation as

$$z(\tilde{t}) = c_1 e^{-i\tilde{\omega}_1 \tilde{t}} + c_2 e^{i\tilde{\omega}_2 \tilde{t}}, \tag{4.1}$$

where $z(\tilde{t}) = \tilde{U}(\tilde{t}) + i\tilde{V}(\tilde{t})$, and require that the ratio $\tilde{\omega}_2/\tilde{\omega}_1$ is an integer connected to the degree of rotational symmetry of the polygon. This detemines the required value of $\varGamma$. Appropriate initial conditions may be found such that the complex constants c_1 and c_2 in (4.1) correspond to the Fourier coefficients for the particular polygon (see, for example, [18]).

Two examples are shown in Fig. 6 where a prescribed (a) hexagon and (b) pentagon may be seen together with their approximating trajectories. The Fourier coefficients in these two cases are (a) $c_1 = -0.91189i$ and $c_2 = -0.03648i$, and (b) $c_1 = -0.51439 - 0.70800i$ and $c_2 = -0.03215 - 0.04425i$. The resulting values of the parameter $\varGamma$ together with the initial conditions are shown in the figure caption. There is very good agreement between the two prescribed shapes and their approximating trajectories considering a truncated Fourier series is used from the analytical solution. In addition, it is also noted that the ratio $|c_2|/|c_1|$ is small in both cases, highlighting how these polygonal shapes represent a perturbation of the basic circular motion.

Interestingly, an inverted gyropendulum does not follow the trajectories shown in Fig. 6. For the inverted gyropendulum of sufficiently high gyricity, an attempt to simulate a path, which loosely resembles a regular hexagon or a pentagon, results in the class 3 trajectories shown in Fig. 7. Of course, such an "approximation" is different from the path shown in Fig. 6.

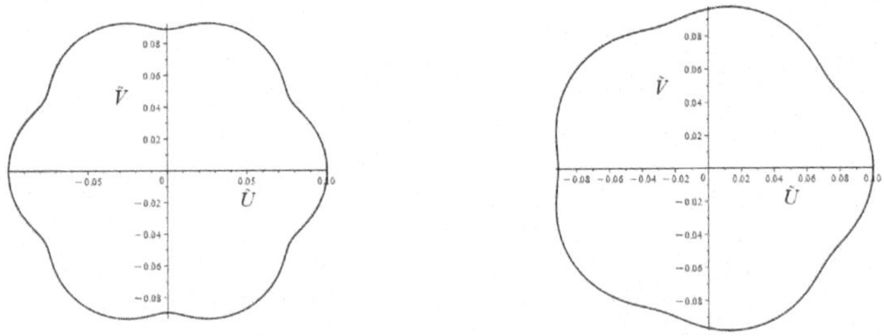

(a) (b)

Fig. 7 The class 3 orbits for the inverted gyropendulum. There is a resemblance of polygonal shapes, for the set of initial conditions: $\tilde{U}_0 = 0.1$, $\tilde{V}_0 = 0$, $\dot{\tilde{U}}_0 = 0$, $\dot{\tilde{V}}_0 = 0.05$, and the following values of Γ: (**a**) $\Gamma = 8/\sqrt{7}$, and (**b**) $\Gamma = 7/\sqrt{6}$

5 Angular Momentum Method of Classification of the Bounded Transient Motion for an Inverted Gyropendulum

In the paper [18], we introduced a method of classification for the trajectories of the gyropendulum, based on the initial value of the normalised angular momentum defined by

$$H_\gamma(\tilde{t})\mathbf{k} = (\tilde{U}(\tilde{t})\dot{\tilde{V}}(\tilde{t}) - \tilde{V}(\tilde{t})\dot{\tilde{U}}(\tilde{t}))\mathbf{k} = \tilde{\mathbf{Q}}(\tilde{t}) \times \dot{\tilde{\mathbf{Q}}}(\tilde{t}),$$

where $\mathbf{k}$ is the basis vector along the z-axis and $\tilde{\mathbf{Q}}(\tilde{t}) = (\tilde{U}(\tilde{t}), \tilde{V}(\tilde{t}))$. This has been explored in [18] for the case of $\gamma = 1$ (standard gyropendulum).

In this section, we consider this method for the case of $\gamma = -1$ (inverted gyropendulum), subject to an additional assumption of $|\Gamma| > 2$. It is worth mentioning that for any transient (non-time-harmonic) bounded motion of the gyropendulum, the function $H_\gamma(\tilde{t})$ is always time-harmonic:

$$H_\gamma(\tilde{t}) = \frac{[|\dot{\tilde{\mathbf{Q}}}(0)|^2 - \gamma|\tilde{\mathbf{Q}}(0)|^2]\Gamma + 4\gamma H_\gamma(0)}{\Xi^2} + \frac{\sqrt{P^2 + R^2}}{\Xi^2}\cos\left(\Xi\tilde{t} - \chi\right), \tag{5.1}$$

where

$$\Xi = \sqrt{\Gamma^2 + 4\gamma},\ P = \Gamma(\Xi\tilde{\mathbf{Q}}(0)\cdot\dot{\tilde{\mathbf{Q}}}(0)^T),\ R = \Gamma(H_\gamma(0)\Gamma + \gamma|\tilde{\mathbf{Q}}(0)|^2 - |\dot{\tilde{\mathbf{Q}}}(0)|^2),$$
$$\chi = \tan^{-1}\left(\tfrac{P}{R}\right) \text{ for } R > 0,\quad \chi = \pi + \tan^{-1}\left(\tfrac{P}{R}\right) \text{ for } R < 0,$$
$$\chi = \mathrm{sgn}(P)\tfrac{\pi}{2} \text{ for } R = 0. \tag{5.2}$$

For convenience, we use the notation $H_-(\tilde{t}) = H_{-1}(\tilde{t})$. The product of the maximum and minimum values of $H_-(\tilde{t})$ for $|\Gamma| > 2$ and $\gamma = -1$ is given by

$$-\frac{(\Gamma|\tilde{\mathbf{Q}}(0)|^2 - 2H_-(0))(\Gamma|\dot{\tilde{\mathbf{Q}}}(0)|^2 - 2H_-(0))}{\Gamma^2 - 4}, \tag{5.3}$$

and the sign of (5.3) is important for determining the trajectories. The classification of the inverted gyropendulum trajectories is connected to the initial value $H_-(0)$ and to the open interval[6] Υ between the values M_- and N_-, where

$$\begin{gathered} M_- = \Gamma|\dot{\tilde{\mathbf{Q}}}(0)|^2/2, \quad N_- = \Gamma|\tilde{\mathbf{Q}}(0)|^2/2, \\ \text{for } \{|\dot{\tilde{\mathbf{Q}}}(0)| < |\tilde{\mathbf{Q}}(0)|\} \cap \{\Gamma > 2\} \text{ or } \{|\dot{\tilde{\mathbf{Q}}}(0)| > |\tilde{\mathbf{Q}}(0)|\} \cap \{\Gamma < -2\}, \end{gathered} \tag{5.4}$$

and

$$\begin{gathered} N_- = \Gamma|\tilde{\mathbf{Q}}(0)|^2/2, \quad M_- = \Gamma|\dot{\tilde{\mathbf{Q}}}(0)|^2/2, \\ \text{for } \{|\dot{\tilde{\mathbf{Q}}}(0)| > |\tilde{\mathbf{Q}}(0)|\} \cap \{\Gamma > 2\} \text{ or } \{|\dot{\tilde{\mathbf{Q}}}(0)| < |\tilde{\mathbf{Q}}(0)|\} \cap \{\Gamma < -2\}. \end{gathered} \tag{5.5}$$

In full consistency with the earlier Sect. 3.1.2, the three classes of inverted gyropendulum trajectories are characterised as follows:

Class 1: Self-intersecting loops[7] of variable orientation occur when $H_-(0) \notin \overline{\Upsilon}$.

Class 2: For this class, the quantity (5.3) vanishes, i.e. $H_-(0) \in \partial\Upsilon$, and we have two sub-classes:

(**a**) Trajectories have cusps[8] when the motion does not change its orientation and $H_-(0) = M_-$.

(**b**) Loops passing through the origin[9] occur when the orientation of the motion does not change and $H_-(0) = N_-$.

Class 3: Smooth curves, which do not pass through the origin, and do not change their orientation, are characterised by the case when $H_-(0) \in \Upsilon$.

6 Concluding Remarks

In the present paper, we have introduced a geometrical method for the classification for the motion of chiral gravitational structures, including gyropendulums and inverted gyropendulums. This method is universal, and it has been shown that

---

[6] When $|\dot{\tilde{\mathbf{Q}}}(0)| = |\tilde{\mathbf{Q}}(0)|$ and $|\Gamma| > 2$, there are no non-trivial solutions such that $H_-(0) = M_- = N_-$, i.e. in this case all compatible orbits will correspond to Class 1 motions.

[7] The trajectories do not contain cusps or loops passing through the origin. The orientation of the curves alternates between a clockwise and anticlockwise motion relative to the origin in accordance with the sign changes of the function $H_-(\tilde{t})$.

[8] The trajectory does not pass through the origin. In this case, the function $H_-(\tilde{t})$ vanishes only at the points corresponding to the cusps.

[9] In this sub-class, there are no cusps, and the function $H_-(\tilde{t})$ equals zero only at the points corresponding to the origin.

although the dynamic features of the gyropendulums and inverted gyropendulums differ, the geometric descriptions of the motion of both systems are closely related.

Connections with the normalised angular momentum method have been discussed with the focus on the motion of the inverted gyropendulum. Applications have been noted in the gyroscopic motion of planetary systems, with special attention to orbits of approximate polygonal shapes.

The study of the gyropendulums and inverted gyropendulums provides insight into the dynamics of mechanically chiral structures, that takes into account the combined effects of gyroscopic forces and gravity. This is especially important in modelling chiral metamaterials in the presence of gravity.

Acknowledgements A.K. gratefully acknowledges the financial support of the EPSRC through the Maths DTP grant EP/V52007X/1, project reference 2599756. I.S.J. is grateful to the Department of Mathematical Sciences, University of Liverpool and to the Liverpool Research Centre for Mathematics and Modelling for provision of the research infrastructure and computational facilities.

References

1. Nash, L.M., Kleckner, D., Read, A., Vitelli, V., Turner, A.M., Irvine, W.T.M.: Topological mechanics of gyroscopic metamaterials. PNAS **112**(47), 14495–14500 (2015)
2. Brun, M., Jones, I.S., Movchan, A.B.: Vortex-type elastic structured media and dynamic shielding. Proc. R. Soc. A **468**, 3027–3046 (2012)
3. Carta, G., Jones, I., Movchan, N., Movchan, A., Nieves, M.: "Deflecting elastic prism" and unidirectional localisation for waves in chiral elastic systems. Sci. Rep. **7**, 26 (2017)
4. Carta, G., Jones, I., Movchan, N., Movchan, A., Nieves, M.: Gyro-elastic beams for the vibration reduction of long flexural systems. Proc. R. Soc. A: Math. Phys. Eng. Sci. **473**, 20170136 (2017)
5. Carta, G., Nieves, M.J., Jones, I.S., Movchan, N.V., Movchan, A.B.: Elastic chiral waveguides with gyro-hinges. Quart. J. Mech. Appl. Math. **71**, 157–185 (2018)
6. Nieves, M.J., Carta, G., Jones, I.S., Movchan, A.B., Movchan, N.V.: Vibrations and elastic waves in chiral multi-structures. J. Mech. Phys. Solids **121**, 387–408 (2018)
7. Carta, G., Jones, I.S., Movchan, N.V., Movchan, A.B.: Wave characterisation in a dynamic elastic lattice: lattice flux and circulation. Phys. Mesomech. **22**, 152–163 (2019)
8. Carta, G., Jones, I.S., Movchan, N.V., Movchan, A.B.: Wave polarization and dynamic degeneracy in a chiral elastic lattice. Proc. R. Soc. Lond. A **475**, 20190313 (2019)
9. Carta, G., Colquitt, D., Movchan, A., Movchan, N., Jones, I.: Chiral flexural waves in structured plates: directional localisation and control. J. Mech. Phys. Solids **137**, 103866 (2020)
10. Carta, G., Colquitt, D., Movchan, A., Movchan, N., Jones, I.: One-way interfacial waves in a flexural plate with chiral double resonators. Philos. Trans. R. Soc. A **378**, 20190350 (2020)
11. Movchan, A.B., Movchan, N.V., Jones, I.S., Milton, G.W., Nguyen, H.-M.: Frontal waves and transmissions for temporal laminates and imperfect chiral interfaces. Phil. Trans. R. Soc. A **380**, 20210385 (2022)
12. Movchan, A., Movchan, N., Jones, I.: On waves in multi-scale chiral elastic systems. Math. Mech. Solids **27**, 1855–1868 (2022)
13. Carta, G., Brun, M., Movchan, A.B., Movchan, N.V., Jones, I.S.: Dispersion properties of vortex-type monatomic lattices. Int. J. Solids Struct. **51**, 2213–2225 (2014)
14. Garau, M., Carta, G., Nieves, M., Jones, I., Movchan, N., Movchan, A.: Interfacial waveforms in chiral lattices with gyroscopic spinners. Proc. R. Soc. A: Math. Phys. Eng. Sci. **474**, 20180132 (2018)

15. Carta, G., Nieves, M.J., Jones, I.S., Movchan, N.V., Movchan, A.B.: Flexural vibration systems with gyroscopic spinners. Philos. Trans. R. Soc. A. **377**, 20190154 (2019)
16. Jones, I.S., Movchan, N.V., Movchan, A.B.: Two-dimensional waves in a chiral elastic chain: dynamic Green's matrices and localised defect modes. Quart. J. Mech. Appl. Math. **73**, 305–328 (2020)
17. Jones, I.S., Movchan, N.V., Movchan, A.B.: Chiral waves in structured elastic systems: dynamics of a meta-waveguide. Quart. J. Mech. Appl. Math. **75**, 63–89 (2022)
18. Kandiah, A., Jones, I.S., Movchan, N.V., Movchan, A.B.: Effect of gravity on the dispersion and wave localisation in gyroscopic elastic systems. In: Altenbach, H., Bruno, G., Eremeyev, V.A., Gutkin, M.Y., Müller, W.H. (eds.) Mechanics of Heterogeneous Materials. Advanced Structured Materials, vol. 195. Springer, Switzerland (2023)
19. Webster, A.G.: The Dynamics of Particles and of Rigid, Elastic, and Fluid Bodies. Lectures on Mathematical Physics. B. G. Teubner, Leipzig (1904)
20. Gray, A.: A Treatise on Gyrostatics and Rotational Motion: Theory and Applications. Macmillan and Co., Ltd, London (1918)
21. Goldstein, H., Safko, J., Poole, C.: Classical Mechanics. Addison-Wesley, San Francisco, Boston, New York (1951)
22. Kelvin, L.: The Molecular Tactics of a Crystal. Clarendon Press, Oxford (1894)
23. Roitenberg, I.N.: Gyroscopes. Nauka, Moscow (1975)
24. Andreev, V.D.: Teoriya Inertsial'noy Navigatsii. (Theory of Inertial Navigation). Nauka, Moscow (1966)
25. Ishlinskii, A.Y.: Orientation, Gyroscopes, and Inertial Navigation. Nauka, Moscow (1976)
26. Ishlinskii, A.Y.: Mechanics of Gyroscopic Systems. Izdatel'stvo Akademii Nauk SSSR, Moscow (1963)
27. Scarborough, J.B.: The Gyroscope Theory and Applications. Interscience Publishers Inc, New York (1958)
28. Wrigley, W., Hollister, W.M., Denhard, W.G.: Gyroscopic Theory, Design, and Instrumentation. M.I.T. Press, Cambridge, Massachusetts (1969)
29. D'Eleuterio, G.M.T., Hughes, P.C.: Dynamics of gyro-elastic continua. In: 24th Structures, Structural Dynamics and Materials Conference, 02 May 1983–04 May 1983, Lake Tahoe, NV, U.S.A. https://doi.org/10.2514/6.1983-826
30. D'Eleuterio, G.M.T., Hughes, P.C.: Dynamics of gyroelastic spacecraft. J. Guid. Control. Dyn. **10**, 401–405 (1987)
31. Kelvin, L., Tait, P.G.: Treatise on Natural Philosophy. Clarendon Press, Oxford (1867)
32. Foucault, J.B.L.: Démonstration physique du mouvement de rotation de la terre au moyen du pendule. Comptes Rendus des Séances de L'Académie des Sciences **32**, 135–138 (1851)
33. Shepherd, T.G.: Rossby waves and two-dimensional turbulence in a large-scale zonal jet. J. Fluid Mech. **183**, 467–509 (1987)
34. Ingersoll, A.P.: Cassini exploration of the planet saturn: a comprehensive review. Space Sci. Rev. **216**, 122 (2020). https://doi.org/10.1007/s11214-020-00751-1
35. Tabataba-Vakili, F., Rogers, J.H., Eichstädt, G., Orton, G.S., Hansen, C.J., Momary, T.W., Sinclair, J.A., Giles, R.S., Caplinger, M.A., Ravine, M.A., Bolton, S.J.: Long-term tracking of circumpolar cyclones on Jupiter from polar observations with JunoCam. ICARUS **335**, 113405 (2020). https://doi.org/10.1016/j.icarus.2019.113405
36. Sommeria, J.: Foucault and the rotation of the Earth. C R Phys. **18**, 520–525 (2017)
37. Slade, B.: Classical Symmetric Top in a Gravitational Field (2012)
38. Chelnokov, I.N.: On the theory of the gyropendulum. J. Appl. Math. Mech. **44**, 703–708 (1980)
39. Lorentz, H.A.: Versuch einer Theorie der Electrischen und Optischen Erscheinungen in Bewegten Körpern. B. G. Teubner, Leipzig (1906)
40. Bromwich, T.: On the theory of Foucault's pendulum, and of the gyrostatic pendulum. Proc. Lond. Math. Soc. **2**, 222–235 (1914)
41. Kapitza, P.: Dynamic stability of the pendulum with vibrating suspension point. Sov. Phys. JETP. **21**, 588–597 (1951)

42. Maryuta, A.: Dynamics of an inverted pendulum with a vibrating suspension. Int. Appl. Mech. **29**, 1025–1031 (1993)
43. Godfrey, D.A.: A hexagonal feature around Saturn's north pole. Icarus **76**, 335–356 (1988)
44. Allison, M., Godfrey, D.A., Beebe, R.F.: A wave dynamical interpretation of Saturn's polar hexagon. Science **247**, 1061–1063 (1990)

Manipulation of the Optical Properties of a Plasmonic Metamaterial Film by an External Transverse Voltage

Yakov M. Strelniker and David J. Bergman

Abstract The macroscopic permittivity tensors of a film-shaped metamaterial medium are calculated for the case when the charge carrier concentration N_0 of each constituent of the composite metamaterial varies across the film thickness. The influence of an applied transverse voltage on the surface plasmon resonances and on the optical properties of such a system are studied. It is shown that the macroscopic effective permittivity tensor, as well as other optical properties of the metamaterial, are extremely sensitive to the applied transverse voltage. This can be used to construct fast switches and other optical devices.

Keywords Plasmonics · Extraordinary light transmission · Meta-materials · Composite materials · Magneto-optics

1 Introduction

In our previous publications [1–11] we considered the possibility of manipulating the values of the itinerant [10, 11] and localized [6–9] surface plasmon resonances by application of a strong static magnetic field. This leads to the possibility of also manipulating the optical properties of this system. Until now our suggestions [1–4] were realized only for the case of magnetoresistance (i.e., for the case of *dc* electric fields) [3, 5] but not in optical experiments or devices (i.e., in the case of *ac* fields) due to technical difficulties of implementing a strong *dc* magnetic field in optical experiments. The application of a transverse voltage is a technically easier way to manipulate the optical properties of the above plasmonic systems [12–14].

The permittivity tensor of the i-th constituent of the composite metamaterial can be written for simplicity in the form of the quasistatic regime [1–11]

Y. M. Strelniker (✉)
Department of Physics, Bar-Ilan University, IL-52900 Ramat-Gan, Israel
e-mail: strelnik@biu.ac.il

D. J. Bergman
Faculty of Exact Sciences, Tel Aviv University, IL-69978 Tel Aviv, Israel
e-mail: bergman@tauex.tau.ac.il

F. Willot et al. (eds.), *Continuum Models and Discrete Systems*,
Springer Proceedings in Mathematics & Statistics 457,
https://doi.org/10.1007/978-3-031-58665-1_14

$$\hat{\varepsilon}_i = \varepsilon_{i,0} \cdot \hat{I} + i\frac{4\pi}{\omega}\hat{\sigma}_i = \left(\varepsilon_{i,0} + \frac{i\omega_p^2\tau}{\omega(1-i\omega\tau)}\right) \cdot \hat{I} \simeq \left(\varepsilon_{i,0} - \frac{\omega_p^2}{\omega^2}\right) \cdot \hat{I}, \quad (1)$$

where the conductivity tensor $\hat{\sigma}_i$ is taken in the free-electron Drude approximation, ε_0 is the scalar dielectric constant of the background ionic lattice, which we take to be 1 in this article, $\hat{I}$ is the unit tensor, $\omega_p = \left(4\pi e^2 N_0/m\right)^{1/2}$ is the plasma frequency, N_0 is the charge carrier concentration, m is the effective mass of the charge carriers [1–11]. Both σ and ε of the itinerant charge carriers are proportional to the charge carrier concentration N_0. This opens a way to manipulate the value of the surface plasmon resonance as well as the optical properties by changing the value of N_0. If N_0 is uniform across the entire thickness of the conducting constituent then everything is simple and the result can be obtained by changing the value of N_0 in the expression for the surface plasmon resonance frequency $\omega_{\text{res}} = \omega(N_0)$. Here we consider the more complicated case where N_0 varies along the film thickness. This case necessitates numerical computations and the results cannot be obtained by simple substitution of the local value of N_0 in the corresponding analytical expressions.

The theory developed in this paper can also be applied for calculations of samples with impurity centers creating the coordinate-dependent densities of free charges.

2 Theory

We consider a conducting film with a square array of holes. A monochromatic light beam of angular frequency ω impinges upon this film along the perpendicular axis y, with linear polarization along the principal axis x of the array (see Fig. 1). A film with holes is placed between metal plates to which a potential difference is applied, i.e., an electrical voltage perpendicular to the surface of the film (in y-direction). The plates are electrically insulated from the film so that electric current cannot penetrate the film from the plates. The plates are thin enough for the reasonable transmission of the electromagnetic waves and are made of a different material so that the surface plasmon resonance frequency differs significantly from the corresponding resonant frequency of the film. Thus, the plates can affect optical properties only through a change in the distribution of charge carrier density across the film thickness (i.e., along the y-axis). Both, the conducting perforated film and the metallic plates are assumed to be infinitely long in the x and z directions in order to make change of the charge density (due to the applied transverse voltage) along the y-axis only (see Appendix below).

Following Refs. [6–9] we consider the holes as dielectric inclusions embedded in a conducting host and treat the system in the quasi-static approximation. This description is suitable, provided the lattice constant is much smaller than the wavelength, i.e., when the film is homogeneous on the scale of the electromagnetic wavelength. In this approach, the local electric potential $\phi^{(\alpha)}(\mathbf{r})$ is the solution of a boundary value problem based upon the Laplace partial differential equation (see below).

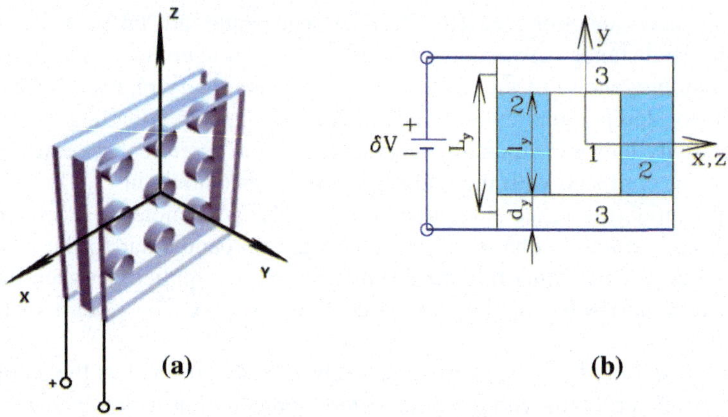

Fig. 1 **a** Schematic drawing of a conducting film with a periodic square array of holes. The coordinate axes x and z are always directed along the principal axes of the simple square lattice. The incident light beam is normal to the film surface, i.e., the *ac* electric field $\mathbf{E}$ is parallel to the film plane, while the wave vector is normal to it (i.e., $\mathbf{k} \parallel y$). Voltage is applied between two plates. **b** Side view of the unit cell used for calculation of the conducting film (denoted 2 and described by a θ_2 characteristic function) with a periodic arrangement of circular or square cylindrical holes (denoted as 1 and described by another θ_1 characteristic function). The symmetry axes of the cylindrical holes (inclusions) are directed along the y-axis. d_y is the thickness of the rectangular prism on the top and bottom of the unit cell, l_y is the thickness of the film and the length of the cylindrical hole

2.1 *Numerical Scheme*

The position-dependent local permittivity tensors $\hat{\varepsilon}(\mathbf{r})$ of the metamaterial medium are written as

$$\hat{\varepsilon} = \left[\hat{\varepsilon}_1 + \Delta\hat{\epsilon}_1(\mathbf{r})\right]\theta_1 + \left[\hat{\varepsilon}_2 + \Delta\hat{\epsilon}_2(\mathbf{r})\right]\theta_2 = \hat{\varepsilon}_2 - \delta\hat{\varepsilon}\theta_1 + \Delta\hat{\epsilon}_1(\mathbf{r})\theta_1 + \Delta\hat{\epsilon}_2(\mathbf{r})\theta_2, \quad (2)$$

where $\hat{\varepsilon}_1$, $\hat{\varepsilon}_2$, and $\delta\hat{\varepsilon} \equiv \hat{\varepsilon}_2 - \hat{\varepsilon}_1$ are position-independent parts of the permittivity tensors while $\Delta\hat{\epsilon}_1(\mathbf{r})$ and $\Delta\hat{\epsilon}_2(\mathbf{r})$ represent their position dependent parts (i.e., parts depending on the variation of the charge carrier concentration N_0 across the film thickness). Note that $\hat{\varepsilon}(\mathbf{r})$ is a tensorial quantity in general. Here we have used the characteristic or indicator function of the i-constituent [$\theta_i(\mathbf{r}) = 1$ for $\mathbf{r}$ inside the i-constituent, and 0 elsewhere].

The bulk effective permittivity tensor of the metamaterial $\hat{\varepsilon}_e$ is defined as providing the linear relationship between the volume averages of the electric field $\nabla\phi(\mathbf{r})$ and the electrical displacement $\mathbf{D}(\mathbf{r})$: $\langle\mathbf{D}\rangle \equiv \hat{\varepsilon}_e \cdot \langle\nabla\phi(\mathbf{r})\rangle$, where $\langle\ldots\rangle \equiv \frac{1}{V_a}\int dV \ldots$ is an average over the entire system volume. The final result for $\hat{\varepsilon}_e$ will be independent of the precise way in which the local electric potential is created, provided that the composite nanostructure, as well as the boundary conditions that determine $\phi(\mathbf{r})$, are macroscopically homogeneous. We choose a scheme where the metamaterial medium occupies the entire volume in-between the infinitely conducting plates of a parallel plate capacitor (see Refs. [1, 2]). The plates are taken to be infinitely large, and the distance between them is taken to be finite but large compared to any scale of inhomogeneity of the system. These are imaginary plates of an imaginary capacitor (located perpendicular to the direction of the corresponding component of the permittivity tensor), which are necessary only for numerical scheme and which should not be confused with the real capacitor plates (shown in Fig. 1) to which an electric potential difference is applied [1, 2, 6–11]. We denote by $\phi^{(\alpha)}(\mathbf{r})$ the local potential field which results when the plates are perpendicular to the r_α-axis and a potential difference equal to their distance apart L is applied between them. The volume averaged electric field is then $\langle\nabla\phi^{(\alpha)}\rangle = \nabla r_\alpha = \mathbf{e}_\alpha$, i.e., a unit vector in the r_α-direction.

The potential field $\phi^{(\alpha)}(\mathbf{r})$ is usually taken to be determined as the solution of a boundary value problem with a partial differential equation which follows from the Maxwell equation $\nabla\cdot\mathbf{D}(\mathbf{r}) = 0$, namely $\nabla\cdot\hat{\varepsilon}(\mathbf{r})\cdot\nabla\phi^{(\alpha)}(\mathbf{r}) = 0$ inside the capacitor and $\phi^{(\alpha)}(\mathbf{r}) = 0, L$ at the plates. We find it more convenient to base our discussion on an integral equation for $\phi^{(\alpha)}(\mathbf{r})$. In order to obtain such a formulation, we first use Eq. (2) to get:

$$\nabla\cdot\hat{\varepsilon}_2\cdot\nabla\phi^{(\alpha)} = \nabla\cdot\theta_1\delta\hat{\varepsilon}\cdot\nabla\phi^{(\alpha)} - \nabla\cdot\theta_1\Delta\hat{\epsilon}_1(\mathbf{r})\cdot\nabla\phi^{(\alpha)} - \nabla\cdot\theta_2\Delta\hat{\epsilon}_2(\mathbf{r})\cdot\nabla\phi^{(\alpha)}. \quad (3)$$

We then define a Green function $G(\mathbf{r},\mathbf{r}'|\hat{\varepsilon}_2)$ as the solution of the following boundary value problem: $\nabla\cdot\hat{\varepsilon}_2\cdot\nabla G = -\delta^3(\mathbf{r}-\mathbf{r}')$ inside the capacitor while $G = 0$ at the plates and use it to "solve" Eq. (3), treating the r.h.s. of that equation as though it were given. In this way we obtain

$$\phi^{(\alpha)}(\mathbf{r}) = r_\alpha + \hat{\Gamma}\phi^{(\alpha)} - \hat{\Gamma}_1\phi^{(\alpha)} - \hat{\Gamma}_2\phi^{(\alpha)}, \quad (4)$$

where we used integration by parts to transform the equation and where the linear integro-differential operators $\hat{\Gamma}$, $\hat{\Gamma}_1$ and $\hat{\Gamma}_2$ are defined by

$$\hat{\Gamma}\phi \equiv \int dV'\theta_1(\mathbf{r}')\nabla' G(\mathbf{r},\mathbf{r}'|\hat{\varepsilon}_2)\cdot\delta\hat{\varepsilon}\cdot\nabla'\phi(\mathbf{r}'), \quad (5)$$

$$\hat{\Gamma}_k\phi \equiv \int dV'\theta_k(\mathbf{r}')\nabla' G(\mathbf{r},\mathbf{r}'|\hat{\varepsilon}_2)\cdot\Delta\hat{\epsilon}_k(\mathbf{r}')\cdot\nabla'\phi(\mathbf{r}'), \quad (6)$$

where $k = 1, 2$. Although G, and hence also $\hat{\Gamma}$, depend on the orientation of the capacitor plates, this dependence is only important if at least one of $\mathbf{r}$, $\mathbf{r}'$ is near the plates. For most purposes G can be approximated by the Green function that vanishes at infinity, and in that case the only dependence of (4) on the boundary conditions comes from the term r_α, which ensures that $\phi^{(\alpha)}$ satisfies them. The nanostructure enters into $\hat{\Gamma}$, $\hat{\Gamma}_1$, and $\hat{\Gamma}_2$ through the characteristic functions $\theta_1(\mathbf{r})$ and $\theta_2(\mathbf{r})$. The Green function $G(\mathbf{r},\mathbf{r}'|\hat{\varepsilon}_2)$ that vanishes at infinity depends on $\mathbf{r}$, $\mathbf{r}'$ only through their difference $\mathbf{r}-\mathbf{r}'$. In the case where $\hat{\varepsilon}_2$ is a diagonal tensor, this Green function has a relatively simple closed form presented in Refs. [1, 2]. The Fourier transform of G has an even simpler form, namely

$$\int dVe^{-i\mathbf{k}\cdot(\mathbf{r}-\mathbf{r}')}G(\mathbf{r},\mathbf{r}'|\hat{\varepsilon}_{2s}) = \frac{1}{\mathbf{k}\cdot\hat{\varepsilon}_{2s}\cdot\mathbf{k}}, \tag{7}$$

where $\hat{\varepsilon}_{2s}$ is the symmetric part of $\hat{\varepsilon}_2$. From the definition of $\hat{\varepsilon}_e$ and Eq. (2) we can write:

$$\begin{aligned}&\left(\hat{\varepsilon}_2 - p_1\delta\hat{\varepsilon} - \hat{\varepsilon}_e + \langle\Delta\epsilon_1(\mathbf{r})\theta_1\rangle + \langle\Delta\epsilon_2(\mathbf{r})\theta_2\rangle\right)_{\alpha\beta}\\ &\quad = \delta\hat{\varepsilon}_{\alpha\gamma}\langle\theta_1\nabla_\gamma\psi^{(\beta)}\rangle - \langle\Delta\hat{\epsilon}_{1,\alpha\gamma}(\mathbf{r}')\theta_1\nabla_\gamma\psi^{(\beta)}\rangle - \langle\Delta\hat{\epsilon}_{2,\alpha\gamma}(\mathbf{r}')\theta_2\nabla_\gamma\psi^{(\beta)}\rangle,\end{aligned} \tag{8}$$

where $p_1 = \langle\theta_1\rangle$ is the volume fraction of the No. 1 constituent and $\psi^{(\alpha)}(\mathbf{r}) \equiv \phi^{(\alpha)}(\mathbf{r}) - r_\alpha$. Here and in the rest of this article we use the convention that a summation over repeated tensor indices is always implied.

2.2 *Periodic Composites*

When the composite medium has a periodic nanostructure, great simplifications ensue due to the fact that $\theta_1(\mathbf{r})$ is now a periodic function. From the fact that, away from the external boundaries, G depends on $\mathbf{r}, \mathbf{r}'$ only through their difference $\mathbf{r} - \mathbf{r}'$, we get that for any periodic or linear function $f(\mathbf{r})$, $\hat{\Gamma}f$ is also periodic except near those boundaries. Hence $\psi^{(\alpha)}(\mathbf{r})$ and $\theta_1(\mathbf{r})$ are periodic and can be expanded in a Fourier series

$$\psi^{(\alpha)}(\mathbf{r}) = \sum_{\mathbf{g}}\psi_{\mathbf{g}}^{(\alpha)}e^{i\mathbf{g}\cdot\mathbf{r}}, \quad \psi_{\mathbf{g}}^{(\alpha)} = \frac{1}{V_a}\int_{V_a} dVe^{-i\mathbf{g}\cdot\mathbf{r}}\psi^{(\alpha)}(\mathbf{r}), \tag{9}$$

$$\theta(\mathbf{r}) = \sum_{\mathbf{g}}\theta_{\mathbf{g}}e^{i\mathbf{g}\cdot\mathbf{r}}, \quad \theta_{\mathbf{g}} = \frac{1}{V_a}\int_{V_a} dVe^{-i\mathbf{g}\cdot\mathbf{r}}\theta_1(\mathbf{r}), \tag{10}$$

where the sums are over all the vectors $\mathbf{g}$ of the appropriate reciprocal lattice. Any other periodic function can also be expanded in this fashion.

The parts $\Delta\hat{\epsilon}_1(\mathbf{r})$ and $\Delta\hat{\epsilon}_2(\mathbf{r})$, which depend on variation of the charge carriers concentration, appear due to application of the transverse voltage δV. It is convenient to represent each of them as a product:

$$\Delta\hat{\epsilon}_1(\mathbf{r}) = \delta V\hat{F}_1(\mathbf{r}), \quad \Delta\hat{\epsilon}_2(\mathbf{r}) = \delta V\hat{F}_2(\mathbf{r}), \tag{11}$$

where we put all coordinate dependences into the tensorial functions $\hat{F}_1(\mathbf{r})$ and $\hat{F}_2(\mathbf{r})$, and where $\delta V = eEl_y/k_BT$ is the dimensionless transverse voltage (see the Appendix). It is convenient to consider the coordinate dependent functions $\hat{F}_1(\mathbf{r})$

and $\hat{F}_2(\mathbf{r})$ together with coordinate dependent functions $\theta_1(\mathbf{r})$ and $\theta_2(\mathbf{r})$. Then it is possible to expand the products $\theta_1(\mathbf{r})\hat{F}_1(\mathbf{r})$ and $\theta_2(\mathbf{r})\hat{F}_2(\mathbf{r})$ into a Fourier series with coefficients

$$\hat{\theta}_{1,\mathbf{g}} = \frac{1}{V_a}\int_{V_a} e^{-i\mathbf{g}\cdot\mathbf{r}}\hat{F}_1(\mathbf{r})\theta_1(\mathbf{r})dV, \quad \hat{\theta}_{2,\mathbf{g}} = \frac{1}{V_a}\int_{V_a} e^{-i\mathbf{g}\cdot\mathbf{r}}\hat{F}_2(\mathbf{r})\theta_2(\mathbf{r})dV. \tag{12}$$

Note again that here and everywhere below $\hat{\theta}_{1,\mathbf{g}}$ and $\hat{\theta}_{2,\mathbf{g}}$ should be understood not as the Fourier coefficients of θ_1 and θ_2 but as Fourier coefficients of the products $\theta_1(\mathbf{r})\hat{F}_1(\mathbf{r})$ and $\theta_2(\mathbf{r})\hat{F}_2(\mathbf{r})$ given by Eq. (12).

It is convenient to multiply each of the Fourier coefficients by the factor $i(\mathbf{g}\cdot\hat{\varepsilon}_{2s}\cdot\mathbf{g})$. Some straightforward calculations then lead to the following results for the Fourier coefficients:

$$r_{\mathbf{g}}^{(\alpha)} \equiv i(\mathbf{g}\cdot\hat{\varepsilon}_{2s}\cdot\mathbf{g})^{1/2}(\hat{\Gamma}r_\alpha)_{\mathbf{g}} = \frac{g_\beta\delta\varepsilon_{\beta\alpha}\theta_{\mathbf{g}}}{(\mathbf{g}\cdot\hat{\varepsilon}_{2s}\cdot\mathbf{g})^{1/2}}, \tag{13}$$

$$i(\mathbf{g}\cdot\hat{\varepsilon}_{2s}\cdot\mathbf{g})^{1/2}(\hat{\Gamma}f)_{\mathbf{g}} = \sum_{\mathbf{g}'\neq 0}\Gamma_{\mathbf{g}\mathbf{g}'}i(\mathbf{g}'\cdot\hat{\varepsilon}_{2s}\cdot\mathbf{g}')^{1/2}f_{\mathbf{g}'}, \tag{14}$$

$$\Gamma_{\mathbf{g}\mathbf{g}'} \equiv \frac{(\mathbf{g}\cdot\delta\hat{\varepsilon}\cdot\mathbf{g}')}{(\mathbf{g}\cdot\hat{\varepsilon}_{2s}\cdot\mathbf{g})^{1/2}(\mathbf{g}'\cdot\hat{\varepsilon}_{2s}\cdot\mathbf{g}')^{1/2}}\theta_{\mathbf{g}-\mathbf{g}'}, \tag{15}$$

$$r_{k,\mathbf{g}}^{(\alpha)} \equiv i(\mathbf{g}\cdot\hat{\varepsilon}_{2s}\cdot\mathbf{g})^{1/2}(\hat{\Gamma}_k r_\alpha)_{\mathbf{g}} = \frac{\delta V g_\beta(\hat{\theta}_{k,\mathbf{g}})_{\beta\alpha}}{(\mathbf{g}\cdot\hat{\varepsilon}_{2s}\cdot\mathbf{g})^{1/2}}, \tag{16}$$

$$i(\mathbf{g}\cdot\hat{\varepsilon}_{2s}\cdot\mathbf{g})^{1/2}(\hat{\Gamma}_k f)_{\mathbf{g}} = \sum_{\mathbf{g}'\neq 0}\Gamma_{k,\mathbf{g}\mathbf{g}'}i(\mathbf{g}'\cdot\hat{\varepsilon}_{2s}\cdot\mathbf{g}')^{1/2}f_{\mathbf{g}'}, \tag{17}$$

$$\Gamma_{k,\mathbf{g}\mathbf{g}'} \equiv \frac{\delta V(\mathbf{g}\cdot\hat{\theta}_{k,\mathbf{g}-\mathbf{g}'}\cdot\mathbf{g}')}{(\mathbf{g}\cdot\hat{\varepsilon}_{2s}\cdot\mathbf{g})^{1/2}(\mathbf{g}'\cdot\hat{\varepsilon}_{2s}\cdot\mathbf{g}')^{1/2}}, \tag{18}$$

where $k = 1, 2$ and $f(\mathbf{r})$ is an arbitrary periodic function. We recall here that $\theta_{\mathbf{g}}$ is the Fourier component of $\theta_1(\mathbf{r})$ [given by Eq. (10) above] while $\hat{\theta}_{1,\mathbf{g}}$ and $\hat{\theta}_{2,\mathbf{g}}$ are the Fourier components of the products $\hat{F}_1(\mathbf{r})\theta_1(\mathbf{r})$ and $\hat{F}_2(\mathbf{r})\theta_2(\mathbf{r})$, respectively [see Eq. (11)]. Clearly, the matrices defined in Eqs. (15), (18) represent the integro-differential operators $\hat{\Gamma}$, $\hat{\Gamma}_1$ and $\hat{\Gamma}_2$ in this scheme. We now translate the integro-differential equation for $\psi^{(\alpha)}(\mathbf{r})$ (see Refs. [1, 2, 6–11]) into an infinite set of linear algebraic equations for the Fourier coefficients $\psi_{\mathbf{g}}^{(\alpha)}$, which are represented by $a_{\mathbf{g}}^{(\alpha)} \equiv i(\mathbf{g}\cdot\hat{\varepsilon}_{2s}\cdot\mathbf{g})^{1/2}\psi_{\mathbf{g}}^{(\alpha)}$ for $\mathbf{g}\neq 0$. From Eqs. (13)–(18) it follows that

$$a_{\mathbf{g}}^{(\alpha)} = r_{\mathbf{g}}^{(\alpha)} - r_{1,\mathbf{g}}^{(\alpha)} - r_{2,\mathbf{g}}^{(\alpha)} + \sum_{\mathbf{g}'\neq 0}\left(\Gamma_{\mathbf{g}\mathbf{g}'} - \Gamma_{1,\mathbf{g}\mathbf{g}'} - \Gamma_{2,\mathbf{g}\mathbf{g}'}\right)a_{\mathbf{g}'}^{(\alpha)} \quad \text{for} \quad \mathbf{g}\neq 0. \tag{19}$$

Substituting the Fourier expansion of $\psi^{(\alpha)}(\mathbf{r})$ in Eq. (8), we now get

$$\left[\hat{\varepsilon}_2 - p_1\delta\hat{\varepsilon} - \hat{\varepsilon}_e + \delta V\left(\hat{\theta}_{1,0} + \hat{\theta}_{2,0}\right)\right]_{\alpha\beta} =$$

$$\delta\varepsilon_{\alpha\gamma}\sum_{\mathbf{g}} ig_\gamma\psi_{\mathbf{g}}^{(\beta)}\theta_{-\mathbf{g}} - \delta V\sum_{\mathbf{g}} i(\theta_{1,-\mathbf{g}})_{\alpha\gamma}g_\gamma\psi_{\mathbf{g}}^{(\beta)} - \delta V\sum_{\mathbf{g}} i(\theta_{2,-\mathbf{g}})_{\alpha\gamma}g_\gamma\psi_{\mathbf{g}}^{(\beta)} =$$

$$\sum_{\mathbf{g}\neq 0}\frac{\left[\delta\varepsilon_{\alpha\gamma}\theta_{-\mathbf{g}} - \delta V(\theta_{1,-\mathbf{g}})_{\alpha\gamma} - \delta V(\theta_{2,-\mathbf{g}})_{\alpha\gamma}\right]g_\gamma}{(\mathbf{g}\cdot\hat{\varepsilon}_{2s}\cdot\mathbf{g})^{1/2}}a_{\mathbf{g}}^{(\beta)}. \tag{20}$$

One way to proceed is to solve Eq. (19) numerically for the $a_{\mathbf{g}}^{(\alpha)}$, using a truncated version of the $\Gamma_{\mathbf{gg}'}$, $\Gamma_{1,\mathbf{gg}'}$, and $\Gamma_{2,\mathbf{gg}'}$ matrices of Eqs. (15), (18), and then use those coefficients in Eq. (20) in order to evaluate $\hat{\varepsilon}_e$.

2.3 *Non-uniformity of the Permittivity Tensor*

In contrast with Refs. [1–11] where the local permittivity tensor (1) was assumed to have a constant value within each constituent, we now assume that in each constituent these tensors vary with the coordinates. The position dependence of the local permittivity tensor, even in a homogeneous constituent, is due to the position dependence of the local density of free charges: When a potential difference is applied between the two film surfaces the negative charge carriers are attracted to one surface while the positive charge carriers are attracted to the other surface. But the free charges of one kind cannot concentrate entirely at the surfaces due to their mutual Coulomb repulsion. This repulsion is most effective when the local electric conductivity is low, as when the conducting constituent is a semiconductor rather than a metal. For simplicity we consider the case when this non-uniform charge density depends linearly on the perpendicular coordinate y. We thus assume that $F_2(\mathbf{r}) = y/l_y$, where l_y is the thickness of the conducting film and the length of the cylinder (see Fig. 1). Evaluation of the Fourier transform (12) for this case gives:

$$\theta_{2,\mathbf{g}} = \frac{1}{a^3}\int_{V_a} dxdzdy(1-\theta_1)\left(\frac{y}{l_y}\right)e^{-ig_x\cdot x - ig_z\cdot z - ig_y\cdot y}$$

$$= \frac{1}{a^3}\int_{V_a} dxdzdy\left(\frac{y}{l_y}\right)e^{-ig_x\cdot x - ig_z\cdot z - ig_y\cdot y} \tag{21}$$

$$-\frac{1}{a^3}\int_{V_a} dxdz\, dy\theta_1\left(\frac{y}{l_y}\right)e^{-ig_x\cdot x - ig_z\cdot z - ig_y\cdot y}. \tag{22}$$

We recall that a is the size of the unit cell. The integration in expression (21) is over the entire unit cell. It yields

$$\frac{2i\left(g_y\frac{a}{2}\cos g_y\frac{a}{2} - \sin g_y\frac{a}{2}\right)}{al_yg_y^2}\delta_{g_x,0}\delta_{g_z,0}. \tag{23}$$

The integration in expression (22) depends on the geometry of a unit cell. For the geometry where the hole in the film has the shape of a circular cylinder (see Fig. 1) this expression takes the form

$$-\frac{2\pi}{g_\perp a^2} R J_1(Rg_\perp) \frac{2i\left(g_y \frac{l_y}{2}\cos g_y \frac{l_y}{2} - \sin g_y \frac{l_y}{2}\right)}{a l_y g_y^2} \tag{24}$$

$$-2\cos\left(g_y \frac{L_y}{2}\right) \frac{2i\left(g_y \frac{d_y}{2}\cos g_y \frac{d_y}{2} - \sin g_y \frac{d_y}{2}\right)}{a l_y g_y^2} \delta_{g_x,0}\delta_{g_z,0}, \tag{25}$$

where the expression (24) is obtained from the cylindrical hole, while the second expression (25) is obtained from the two prisms (on the bottom and top of the unit cell, see Fig. 1), d_y is the thickness of these prisms, L_y is the distance between the centers of these prisms, R is the radius of the cylinder, J_1 is a Bessel function and $g_\perp = \sqrt{g_x^2 + g_z^2}$ (see Refs. [1, 2]). It is easy to see that $\theta_{2,g} = 0$ for $g_y = 0$ and the factor $\frac{J_1(Rg_\perp)}{g_\perp}$ simplifies to $R/2$ for $g_\perp = 0$.

2.4 Linear Dependence Along the Sphere Radius

The theory described above can also be used for calculations of permittivity tensors of samples with impurity centers which create spherical volumes where the density of free charges changes along the radii of these spheres.

Let us consider the case $\hat{\Gamma}_2 = 0$ while $\hat{\Gamma}_1 \neq 0$. The simplest case is the linear dependence of the free charge density N_0 along the sphere radius:

$$\varepsilon_1 = \varepsilon_{1,0} + \Delta\epsilon_1 F_1(\mathbf{r}), \tag{26}$$

where $F_1(\mathbf{r}) = -\frac{\rho}{R}$ and ρ is the distance from the sphere center while R is the radius of the sphere. The evaluation of the Fourier transform for this case gives:

$$\theta_{1,\mathbf{g}} = \frac{1}{V_a}\int \theta(\mathbf{r})\rho e^{-i\mathbf{g}\cdot\mathbf{r}} dV = \frac{1}{a^3}\int_0^{2\pi} d\phi \int_0^R \theta(\rho)\rho^3 d\rho \int_0^\pi \sin\vartheta e^{-ig\rho\cos\vartheta} d\vartheta$$
$$= \frac{4\pi}{a^3}\int_0^R \theta(\rho)\rho^3 \sin(g\rho) d\rho = \frac{4\pi}{a^3 g^4}\left[-2 + (2 - g^2R^2)\cos(gR) + 2gR\sin(gR)\right]. \tag{27}$$

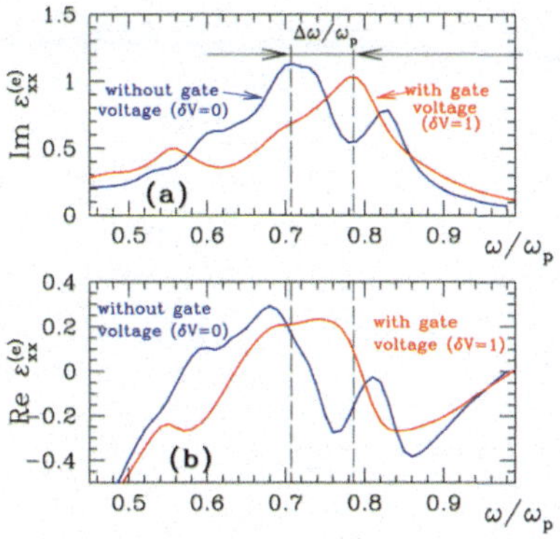

Fig. 2 Imaginary **a** and real **b** parts of the in-plane component $\varepsilon_{xx}^{(e)} = \varepsilon_{zz}^{(e)}$ of the macroscopic electric permittivity tensor of the metamaterial with periodic array of holes shown in Fig. 1 versus ω/ω_p. Cases without transverse voltage ($\delta V = 0$, blue curves) and with applied transverse voltage ($\delta V = 1$, red curves). The frequency of the localized surface plasmon resonance [6–12] $\omega = \omega_p/\sqrt{2}$ shifts by $\Delta\omega \sim 0.08\omega_p$ when the transverse voltage $\delta V = 1$ is applied. The vertical dashed lines show the resonance frequences. The holes radii are $R = 0.3a$ (where a is the lattice constant and the unit cell size), $l_y = 0.8a$, $d_y = 0.1a$ [see Fig. 1 and Eq. (23)–(25) below]. The insulator permittivity $\hat{\varepsilon}_1 = 1$, while the permittivity $\hat{\varepsilon}_2$ has the form of Eq. (1) with $\omega_p\tau = 40$

3 Results

We have solved the system of linear algebraic Eq. (19) for the $a_{\mathbf{g}}^{(\alpha)}$ coefficients and used them in Eq. (20) to obtain $\hat{\varepsilon}_e$. The results show evident sensitivity of the permittivity tensor and therefore of all the optical properties to the transverse voltage applied perpendicular to the metamaterial film plane. In Fig. 2 we show the imaginary and real parts of the in-plane component $\varepsilon_{xx}^{(e)}$ of the macroscopic electric permittivity tensor of the metamaterial with periodic array of holes shown in Fig. 1 versus ω/ω_p. Shown are the cases with zero transverse voltage ($\delta V = 0$, blue curves) and with nonzero transverse voltage ($\delta V = 1$, red curves). It is shown that the frequency of the surface plasmon resonance $\omega = \omega_p/\sqrt{2}$ shifts by $\Delta\omega \sim 0.08\omega_p$ when the dimesionless voltage $\delta V = 1$ is applied.

In Fig. 3a and b we show respectively the reflectivity $R = |(n-1)/(n+1)|^2$ and the transmissivity $T = 1 - R$ coefficients of the same sample as in Fig. 2, computed from the values of $\hat{\varepsilon}_e$. The factors R and T are defined so as to sum up to 1. The absorption is neglected because it is expected to be very small: In the insulating constituent there are no dissipative mechanisms. In the conducting constituent there are dissipative mechanisms but the incident wave almost does not penetrate there so there is very little dissipation. This expected behavior is in accordance with the experimental observations of the so-called *extraordinary light transmission* effect [15] at the frequency of the localized surface plasmon resonance [6, 7].

In Fig. 4 we show the real and imaginary parts of of the in-plane component $\varepsilon_{xx}^{(e)}$ of the effective volume averaged permittivity tensor of the insulating host medium with

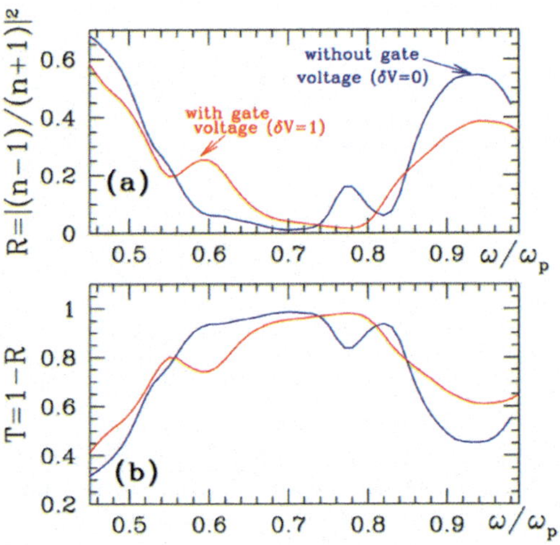

Fig. 3 Reflectivity $R = |(n-1)/(n+1)|^2$ and transmissivity $T = 1 - R$ coefficients of the same sample as in Fig. 2 computed from the values of ε_e shown in Fig. 2. By the blue curve is shown case of zero applied transverse voltage $\delta V = 0$ while by the red curve is shown the case of nonzero transverse voltage $\delta V = 1$. $n = \sqrt{\varepsilon_{zz}^{(e)}(\omega)}$

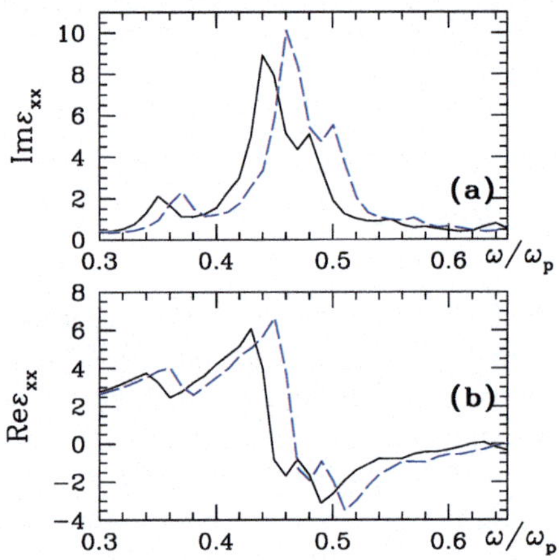

Fig. 4 Imaginary and real parts of the in-plane component $\varepsilon_{xx}^{(e)}$ of the macroscopic electric permittivity tensor of the metamaterial with periodic array of spheres conductivity of which are radius-dependent functions (blue curves). By black curves are shown the same but for the case when the conductivity of the spheres does not depend on coordinate. The spheres are placed into an insulating host

periodic array of spheres conductivity of which are radius-depended function [see Eq. (27)]. As in case of the applied transverse voltage the macroscopic permittivity tensor depends strongly on the free charges density space distribution.

4 Discussion and Summary

In our calculations we considered perforated conductor-insulator metamaterial films with the simple micro-geometry shown in Fig. 1. The insulator permittivity $\hat{\varepsilon}_1$ is assumed to be a diagonal tensor equal to 1 (air), while the permittivity of the conductor $\hat{\varepsilon}_2$ is used in the form of Eq. (1) with $\omega_p\tau$ taken as typical for metals and consistent with semi-metals like bismuth [5] and highly doped semiconductors [3]: $\omega_p \tau = 40$ [see Eq. (1)]. The sample was considered to be purely classical. This means that in order to test our predictions an experiment should be done on a metamaterial conducting medium in which the inclusions are large compared to the electron mean-free path [1–9]. Therefore, the sample for a real experiment can be easily made as a metal-insulator or semiconductor-insulator film on the top and bottom of which the voltage electrodes are situated.

We showed numerically that the macroscopic permittivity tensor, as well as optical properties, of a metamaterial strongly depend on the spatial distribution $N_0(\mathbf{r})$ of the free charges. The charge distribution can be altered by applying a transverse voltage. When the dimensionless voltage is $\delta V = eEl_y/k_B T = 1$ then for a film thickness $l_y = 100$ nm the physical electric field will be $E = 10^5$ V/m. As a result the optical properties of the metamaterial can be manipulated and can be used for the construction of fast switches and other optical devices. Qualitatively, this can be explained by analogy with an applied external longitudinal (along the x or z axis) magnetic field [7, 9], which leads to the appearance of a Hall electric field across the film. In our case, the role of the Hall field is played by the applied external transverse electric field (along y-axis). This field (as in case of magnetic field) leads to distortion of the current trajectories, which affects the macroscopic permittivity ε_e and, as a consequence, the optical properties of the film.

We have used the Drude approximation for the local permittivity tensor since it allows us to perform analytical evaluations and to predict the optical behavior and features of the considered film system. For the case of real experiment it is possible to take not the Drude approximation but measured data for the local permittivity tensor. Then the effective permittivity tensor can be calculated numerically.

The phenomenon discussed in this paper can find different technical applications. These can be, for example, the *fast optical switch* and *stealth technology*. Another possible technical application can be *optical frequency modulation*, similar to radio frequency modulation, which can be based on *extraordinary light transmission* and transverse voltage manipulation. If white light is directed at a film with holes then only light of a certain frequency will pass. On the same basis can also be built a *color generator* and color TV. It would be interesting to perform experiments on perforated metal-dielectric film nanostructures, for which we have now presented exact theoretical predictions.

Appendix: Justification of Eq. (11)

In order to justify Eq. (11) let us write down the expression for electrical conductivity obtained in the framework of Boltzmann kinetic equation

$$\sigma_{ij} = -\frac{e^2}{4\pi^3} \int \tau(\mathbf{k}) \frac{\partial f_0}{\partial \varepsilon_K} v_i v_j d^3k, \quad f_0 = \left[\exp\left(\frac{\mathcal{E} - \mu}{k_B T} \right) + 1 \right]^{-1}, \tag{28}$$

where f_0 is the Fermi-Dirac electron distribution function in a state of statistical equilibrium, $\partial f_0/\partial \varepsilon_K$ is its derivative with respect to the kinetic energy ε_K, μ is the chemical potential of the electrons, k_B is the Boltzmann constant, $\tau(\mathbf{k})$ is the wave vector **k** dependent conductivity relaxation time, $\mathcal{E} = \varepsilon_K + \mathcal{U}$ is a sum of the kinetic ε_K and potential energies $\mathcal{U}$. In the case where the applied transverse electrical field E is perpendicular to the film surface (i.e., along y-axis, see Fig. 1), the potential energy $\mathcal{U}$ can be written as $\mathcal{U} = eEy$, where y is the coordinate across the film thickness. If $|eEl_y| \ll |\varepsilon_k - \mu|$ and $k_B T \ll \mu$, then Eq. (28) can be expanded into a series in powers of $eEy/k_B T$, so that only the first member is retained. After substitution of this into Eq. (28), some assumptions and algebra, one can write an approximate expression for the conductivity tensor in the presence of the applied voltage and the resulting non-uniformity of the charge density distribution across the film thickness:

$$\sigma_{ij} \simeq \sigma_{ij0}\left(1 + \delta V \frac{y}{l_y}\right), \tag{29}$$

where l_y is the film thickness, σ_{ij0} is the conductivity tensor in the absence of an applied transverse voltage, and $\delta V \equiv \frac{eEl_y}{k_B T}$ is the dimensionless voltage.

References

1. Bergman, D.J., Strelniker, Y.M.: Phys. Rev. B **49**, 16256 (1994)
2. Strelniker, Y.M., Bergman, D.J.: Phys. Rev. B **50**, 14001 (1994)
3. Tornow, M., Weiss, D., von Klitzing, K., Eberl, K., Bergman, D.J., Strelniker, Y.M.: Phys. Rev. Lett. **77**, 147 (1996)
4. Bergman, D.J., Strelniker, Y.M.: Phys. Rev. B **60**, 13016 (1999)
5. Strijkers, G.J., Yang, F.Y., Reich, D.H., Chien, C.L., Searson, P.C., Strelniker, Y.M., Bergman, D.J.: IEEE Trans. Magn. **37**, 2067 (2001)
6. Bergman, D.J., Strelniker, Y.M.: Phys. Rev. Lett. **80**, 857 (1998)
7. Strelniker, Y.M., Bergman, D.J.: Phys. Rev. B **59**, R12763 (1999)
8. Strelniker, Y.M.: Phys. Rev. B **76**, 085409 (2007)
9. Strelniker, Y.M., Bergman, D.J.: Phys. Rev. B **77**, 205113 (2008)
10. Strelniker, Y.M., Bergman, D.J.: Phys. Rev. B **102**, 035302 (2020)
11. Strelniker, Y.M., Bergman, D.J.: Phys. Rev. B **103**, 205302 (2021)
12. Fei, Z., Rodin, A.S., Andreev, G.O., Bao, W., McLeod, A.S., Wagner, M., Zhang, L.M., Zhao, Z., Dominguez, G., Thiemens, M., Fogler, M.M., Castro-Neto, A.H., Lau, C.N., Keilmann, F., Basov, D.N.: Nature **487**, 82 (2012)
13. Strelniker, Y.M., Bergman, D.J.: Proceedings of SPIE. **12197**, Plasmonics: Design, Materials, Fabrication, Characterization, and Applications XX, 1219706 (2022)
14. Elhabbash, M.K.M., Halim, M.M., Yoon, T.L., Taya, S.A.: Opt. Quantum Electron. **55**, 214 (2023)
15. Ebbesen, T.W., Lezec, H.J., Ghaemi, H.F., Thio, T., Wolff, P.A.: Nature (London) **391**, 667 (1998)

Variational Principle for a Damped, Quadratically Interacting Particle Chain with Nonconservative Forcing

Amit Acharya and Ambar N. Sengupta

Abstract A method for designing variational principles for the dynamics of a possibly dissipative and non-conservatively forced chain of particles is demonstrated. Some qualitative features of the formulation are discussed.

Keywords Variational principle · Particle system · Duality · FPUT

1 Introduction

Starting from the work of Fermi-Pasta-Ulam-Tsingou [5], much has been understood about non-generic, non-thermalized behavior in the mechanics of discrete particle chains, e.g. [6, 11, 12, 16, 19]-[4, and related Parts II, III, IV]-[14]-[3, and Part II], [8, 9, 13, 15, 17, 18]. Such mechanical systems are considered to be idealized problems for understanding nonlinear dynamics of real solids. A variational description of such chains can help in the identification of interesting solutions like periodic orbits and solitary waves, and their stability, e.g. [8, 13–15]. Our recent work [2] outlines a general setup for constructing variational principles for general anholonomically constrained, possibly dissipative Newtonian particle systems. Here, we apply those ideas to the problem of a special class of particle chains to enable future work on the subject from this point of view. We have chosen the case of at most quadratic force interactions in order to explicitly write out the action functional involved.

We note here that the existence of variational action principles for systems with dissipation and non-conservative forces is a surprising and novel result; it has generally been believed that dissipative systems do not admit variational action principles. Moreover, Hamilton's principle does not yield an initial value problem as its Euler

A. Acharya (✉)
Department of Civil & Environmental Engineering, and Center for Nonlinear Analysis, Carnegie Mellon University, Pittsburgh, PA 15213, USA
e-mail: acharyaamit@cmu.edu

A. N. Sengupta
Department of Mathematics, University of Connecticut, Storrs, CT 06269, USA

F. Willot et al. (eds.), *Continuum Models and Discrete Systems*,
Springer Proceedings in Mathematics & Statistics 457,
https://doi.org/10.1007/978-3-031-58665-1_15

Lagrange equations and side conditions but a boundary value problem whose final-time boundary condition is unknown for solving initial value problems—see [7] for a discussion. In this connection, we also note that constructing an action principle whose first variation gives a certain set of dissipative equations is not the same as incorporating a Rayleigh dissipation function to the Lagrangian of an associated conservative system and obtaining the Euler-Lagrange equations of the resulting functional; a typical dissipative force in the Rayleigh formalism with Rayleigh function $\mathcal{R}(\dot{q})$ is given by $-\frac{\partial \mathcal{R}}{\partial \dot{q}}$ which is not the same as the variational derivative $\frac{\delta \int_0^T \mathcal{R}(\dot{q}(t))\,dt}{\delta q}$.

2 Dual Variational Principles for ODE Systems: The General Formalism

We briefly review our strategy for developing action principles for particle systems [2]. Consider a (primal) system of ODE for a vector of degrees of freedom $t \mapsto U(t) \in \mathbb{R}^s$ for some $s \geq 1$, and F is a given function of its arguments with t being time:

$$\mathcal{A}\dot{U} - F(U,t) = 0; \qquad U(0) = U^{(0)}, \tag{1}$$

and $\mathcal{A}$ is a constant matrix. Treating the above system as constraints for extremizing an arbitrarily chosen objective functional $\int_0^T H(U,t)\,dt$ and introducing *dual* Lagrange multiplier functions $t \mapsto D(t) \in \mathbb{R}^s$, a pre-dual functional is defined as

$$\widehat{S}_H[U,D] = \int_0^T \mathcal{L}_H(U,\mathcal{D},t)\,dt - \mathcal{A}U^{(0)} \cdot D(0); \qquad \mathcal{D} := (D,\dot{D});$$
$$\mathcal{L}_H(U,\mathcal{D},t) := -\mathcal{A}U \cdot \dot{D} - D \cdot F(U,t) + H(U,t),$$

where we include only those 'boundary' terms that allow inclusion of information from the primal problem. We now require the function H to be such that it allows for solving the algebraic equation $\frac{\partial \mathcal{L}_H}{\partial U}(U,\mathcal{D},t) = 0$ for U in terms of $(\mathcal{D},t)$, i.e. it allows the definition of a function, the *dual-to-primal* (DtP) mapping, $U = U^{(H)}(\mathcal{D},t)$ such that

$$\frac{\partial \mathcal{L}_H}{\partial U}\left(U^{(H)}(\mathcal{D},t),\mathcal{D},t\right) = 0 \tag{2}$$

is satisfied in some neighborhood of the space of $(\mathcal{D},t)$.

The desired action functional, *for any H with the above property*, is then defined by substituting the function $U^{(H)}$ in the pre-dual functional $\widehat{S}$ for U:

$$S_H[D] = \int_0^T \mathcal{L}_H\left(U^{(H)}(\mathcal{D}(t),t),\mathcal{D}(t),t\right)\,dt - \mathcal{A}U^{(0)} \cdot D(0),$$
$$D \text{ arbitrarily specified at } t = T,$$

and it can be checked that its Euler-Lagrange (E-L) equations and implied b.c. are simply the primal system (1) with U replaced by $U^{(H)}(\mathcal{D}, t)$.

3 A Dual Action Functional for a Forced Particle Chain with Damping and Quadratic Interaction Forces

We apply this machinery to a 1-d chain of N particles that have the same mass $m > 0$, $d \geq 0$ is the damping coefficient, $x \mapsto K(x) \in \mathbb{R}^N$ is a quadratic function (allowing for long-range interactions) characterized by the given constants $C \in \mathbb{R}^N$, $A \in \mathbb{R}^{N\times N}$, and $B \in \mathbb{R}^{N\times N\times N}$, B symmetric in the last two indices, and $t \mapsto f(t) \in \mathbb{R}^N$ is a prescribed forcing function of time (not necessarily the gradient of a scalar potential in x):

$$\begin{aligned}
& m\dot{v}_i + dv_i + K_i(x) - f_i(t) = 0 \\
& \dot{x}_i - v_i = 0 \\
& \dot{x}_i(0) = x_i^{(0)}; \qquad \dot{v}_i(0) = v_i^{(0)} \\
& K_j(x) := K_j(\bar{x}) + A_{jr}(x_r - \bar{x}_r) + \frac{1}{2}B_{jrs}(x_r - \bar{x}_r)(x_s - \bar{x}_s),
\end{aligned} \tag{3}$$

where the interaction force $K(x)$ is at most a quadratic polynomial in the particle positions, expanded about an arbitrarily chosen 'base' configuration $\bar{x}$ to be discussed further below.

Now consider a pre-dual functional with a shifted quadratic form for the auxiliary potential H:

$$\begin{aligned}
\widehat{S}[x, v, \gamma, \lambda] = \int_0^T & -v_i m\dot{\lambda}_i + \lambda_i d v_i + \lambda_i K_i(x) - \lambda_i f_i - x_I \dot{\gamma}_i - \gamma_i v_i \\
& + \frac{1}{2}c_x|x - \bar{x}|^2 + \frac{1}{2}c_v|v - \bar{v}|^2\, dt - \lambda_i(0) m v_i^{(0)} - \gamma_i(0) x_i^{(0)},
\end{aligned} \tag{4}$$

where the shift is by an arbitrarily chosen 'base state' $(t \mapsto (\bar{x}(t), \bar{v}(t))$. Define

$$U = (x, v); \qquad \mathcal{D} = (\gamma, \lambda, \dot{\gamma}, \dot{\lambda}); \qquad \mathcal{L}(U, \mathcal{D}; \bar{U}) = \text{integrand of (4)}$$

and solve for U in terms of $(\mathcal{D}, \bar{U})$ from the equations

$$\frac{\partial \mathcal{L}}{\partial U}(U, \mathcal{D}; \bar{U}) = 0$$

to obtain the system

$$\frac{\partial \mathcal{L}}{\partial x_i}: \qquad \lambda_j(A_{ji} + B_{jri}(x_r - \bar{x}_r)) - \dot{\gamma}_i + c_x\delta_{ir}(x_r - \bar{x}_r) = 0$$

$$\Longrightarrow \qquad c_x\mathbb{K}_{ir}\big|_\lambda(x_r - \bar{x}_r) = \dot{\gamma}_i - \lambda_j A_{ji}; \qquad \text{where } \mathbb{K}_{ir}\big|_\lambda := \delta_{ir} + \frac{1}{c_x}\lambda_j B_{jir} \tag{5a}$$

$$\frac{\partial \mathcal{L}}{\partial v_i}: \qquad -m\dot{\lambda}_i + d\lambda_i - \gamma_i + c_v(v_i - \bar{v}_i) = 0$$

$$\Longrightarrow \qquad c_v(v_i - \bar{v}_i) = \gamma_i + m\dot{\lambda}_i - d\lambda_i \tag{5b}$$

Substituting (5) in (4), we obtain

$$\begin{aligned}
\widehat{S}[x, v, \gamma, \lambda] = & \int_0^T (v_i - \bar{v}_i)(-m\dot{\lambda}_i + d\lambda_i - \gamma_i) + \frac{1}{2}c_v(v_i - \bar{v}_i)(v_i - \bar{v}_i)\, dt \\
& + \int_0^T \bar{v}_i(-m\dot{\lambda}_i + d\lambda_i - \gamma_i) - \bar{x}_r\dot{\gamma}_r + \lambda_i K_i\big|_{\bar{x}} - \lambda_i f_i\, dt \\
& + \int_0^T \frac{1}{2}\lambda_j B_{jri}(x_r - \bar{x}_r)(x_i - \bar{x}_i) + \frac{1}{2}c_x\delta_{ri}(x_r - \bar{x}_r)(x_i - \bar{x}_i) \\
& - (x_i - \bar{x}_i)\left(\dot{\gamma}_i - \lambda_j A_{ji}\right) dt - \lambda_i(0)mv_i^{(0)} - \gamma_i(0)x_i^{(0)} \\
= & \int_0^T -\frac{1}{2}c_v(v_i - \bar{v}_i)(v_i - \bar{v}_i) - \frac{1}{2}c_x\mathbb{K}_{ir}(x_r - \bar{x}_r)(x_i - \bar{x}_i)\, dt \\
& + \int_0^T \bar{v}_i(-m\dot{\lambda}_i + d\lambda_i - \gamma_i) - \bar{x}_r\dot{\gamma}_r + \lambda_i K_i\big|_{\bar{x}} - \lambda_i f_i\, dt \\
& - \lambda_i(0)mv_i^{(0)} - \gamma_i(0)x_i^{(0)}.
\end{aligned} \tag{6}$$

Finally, using (5) in (6) we obtain a *dual* functional

$$\begin{aligned}
S[\gamma, \lambda] = & -\frac{1}{2}\int_0^T \frac{1}{c_v}\left|\gamma + m\dot{\lambda} - d\lambda\right|^2 + \frac{1}{c_x}(\dot{\gamma} - A^T\lambda)\cdot \mathbb{K}\big|_\lambda^{-1}\left(\dot{\gamma} - A^T\lambda\right) dt \\
& + \int_0^T -\bar{v}\cdot\left(\gamma + m\dot{\lambda} - d\lambda\right) - \bar{x}\cdot\dot{\gamma} + \lambda\cdot K\big|_{\bar{x}} - \lambda\cdot f\, dt \\
& - \lambda(0)\cdot mv^{(0)} - \gamma(0)\cdot x^{(0)} \\
& \text{with } \lambda(T) = 0, \quad \gamma(T) = 0 \text{ without loss of generality.}
\end{aligned} \tag{7}$$

We make the following observations:

- Since the E-L equations of S are guaranteed to be the system in (3) (with the DtP mapping in place), if the functions $(\bar{x}, \bar{v})$ are actual solutions to (3), then the DtP mapping (5) suggests that $t \mapsto (\lambda(t), \gamma(t)) = (0, 0)$ is an extremal of S. Thus, at least for dual functionals designed with base states that are 'close to' actual (un)stable solutions of the primal problem, it is reasonable to expect solutions (extremals) to exist for the dual functional.

- Why should it be possible to solve an initial-value-problem by prescribing final-time boundary conditions as well in time, especially when the primal problem has a unique solution for prescribed initial conditions? This is because the DtP mapping Eq. (5) show that the mapped primal functions at the final time, $(x(T), v(T))$, depend on the rates $(\dot{\lambda}(T), \dot{\gamma}(T))$, and specifying $(\lambda(T), \gamma(T))$ leaves enough freedom in the dual problem to adjust the rates to satisfy the demands of recovering the correct 'nearby' primal solution - this has been discussed and demonstrated in [1, Sect. 7]-[10].
- The dual E-L equations is a coupled system of 2nd-order, 2-point boundary value problems in the dual variables (λ, γ) whose ellipticity is governed by the matrix

$$\begin{bmatrix} \frac{m^2}{c_v}\mathbb{I} & 0 \\ 0 & \frac{1}{c_x}\mathbb{K}\big|_{\lambda}^{-1} \end{bmatrix}$$

which is positive definite as long as $\mathbb{K}$ remains positive definite. Thus, in a neighborhood of $t \mapsto (\lambda(t), \gamma(t)) = (0, 0)$ corresponding to the primal trajectories $(\bar{x}, \bar{v})$, the dual problem is elliptic and therefore, if $(\bar{x}, \bar{v})$ is a primal solution, then the corresponding dual solution $t \mapsto (\lambda(t), \gamma(t)) = (0, 0)$ is neutrally stable - this can also be seen by calculating the second variation of the dual functional $-S$ about the orbit $t \mapsto (\lambda(t), \gamma(t)) = (0, 0)$ and noticing that it is non-negative (the Lagrangian of the dual action S is concave in $\mathcal{D}$ and hence it is the (local) maximization of S that is relevant for defining primal solutions through the DtP mapping).
- The dual functional results in second order equations for the dual variables. Seeking dual extremals in the class of periodic functions in time—with the period as an additional variable that does not evolve (see [2, Sect. 3, point 5.]) provides a seamless way of seeking periodic solutions in time (e.g. 'breathers') for a nonlinear initial value problem, which is otherwise a non-trivial question even at the formal procedural level and computationally, even more so.
- As is well-known, the existence of constants of motion for nonlinear chains plays a significant role in their analysis and understanding. If the Lagrangian of the dual functional is strictly monotone in $\dot{D}$, then an equivalent Hamiltonian description of the dual problem can be written down *for each chosen* H, thus resulting in the possibility of defining several constants of motion for the dual problem. Since each solution of the dual problem is associated with a solution to the primal problem, this seems to suggest, at least superficially, that such constants of motion can be associated with primal orbits as well (cf. [2]). This feature of our formulation is interesting, and awaits further study.

4 Conclusion

In this paper we have described a novel method, developed more extensively in our recent work [2], to construct a variational principle from a given dynamical equation that could involve nonholonomic constraints and dissipative forces. The variational principle is framed in terms of certain dual variables that are related to the given primal variables and an appropriate new objective function H. We have applied our method to a system of N particles experiencing an applied force that is quadratic in the position coordinates, an additional time-dependent force, and a dissipative force, linear in the velocity, as well. We obtain in (7) an exact form of the variational action functional for this system in terms of the dual variables. We made some observations concerning the dual action functional and possible implications for the existence of constants of motion.

Acknowledgements This work was supported by the Simons Pivot Fellowship grant # 983171 and NSF OIA-DMR grant # 2021019.

References

1. Acharya, A.: Variational principles for nonlinear PDE systems via duality. Q. Appl. Math. LXXXI:127–140, 2023, Article electronically published on September 26, 2022
2. Acharya, A., Sengupta, A.N.: Action principles for dissipative, non-holonomic Newtonian mechanics (2023). arXiv:2306.10616. To appear in Proc. R. Soc. A
3. Balk, A.M., Cherkaev, A.V., Slepyan, L.I.: Dynamics of chains with non-monotone stress–strain relations. i. model and numerical experiments. J. Mech. Phys. Solids **49**(1), 131–148 (2001)
4. Friesecke, G., Pego, R.L.: Solitary waves on FPU lattices: I. Qualitative properties, renormalization and continuum limit. Nonlinearity **12**(6), 1601 (1999)
5. Fermi, E., Pasta, J., Ulam, S., Tsingou, M.: Studies of the nonlinear problems I. Technical report, LA-1940, Los Alamos National Lab. (LANL), Los Alamos, NM (United States) (1955)
6. Gallavotti, G.: The Fermi-Pasta-Ulam problem: a status report, vol. 728. Springer, Berlin (2007)
7. Galley, C.R.: Classical mechanics of nonconservative systems. Phys. Rev. Lett. **110**(17), 174301 (2013)
8. Herrmann, Michael: Unimodal wavetrains and solitons in convex Fermi-Pasta-Ulam chains. J. Mech. Phys. Solids **140**(4), 753–785 (2010)
9. Ingimarson, B., Pego, R.L.: On long waves and solitons in particle lattices with forces of infinite range (2023). arXiv:2310.02271
10. Kouskiya, U., Acharya, A.: Hidden convexity in the heat, linear transport, and Euler's rigid body equations: a computational approach. to appear in Quarterly of Applied Mathematics (2023) arXiv:abs/2304.09418
11. Lax, P.D.: Integrals of nonlinear equations of evolution and solitary waves. Commun. Pure Appl. Math. **21**(5), 467–490 (1968)
12. Moser, J.: Three integrable hamiltonian systems connected with isospectral deformations. In: Surveys in Applied Mathematics, pp. 235–258. Elsevier (1976)
13. Pankov, A.: Traveling waves in Fermi-Pasta-Ulam chains with nonlocal interaction. Discret. Contin. Dyn. Syst.-Ser. S **12**(7) (2019)

14. Smets, D., Willem, M.: Solitary waves with prescribed speed on infinite lattices. J. Funct. Anal. **149**(1), 266–275 (1997)
15. Schwetlick, Hartmut: Zimmer, Johannes: Solitary waves for nonconvex FPU lattices. J. Nonlinear Sci. **17**, 1–12 (2007)
16. Toda, M.: Waves in nonlinear lattice. Prog. Theor. Phys. Suppl. **45**, 174–200 (1970)
17. Truskinovsky, Lev: Vainchtein, Anna: Solitary waves in a nonintegrable Fermi-Pasta-Ulam chain. Phys. Rev. E **90**(4), 042903 (2014)
18. Vainchtein, Anna: Solitary waves in FPU-type lattices. Phys. D **434**, 133252 (2022)
19. Zabusky, N.J., Kruskal, M.D.: Interaction of "solitons" in a collisionless plasma and the recurrence of initial states. Phys. Rev. Lett. **15**(6), 240 (1965)

Optimal Design of Programmable Mechanical Metamaterials

Alexander Leichner, Heiko Andrä, Thomas Palmer, Franziska Wenz, and Chris Eberl

Abstract Programmable metamaterials (PMs) belong to the class of metamaterials offering controllable and variable physical properties. As metamaterials, they are architectured, cellular materials and exhibit exotic behavior, but are more specifically tailored for engineering purposes: PMs can be applied wherever a custom-fit reaction to environmental and operating conditions of a material is required. Instead of a homogeneous layout of unit cells, often considered for common metamaterials, we construct PMs by an individual distribution of varying unit cells. In order to tackle this customization of material response, a computational framework similar to topology or material optimization is proposed. Beyond that, our work is based on a multiscale approach and a surrogate material model, allowing a broad range of applications with different classes of unit cells and target functions under finite strains. In this contribution, we present the complete simulation chain: from a parametrized unit cell to the final model of the programmable material, ready to be manufactured. Numerical results with different unit cells are shown and compared to fully resolved simulations as well as to experiments.

Keywords Multiscale problems · Optimization · Metamaterials · Homogenization · Surrogate model

A. Leichner (✉) · H. Andrä · T. Palmer
Fraunhofer Institute for Industrial Mathematics, Fraunhofer-Platz 1, 67663 Kaiserslautern, Germany
e-mail: alexander.leichner@itwm.fraunhofer.de

H. Andrä
e-mail: heiko.andrae@itwm.fraunhofer.de

T. Palmer
e-mail: thomas.palmer@itwm.fraunhofer.de

F. Wenz · C. Eberl
Fraunhofer Institute for Mechanics of Materials, Wöhlerstrasse 11, 79108 Freiburg, Germany
e-mail: franziska.wenz@iwm.fraunhofer.de

C. Eberl
e-mail: chris.eberl@iwm.fraunhofer.de

F. Willot et al. (eds.), *Continuum Models and Discrete Systems*,
Springer Proceedings in Mathematics & Statistics 457,
https://doi.org/10.1007/978-3-031-58665-1_16

1 Introduction

The special properties of metamaterials result from their small-scale structure, e.g., a pattern of cavities or beams. This pattern repeats periodically and is therefore called a unit cell, analogous to those of crystals. However, biological organisms and structures use specialized cells with spatially and temporally variable properties that allow for optimal adaptation to varying environmental conditions such as wind loads and moisture during growth. As a step toward this ideal, in this paper we consider components in which the properties of unit cells are spatially varied and optimized.

Assuming sufficient scalability of the cells, the final manufactured PM can be considered as continuum material with locally varying effective properties. Through their built-in control, the reaction to environmental conditions is triggered and performed intrinsically, in contrast to systems which are controlled externally by power sources, actuators, sensors and control circuits. As a consequence, the material with programmed behavior is more autonomous from external influences, more unlikely to fail and, hence, more robust in application compared to a complex mechatronic system [4].

In our research, we assume a custom-designed, geometrical parametrization of unit cells in order to ensure their connectivity and manufacturability. Because of the individual architecture of the unit cell, the mapping from the design to the effective mechanical properties of the cells might become non-trivial. We construct therefore a map by computing a structured database consisting of all possible design and load scenarios in advance. After that, we compress this database significantly by tensor decomposition methods and use this reduced representation as interpolation basis. This is also known as surrogate modelling, which is a common technique when certain relations are too difficult to be modelled by standard approaches.

Within the scope of this contribution, we focus to establish a desired shape under given boundary conditions in an array of cells. This is done by optimizing the distribution of design parameters, e.g., beam thicknesses or angles inside an array of cells. Other examples are the optimization of fiber orientations [3] or the layout of different unit-cells inside a component [1]. In contrast to topology optimization, no cells inside the design domain are removed, as we aim at the robustness and manufacturability of PM. Besides, several challenges for the surrogate model and the optimization procedure have to be tackled: At macroscale a boundary value problem of unit cell arrays, undergoing large deformations, has to be solved, where design parameters have strong influence on the solution. Moreover, these units cells exhibit highly nonlinear, load case and design dependent behavior and may have multiple design parameters opening a huge design space. In addition, the computational load might become substantial since the size of these arrays must be sufficiently large, such that it is perceived as a material (not only as a structure).

This contribution is structured as follows: In Sect. 2 we introduce our multiscale approach and couple geometrical parameters of unit cells with mechanical properties of arrays composed of mentioned cells. With this multiscale connection, we describe, how programming of material behaviour can be understood as optimization task. In subsequent Sect. 3, we give some insights of our surrogate model strategy and show its important role in our adjoint based optimization procedure. In the last Sect. 4, we present some examples proving the validity and correctness of our concepts.

2 Using Multiscale Optimization for Programming Materials

In Sect. 1, we described the motivation for PMs and distinguished them from classical metamaterials. Here, we intend to define the geometrical conditions as well as the mathematical and engineering background. We begin with the description: a PM is considered as a domain or array $\hat{\Omega}$ which is the union of finitely many unit cells $\tilde{\Omega}_i(\alpha)$, see Fig. 1. The macroscopic behavior of a PM is mainly determined by the distribution of design variables α inside this array, i.e., each unit cell may have different parameters. Therefore, they are the actual control in the "programming" and allow many different design cases. Due to the given discretization, α is chosen to be a piece-wise constant function over $\hat{\Omega}$:

$$\alpha\Big|_{\tilde{\Omega}_i} = \left[\alpha_1^{(i)} \ldots \alpha_{M_\alpha}^{(i)}\right]^T \in \mathbb{R}^{M_\alpha}, \tag{1}$$

where $\alpha_j^{(i)}$ are constants for $j = 1, \ldots, M_\alpha$. Through this choice of parametrical distribution, the assembly (and manufacture) of the final array of cells is straight forward. As (manufacture) restrictions, we consider thresholds c_j and C_j, such that $c_j \leq \alpha_j^{(i)} \leq C_j$ with $j = 1, \ldots, M_\alpha$ in $\tilde{\Omega}_i(\alpha)$.

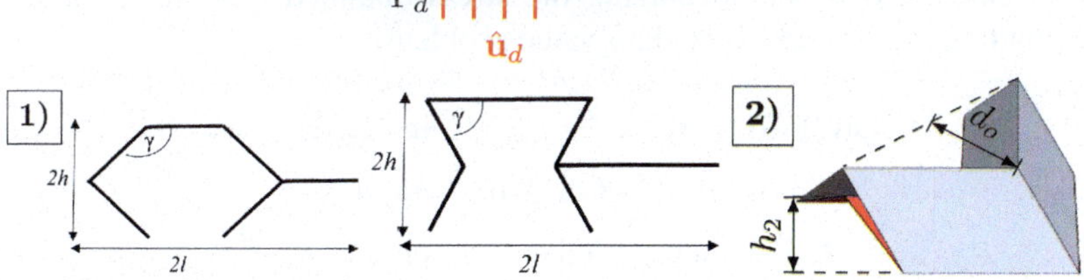

(a) Classical honeycomb (b) Auxetic honeycomb (c) Miura Ori unit cell

Fig. 1 Multiscale optimization problem and examples of unit cells

Recalling the definition of an array as shown in Fig. 1, the multiscale coupling emanates from the decomposition of the domain $\hat{\Omega}$ into unit cell domains $\tilde{\Omega}_i(\alpha)$, $i = 1, \ldots, N$. Therefore, the (macroscopic) displacements $\hat{\mathbf{u}}$ are defined as function of the (microscopic) design:

$$\hat{\mathbf{u}} = \hat{\mathbf{u}}(\alpha). \tag{2}$$

Given a cost functional $\mathcal{J}$ as a function of the displacements and design variables, the material programming is defined as the optimization problem:

$$\begin{aligned} &\min_{\alpha \in \mathcal{D}} \; \mathcal{J}(\hat{\mathbf{u}}(\alpha), \alpha) \\ \text{s.t.} \quad & \\ &\mathrm{Div}(\hat{\mathbf{F}}\,\hat{\mathbf{S}}) = 0 \quad \text{in} \;\; \hat{\Omega}, \\ &\hat{\mathbf{u}} = \hat{\mathbf{u}}_d \quad \text{on} \;\; \hat{\Gamma}_d, \\ &\hat{\mathbf{F}}\,\hat{\mathbf{S}}\,\hat{\mathbf{N}} = \hat{\mathbf{t}}_n \quad \text{on} \;\; \hat{\Gamma}_n, \end{aligned} \tag{3}$$

where $\hat{\mathbf{S}}$ and $\hat{\mathbf{F}}$ is the second Piola-Kirchhoff stress tensor and deformation gradient on the macro-scale, respectively, $\hat{\mathbf{N}}$ is the outward normal on $\hat{\Gamma}_n$ and Div($\bullet$) is the divergence w.r.t. the reference configuration. Since we consider sufficiently large arrays of cells, we can use a hyperelastic constitutive law on the macro-scale. Hyperelasticity assumes the existence of an elastic potential which describes the relation between stresses, the tangential stiffness $\hat{\mathbb{C}}$ and strain state $\hat{\mathbf{E}}$. We add the influence of the design α to this relation such that:

$$\hat{\mathbf{S}} = \hat{\mathbf{S}}(\hat{\mathbf{E}}, \alpha), \quad \hat{\mathbb{C}} = \frac{\partial \hat{\mathbf{S}}(\hat{\mathbf{E}}, \alpha)}{\partial \hat{\mathbf{E}}}. \tag{4}$$

Here, $\hat{\mathbf{E}}$ is the macroscopic Green-Lagrange strain tensor which is defined as

$$\hat{\mathbf{E}} = \frac{1}{2}(\hat{\mathbf{F}}^T\hat{\mathbf{F}} - \mathbf{1}) = \frac{1}{2}(\hat{\mathbf{C}} - \mathbf{1}), \tag{5}$$

where $\hat{\mathbf{C}}$ is the macroscopic right Cauchy-Green tensor. The macroscopic stress in (4) is calculated by homogenizing the stress obtained from the solution of a corresponding microscopic boundary value problem:

$$\begin{aligned} \mathrm{Div}(\tilde{\mathbf{F}}\,\tilde{\mathbf{S}}) &= \mathbf{0} && \text{in} \;\; \tilde{\Omega}_i(\alpha) \subset \hat{\Omega}, \\ \tilde{\mathbf{u}} &= (\hat{\mathbf{U}} - \mathbf{1})\tilde{\mathbf{X}} + \tilde{\mathbf{u}}_p && \text{on} \;\; \tilde{\Gamma}, \\ \tilde{\mathbf{F}}\,\tilde{\mathbf{S}}\,\tilde{\mathbf{N}} & \;\text{anti-periodic} && \text{on} \;\; \tilde{\Gamma}, \end{aligned} \tag{6}$$

where the tilde "∼" labels properties on micro-scale and $\tilde{\mathbf{X}} \in \tilde{\Gamma}$. The transition from macro- to micro-scale is obtained in terms of the design domain $\tilde{\Omega}_i(\alpha)$ and by applying the displacements on the boundary $\tilde{\Gamma}$ of a cell through the macroscopic strain $\hat{\mathbf{U}} = \hat{\mathbf{C}}^{1/2}$, which is also called right stretch tensor. Periodic deformation is ensured, as $\tilde{\mathbf{u}}_p$ is a periodic function over $\tilde{\Gamma}$. Consequently, tractions are anti-periodic over $\tilde{\Gamma}$. In the scope of this work, we choose a hyperelastic material law on both macro- and microscale as all deformations should be fully elastic. Therefore, the constitutive law on the microscale, which yields the material description of the unit cell's base material, can be formulated analogously to (4).

The multiscale coupling defined in (2) allows the cost functional to be reduced to $\check{\mathcal{J}}(\alpha) = \mathcal{J}(\hat{\mathbf{u}}(\alpha), \alpha)$, i.e., it is a function only of the design α. In this contribution, the cost functional is the difference to a desired state, such as a target displacement $\hat{\mathbf{u}}^*$ on the boundary $\hat{\Gamma}^*$ of the array, see Fig. 1:

$$\check{\mathcal{J}}(\alpha) = \mathcal{J}(\hat{\mathbf{u}}(\alpha), \alpha) = \frac{1}{2} \int_{\hat{\Gamma}^*} \|\hat{\mathbf{u}}(\alpha) - \hat{\mathbf{u}}^*\|^2 \, \mathrm{d}\gamma + \frac{\kappa}{2} \int_{\hat{\Omega}} \|\alpha\|^2 + \nabla\alpha \cdot \nabla\alpha \, \mathrm{d}\mathbf{X}. \quad (7)$$

The first integral emphasizes that we consider inverse problems in our work: We track certain distortions or displacements $\hat{\mathbf{u}}^*$ on chosen boundary sides $\hat{\Gamma}^*$ of the macroscopic geometry $\hat{\Omega}$. The second integral is added as Tikhonov based smoothing regulator [3, 9], where $\kappa \geq 0$ penalizes large H^1-norms of the design parameters in the cost functional in (7). This additional term is intended to produce smooth distributions of design parameters due to manufacturing restrictions. Extending the above formulation, the cost functional might also depend on stresses, strains or any other mechanical property.

For the solution method, the constraints in (3) are written as the residual of the mechanical virtual work or the weak form of the boundary value problem:

$$\begin{aligned} \langle \mathcal{R}(\hat{\mathbf{u}}, \alpha), \hat{\mathbf{v}} \rangle &:= \langle C(\hat{\mathbf{u}}, \alpha), \hat{\mathbf{v}} \rangle - \int_{\hat{\Gamma}_n} \hat{\boldsymbol{t}}_n \cdot \hat{\mathbf{v}} \, \mathrm{d}\gamma, \\ \langle C(\hat{\mathbf{u}}, \alpha), \hat{\mathbf{v}} \rangle &:= \int_{\hat{\Omega}} \hat{\mathbf{S}}(\hat{\mathbf{E}}, \alpha) \cdot \delta\hat{\mathbf{E}} \, \mathrm{d}\mathbf{X}, \end{aligned} \quad (8)$$

where $\delta\hat{\mathbf{E}} = (\hat{\mathbf{F}}^T \nabla\hat{\mathbf{v}} + \nabla\hat{\mathbf{v}}^T \hat{\mathbf{F}})/2$ is the variational strain tensor. Here, the functional $C(\hat{\mathbf{u}}, \alpha)$ takes into account the internal mechanical work. The standard trial or solution space is given by $V = \{\mathbf{v} \in H^1(\hat{\Omega}) \mid \mathbf{v} = 0 \text{ on } \hat{\Gamma}_d\}$.

2.1 Optimization Method

To solve the optimization problem (3), we use the standard gradient descent, which presumes the existence of a sufficiently smooth (Frechét) gradient or derivative of $\check{\mathcal{J}}(\alpha)$ w.r.t. the design variable α. In general, the resulting solution procedure can be written in iterative form:

$$\alpha^{(j+1)} = \alpha^{(j)} - \theta D_\alpha \check{\mathcal{J}}\left(\alpha^{(j)}\right), \tag{9}$$

where $j = 1, 2, 3, \ldots$ and the superscript (j) now denotes the iteration of the design α. The parameter θ is the step size which can be adjusted by some scaling criteria.

In [2] we deduce an optimization method which is based on an adjoint solution approach; the central idea is to facilitate the computations by replacing C with an linear alternative $\bar{C}$. We arrive at a simpler, linear formulation which we apply to the nonlinear multiscale optimization problem (3). In the end, we derive the following expression for the adjoint $\mathcal{A}$:

$$\begin{aligned} \langle \mathcal{A}(\hat{\mathbf{u}}, \hat{\boldsymbol{\lambda}}, \alpha), \hat{\mathbf{v}}\rangle &:= \langle \partial_{\hat{\mathbf{u}}} \mathcal{J}(\hat{\mathbf{u}}, \alpha), \hat{\mathbf{v}}\rangle_{V'} + \langle C(\hat{\boldsymbol{\lambda}}, \alpha), \hat{\mathbf{v}}\rangle \\ &= \int_{\hat{\Gamma}^*} (\hat{\mathbf{u}} - \hat{\mathbf{u}}^*) \cdot \hat{\mathbf{v}}\, \mathrm{d}\gamma + \int_{\hat{\Omega}} \hat{\mathbf{S}}(\hat{\Xi}, \alpha) \cdot \delta \hat{\Xi}\, \mathrm{d}\mathbf{X}, \end{aligned} \tag{10}$$

where, analogously to (5), the dual strain tensor $\hat{\Xi}$ and dual deformation gradient $\hat{\Upsilon}$ are defined as

$$\hat{\Xi} = \frac{1}{2}(\hat{\Upsilon}^T \hat{\Upsilon} - \mathbf{1}), \quad \hat{\Upsilon} = \nabla \hat{\boldsymbol{\lambda}} + \mathbf{1}, \tag{11}$$

such that $\delta\hat{\Xi} = (\hat{\Upsilon}^T \nabla \hat{\mathbf{v}} + \nabla \hat{\mathbf{v}}^T \hat{\Upsilon})/2$ (c.f. weak form in (8)). The variational problem for the total gradient in (9) can be obtained like the adjoint problem in (10). The resulting weak form reads:

$$\langle D_\alpha \check{\mathcal{J}}(\alpha), \boldsymbol{\rho}\rangle = \langle \partial_\alpha \mathcal{J}(\hat{\mathbf{u}}, \alpha), \boldsymbol{\rho}\rangle + \langle \partial_\alpha C(\hat{\mathbf{u}}, \hat{\boldsymbol{\lambda}}, \alpha), \boldsymbol{\rho}\rangle \text{ for } \boldsymbol{\rho} \in \left[H^1(\hat{\Omega})\right]^{M_\alpha}. \tag{12}$$

An exact analysis of the last term in (12) results in an integral expression which is too complicated to evaluate efficiently. Therefore, we consider an alternative derivative $\partial_\alpha \mathring{C}$ defined as

$$\langle \partial_\alpha \mathring{C}(\hat{\mathbf{u}}, \hat{\boldsymbol{\lambda}}, \alpha), \boldsymbol{\rho}\rangle := \int_{\hat{\Omega}} \left[\nabla \hat{\boldsymbol{\lambda}} \cdot \partial_\alpha \hat{\mathbb{C}}(\hat{\mathbf{E}}, \alpha) \cdot \nabla \hat{\mathbf{u}}\right] \boldsymbol{\rho}\, \mathrm{d}\mathbf{X}. \tag{13}$$

With the last modification in (13), we arrive at an iterative method resulting from the gradient descent (9) which we apply to the multiscale optimization task from (3). This method requires the solution to three problems, including the primal and adjoint weak forms in (8) and (10), respectively, as well as the design gradient in (12) and (13). Given a feasible, piece-wise constant design distribution α, the method reads:

1. Considering the residual in (8), solve the *primal* problem:
   Find $\hat{\mathbf{u}}_s \in \hat{\mathbf{u}}_d + V$, such that

$$\langle \mathcal{R}(\hat{\mathbf{u}}_s, \alpha), \hat{\mathbf{v}}\rangle = 0 \quad \forall \hat{\mathbf{v}} \in V. \tag{14}$$

2. Considering the adjoint in (10), solve the *adjoint* or *dual* problem:
   Find $\hat{\boldsymbol{\lambda}}_s \in V$, such that

$$\langle \mathcal{A}(\hat{\mathbf{u}}_s, \hat{\boldsymbol{\lambda}}_s, \alpha), \hat{\mathbf{v}} \rangle = 0 \quad \forall \hat{\mathbf{v}} \in V. \tag{15}$$

3. Find the design gradient by solving (12) and update the current design as in (9):
For $i = 1, \ldots, M_\alpha$, find $S_i \in H^1(\hat{\Omega})$ such that for all $\rho \in H^1(\hat{\Omega})$:

$$\int_{\hat{\Omega}} S_i \rho + \nabla S_i \cdot \nabla \rho \, \mathrm{d}\mathbf{X} = \kappa_i \int_{\hat{\Omega}} \alpha_i \rho + \nabla \alpha_i \cdot \nabla \rho \, \mathrm{d}\mathbf{X} + \langle \partial_{\alpha_i} \mathring{C}(\hat{\mathbf{u}}_s, \hat{\boldsymbol{\lambda}}_s, \alpha), \rho \rangle$$
$$\alpha_i \leftarrow \alpha_i - \theta \frac{S_i}{\|S_i\|}. \tag{16}$$

We repeat these three steps until convergence has been reached. Note that we update the design variables α entry-wise. If parameter values occur outside the predefined thresholds, we simply project α_j onto c_j or C_j for $j = 1, \ldots, M_\alpha$. Note that one crucial detail is left to be considered: the evaluation or approximation of the derivative $\partial_\alpha \hat{\mathbb{C}}(\hat{\mathbf{E}}, \alpha)$ from (13). For this issue, we suggest an approach using tensor decomposition in the following section.

3 Surrogate Models and Tensor Decomposition

Generally, it is not trivial to describe the mechanical behavior of arrays consisting of individually designed unit cells. Therefore, we follow a data driven approach, allowing not only to fast and accurately compute stress and stiffness in (4), but also the necessary design derivative $\partial_\alpha \hat{\mathbb{C}}(\hat{\mathbf{E}}, \alpha)$ from (13). This approach will replace the macroscopic constitutive law within an array of cells with a surrogate model, which is constructed on a discrete database. Moreover, this approach benefits our multiscale concept: Instead of solving the microscopic boundary value problem (6) for each cell with different parameters during optimization, the data-driven model is evaluated, which is much less expensive.

In order to establish the database for the surrogate model, the mechanical response of a unit cell $\tilde{\Omega}_i(\alpha)$ is determined for all relevant design and loading possibilities: we traverse all design and strain cases in a canonical way and solve the microscopic boundary value problem (6) to obtain the macroscopic stress tensor $\hat{\mathbf{S}}$ through standard volume averaging. The stress data are then understood as sample points and are ordered on a high-dimensional Cartesian or structured grid. Note that this data sampling can be considered as precomputing part of the material programming and has to be done only once.

Due to the nonlinear behavior of the unit cells, the microscopic boundary value problem (6) has to be solved iteratively. To accelerate the sampling procedure, we organise the incremental solution steps as economically as possible in order to minimize the computational work. After sampling, the obtained data points are collected in a hypercube, whose dimensions correspond to design variables α_j with

$j = 1, \ldots, M_\alpha$ and strain variables $\hat{\mathbf{E}}_i$ with $i = 1, \ldots, M_\mathbf{E}$ (in Voigt notation), where M_α and $M_\mathbf{E}$ are the number of design variables and strain components, respectively. We describe this hypercube as a multidimensional tensor:

$$\mathbf{D} = (\mathbf{D}_I)_{I \in \mathcal{I}} = (\mathbf{D}_{i_1,\ldots,i_M})_{(i_1,\ldots,i_M)\in\mathcal{I}} \in \mathbb{R}^{m_1\times\cdots\times m_M}, \tag{17}$$

where M is the order of the tensor, and m_k is the number of sampling points of the k-th design or strain variable; the multi index $I = (i_1, \ldots, i_M)$ describes the tensor's expansions. Each entry $\mathbf{D}_I$ contains an entry $\hat{\mathbf{S}}_j$ of the homogenized stress tensor $\hat{\mathbf{S}}$ for different design cases at certain strain such that we need between three and six different data tensors in total (depending on the spatial dimension).

To evaluate the macroscopic constitutive law (4) and the design derivative, we need to process the discrete tensor $\mathbf{D}$ further. An interesting approach to this can be found in [5], which shares the same motivation for a surrogate material law and uses the tensor decomposition as low rank approximation:

$$\mathbf{D} \approx \sum_{r=1}^{R} D_1^r \otimes \cdots \otimes D_{M_\alpha}^r \otimes C_1^r \otimes \cdots \otimes C_{M_\mathbf{E}}^r, \tag{18}$$

where R is a given integer. We refer to [6] for an extensive overview of tensor approximation methods. Note that $D_i^r \in \mathbb{R}^{m_i}$ with $i = 1, \ldots, M_\alpha$ are one-rank tensors or vectors (as well as C_j^r with $j = 1, \ldots, M_\mathbf{E}$), such that the data tensor is approximated by a series of outer products of vectors denoted by "$\otimes$". This representation allows the distinction between design and strain variables which is necessary for the optimization method described by (14)–(16).

To access the design and load cases which have not been considered in the precomputing process, we canonically interpolate within $\mathbf{D}$. For the interpolation, we use the vectors $D_i^r \in \mathbb{R}^{m_i}$ as interpolation nodes for $i = 1, \ldots, M_\alpha$. Then, the interpolating function is defined as:

$$\bar{D}_i^r(\alpha_i) := \sum_{n=1}^{m_i} N_n(\alpha_i)\,(D_i^r)_n, \tag{19}$$

where $(D_i^r)_n$ is the n-th component of the vector D_i^r. Analogously, we define the interpolating function $\bar{C}_j^r(\hat{\mathbf{E}}_j)$ for the vectors C_j^r and $j = 1, \ldots, M_\mathbf{E}$. We choose B-Splines as basis interpolating functions N_n since we require smooth interpolation and local influence of data points. The final interpolating function is defined as:

$$\bar{\mathbf{D}}(\alpha_1, \ldots, \alpha_{M_\alpha}, \hat{\mathbf{E}}_1, \ldots, \hat{\mathbf{E}}_{M_\mathbf{E}}) = \sum_{r=1}^{R} \left(\prod_{i=1}^{M_\alpha} \bar{D}_i^r(\alpha_i) \prod_{j=1}^{M_\mathbf{E}} \bar{C}_j^r(\hat{\mathbf{E}}_j) \right). \tag{20}$$

The interpolant in (20) represents the previously mentioned surrogate model which is based on dimensionally separated interpolation as model reduction technique. This variable separation allows now a very simple differentiation w.r.t. to the design, i.e., only $\bar{D}_i^r$ in (20) has to be differentiated for $r = 1, \ldots, R$ and $i = 1, \ldots M_\alpha$. Since we use smooth B-Splines, the evaluation of both the interpolating functions N_n and their derivatives are based on the same efficient algorithm. The differentiation of $\bar{\mathbf{D}}$ w.r.t. the strain variables $\hat{\mathbf{E}}_j$ can be done analogously for $j = 1, \ldots, M_{\mathbf{E}}$. Since $\mathbf{D}$ stores (homogenized) stress data, the differentiation of $\bar{\mathbf{D}}$ w.r.t. $\hat{\mathbf{E}}$ yields the tangential stiffness $\hat{\mathbb{C}}$, see (4). Applying differentiation in both directions, i.e., design and strain, we are able to easily compute the derivative of the stiffness tensor w.r.t. α, i.e., $\partial_\alpha \hat{\mathbb{C}}$ in (13).

4 Results and Discussion

We demonstrate the ability of our optimization framework by "programming" a specimen to obtain a desired shape at the end of a certain mechanical load. The initially rectangular form is divided into 40 × 12 cells in plane. In the example from our recent publication [2], the desired shape is described by a trigonometric function. The exact dimensions, boundary conditions and target functions can be viewed in Fig. 2. In this figure, the specimen is clamped on one side and stretched by $\Delta\mathbf{u}$ on the other side. The desired shape change is expected to emerge on the boundary sides Z_1 and Z_2 and is described analytically on Z_1 by

$$f_1(x) = p_1 \sin(p_2(x + p_3)) + p_4(x + p_5) \quad \text{on } Z_1 \tag{21}$$

with $p_1 = 1.87$, $p_2 = 0.0187$, $p_3 = 9.04$, $p_4 = -2.22 \times 10^{-3}$, $p_5 = 329.1$ and on Z_2 correspondingly $f_2(x) = -f_1(-x + 400)$. We choose the unit cells shown in Fig. 1: the honeycomb cell [7] (Fig. 1a and b) and Miura-Ori or Origami cell [8] (Fig. 1c). We use the tensor decomposition from Sect. 3 for the Origami cell and an analytical description [7] for the honeycomb. As depicted in Fig. 1, the Origami cell is parametrized by the offset d_o and the height h_2 controlling the auxetic behaviour and the stiffness change, respectively. The honeycomb is simply parametrized by the

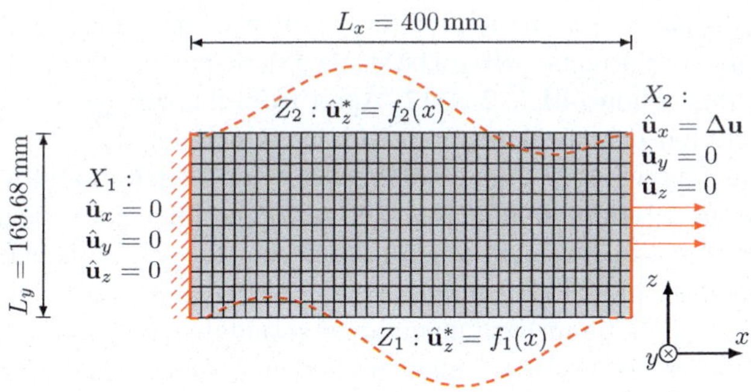

Fig. 2 Array of 40 × 12 cells under tension and targetd shape on Z_1 and Z_2

angle γ influencing its auxetic characteristics and elastic behavior. To precompute the database of the Origami cell, we used the discretization from [2], in particular almost 41 million data points: we considered 35 design combinations of the offset d_o and height h_2, for each of which we computed all 6 entries of the homogenized stress $\hat{\mathbf{S}}$ for 194 481 variants of the strain tensor $\hat{\mathbf{E}}$. To obtain the representation in (20), we reduce this amount of data to 17 400. Note that, through our sampling strategy, the necessary amount of simulations can also be reduced to a fraction.

We begin with the honeycomb cell and analyze the convergence of presented multiscale approach. For the numerical experiments, we applied an analytical model, which returns the homogenized stiffness tensor $\hat{\mathbb{C}}$ and is scalable by the grid size. Besides, it allows the computation of the parameter derivative $\partial_\alpha \hat{\mathbb{C}}$ in (13). As the analytical model is only valid for small deformations, we consider a tension of 2% ($\Delta \mathbf{u} = 8$ mm in Fig. 2). After optimizing the specimen in Fig. 1, we construct an equivalent, fully resolved model and compare it with the surrogate using analytical homogenization. In Fig. 3a, both geometries are displayed with magnified details to demonstrate the good correspondence between analytical surrogate model and fully resolved structure. Next, we look at convergence w.r.t. the grid size: in Fig. 3d, a sequence of solutions to the same problem is presented, where the grid size decreases from left to the right. The number of cells are adapted such that the dimensions of all specimens are identical. All cases have the same uniform parameter distribution as initial guess for the optimization which used 25 iterations for each. All optimized structures look similar and share the same features, i.e., the same distinctive localization of parameters. The convergence plot in Fig. 3c affirms this impression quantitatively as it depicts the error of the first four results w.r.t. the one with the finest grid size. A similar experiment is presented in Fig. 3b: it shows the error between the analytical and the fully resolved models. Both plots show the reduction of the relative error norm while decreasing grid size: we recognize a sublinear rate in Fig. 3c and a superlinear rate in Fig. 3b.

An interesting question from the manufacturers' perspective is whether the identical programming task could be realized with different cells. For the above problem, we consider the Origami cell and summarize the results for the material programming in Fig. 4 at 5% strain ($\Delta \mathbf{u} = 20$ mm). Convergence and the match between targeted and obtained solutions are shown in Fig. 4a and b, respectively. The actual solution, i.e., the parameter distribution for the offset d_o and height h_2, is depicted in Fig. 4c. Note that the Origami cell can be stacked such that we extended the specimen into 3D with dimensions $40 \times 2 \times 12$. Apart from the surrogate model, the fully resolved structure is displayed next to it with magnified details below. In Fig. 4d the main results are shown: the comparison between our model, the fully resolved and manufactured model. In both numerical and real experiments, the results presented in Fig. 4d show good agreement when considering the target displacement fields $\mathbf{u}_z^*$ on the surrogate model (left, min/max –3.18/3.07 mm), resolved model (middle, min/max –3.28/3.20 mm) and manufactured sample (right, min/max –3.44/3.45 mm). For reference, the target minimal and maximal values were ± 3.14 mm.

We can interpret all results in the following way: to a certain extend, all optimization results at different geometrical resolutions share the same characteristic param-

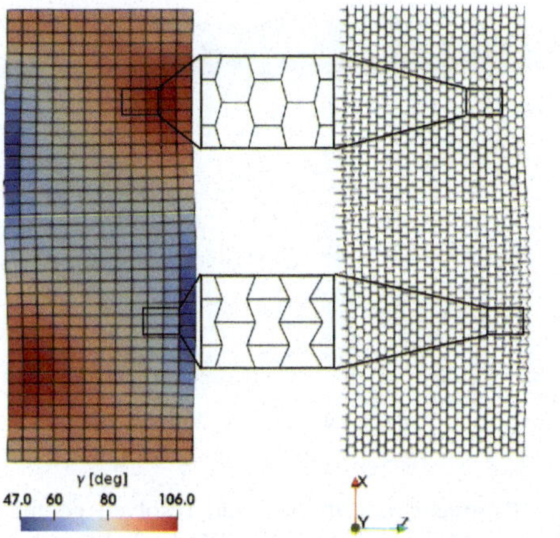

(a) Different models with 40 × 12 cells at 2% strain.

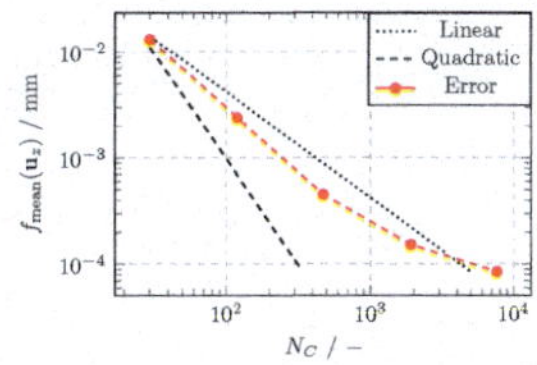

(b) Convergence w.r.t. resolved model

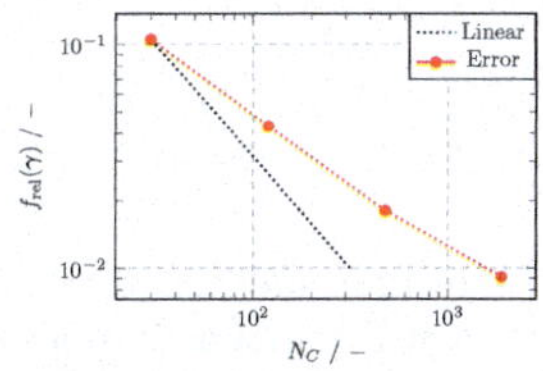

(c) Convergence w.r.t. grid size

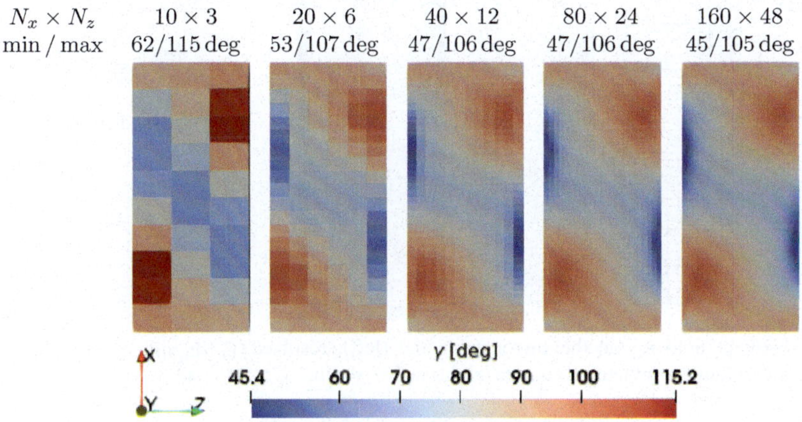

(d) Parameter distribution for different grid sizes

Fig. 3 Optimization of a sinusoidal shape with honeycomb cells

eter distribution. So, the total structure can be viewed as a continuum at a particular grid size, which justifies the notion of programmable *material*. As a final remark, we emphasize the important role of presented multiscale optimization framework which facilitates the design and manufacture of material-like structures for promising purposes in the near future.

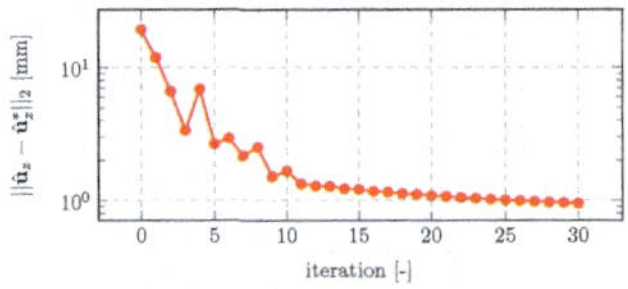

(a) Objective function

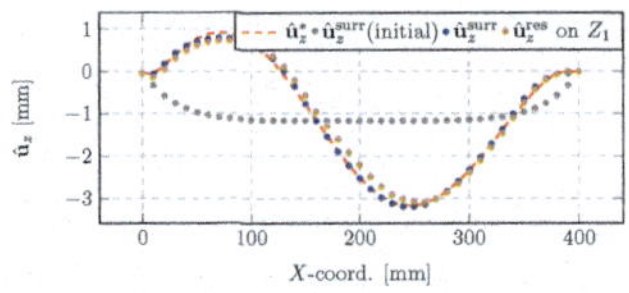

(b) Target function, shape of surrogate model (initial and optimized) and resolved model for the side $\boldsymbol{Z}_1$.

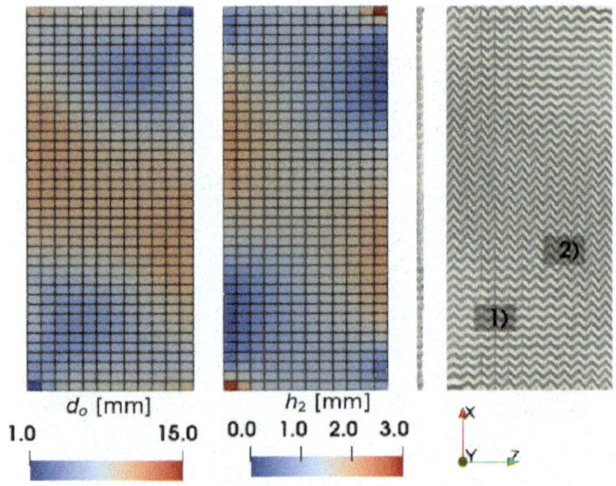

(c) Parameter distribution and resolved geometry. Magnified structures of **1)** and **2)** are below.

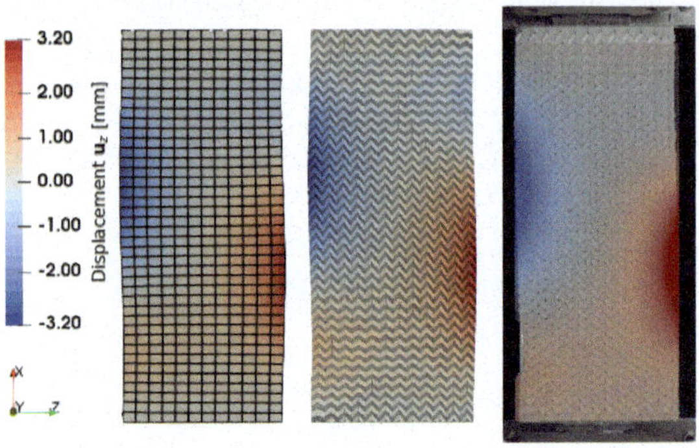

(d) Displacement field $\mathbf{u}_z$ on the surrogate model (left), resolved model (middle) and manufactured sample (right) at 5 % strain.

(e) Magnified structure at **1)** in 4c

(f) Magnified structure at **2)** in 4c

Fig. 4 Optimization of a sinusoidal shape for a Miura-Ori foil stack with two layers

Acknowledgements This contribution was supported by Fraunhofer Cluster of Excellence Programmable Materials (CPM). We appreciate Angela Schwarz and Christof Hübner of the Fraunhofer ICT for the experimental laboratory results displayed in Fig. 4. Furthermore, we thank Tobias Lichti allowing us to present the convergence studies in Fig. 3 from his Ph.D thesis [10].

References

1. Ferrer, A., Cante, J., Hernández, J., Oliver, J.: Two-scale topology optimization in computational material design: an integrated approach. Int. J. Numer. Methods Eng. (2018)
2. Lichti, T., Leichner, A., Andrä, H., Müller, R., Wenz, F., Eberl, C., Schwarz, A., Hübner, C.: Optimal design of shape changing mechanical metamaterials at finite strains (2022). https://doi.org/10.1016/j.ijsolstr.2022.111769

3. Frei, S., Andrä, H., Pinnau, R., Tse, O.: Optimizing fiber orientation in fiber-reinforced materials using efficient upscaling. Comput. Optim. Appl. (2013). https://doi.org/10.1002/nme.5742
4. Weisheit, L., Wenz, F., Eckert, M., Lichti, T., Baumann, S., Hübner, C., Eberl, C., Andrä, H.: Domänenübergreifende workflows zur effizienten Entwicklung Programmierbarer Materialien. Z. Wirtsch. Fabrikbetr. (2020). https://doi.org/10.3139/104.112353/
5. Yvonnet, J., Monteiro, E., He, Q.-C.: Computational homogenization method and reduced database model for hyperelastic heterogeneous structures. Int. J. Multiscale Comput. Eng. (2013). https://doi.org/10.1615/IntJMultCompEng.2013005374
6. Tamara, B.W.B., Kolda, G.: Tensor decompositions and applications. SIAM (2009). https://doi.org/10.1137/07070111X
7. Wenz, F., Schmitt, I., Leichner, A., Lichti, T., Baumann, S., Andrä, H., Eberl, C.: Designing shape morphing behavior through local programming of mechanical metamaterials. Adv. Mater. (2021). https://doi.org/10.1002/adma.202008617
8. Schwarz, A., Lichti, T., Wenz, F., Scheuring, B.M., Hübner, C., Eberl, C., Elsner, P.: Development of a scalable fabrication concept for sustainable, programmable shape morphing metamaterials. Adv. Eng. Mater. (2022). https://doi.org/10.1002/adem.202200386
9. Hinze, M., Pinnau, R., Ulbrich, M., Ulbrich, S.: Optimization with PDE Constraints. Springer, New York (2009)
10. Lichti, T.: Development of methods and tools for the design of programmable materials. Thesis (Ph.D), TU Darmstadt (2023)

Modeling Multi-grain Multi-hole Thin Solid State Films

Amy Novick-Cohen, Daniel Goldberg, Katrine Golubkov, and Rawan Tarabeh

Abstract The modeling of the dynamics of thin solid state films dates back to a series of seminal papers by W.W. Mullins, published in the years 1956–1958. In those years, there was much interest in the stability of bulk bi- and poly-crystalline materials, which can be strongly affected by the dynamics of their exterior surface features. In recent years the focus has been extended to include many new materials, which can enable the manufacture of strong and thin films; however, much of the basic underlying physics remains the same. Thus it is not unreasonable in considering thin film stability, to return and study the implications of the original modeling equations as developed by Mullins, namely surface diffusion and mean curvature motion. We follow this approach, focusing on surface diffusion and mean curvature motions, while neglecting possible anisotropy and elastic effects. We show that even within this somewhat limited framework, it is possible to obtain new criteria for the stability of holes in thin films, based on a new notion of an *effective radius*. This note summarizes results appearing in [7, 18, 19]. Many open questions remain.

Keywords Thin film stability · Surface diffusion · Critical radius · Hole stability · Effective radius

---

A. Novick-Cohen (✉) · D. Goldberg · K. Golubkov · R. Tarabeh
Faculty of Mathematics, Technion–IIT, Haifa, Israel
e-mail: amync@technion.ac.il

D. Goldberg
e-mail: daniel.gold@campus.technion.ac.il

K. Golubkov
e-mail: katringolubkov@gmail.com

R. Tarabeh
e-mail: rawanta@campus.technion.ac.il

F. Willot et al. (eds.), *Continuum Models and Discrete Systems*,
Springer Proceedings in Mathematics & Statistics 457,
https://doi.org/10.1007/978-3-031-58665-1_17

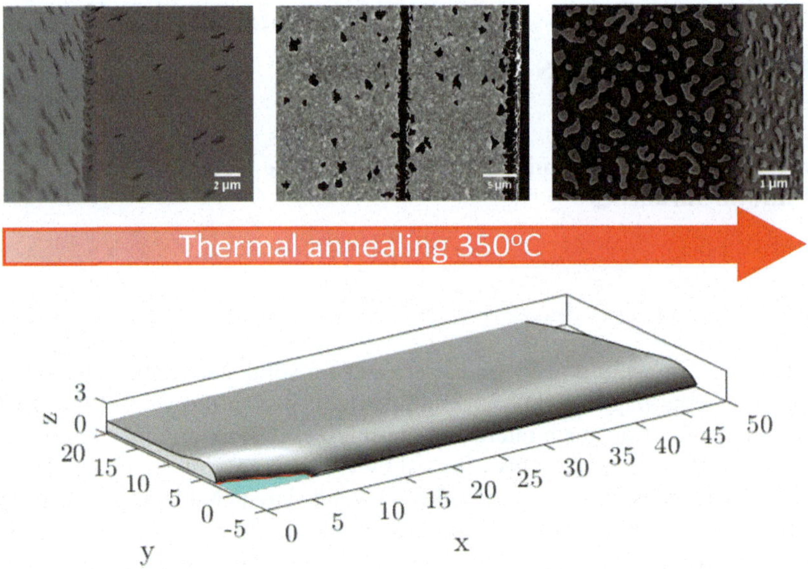

Fig. 1 Upper panels. SEM micrographs of 3 different Au-coated KCl whiskers after annealing at 350° for 3, 10 and 60 min. Lower panel. Numerical simulation of evolution by surface diffusion of a thin solid film on a curved substrate initially containing a hole

1 Introduction

Typically, thin solid state films are metallic and lie on a ceramic substrate. They constitute critical components in a wide range of modern technologies, such as the manufacture of computer chips, thermal sensors, plasmonics, and more. See [19] for further discussion.

Often thin solid state films are polycrystalline, and holes penetrating the film may occur. Hence, to guarantee technical reliability of thin films, a good understanding of their stability is crucial. In this context, it is natural to ask: Should a hole appear, will it close or will it grow, possibly initiating film break up into islands? Classically, thin film stability has focused on avoiding the appearance of holes and on ascertaining conditions under which holes that did form would shrink and disappear. In many applications, such as in computer chip production, one would indeed like to avoid the appearance of holes in order to maintain the integrity of the chip. However, stable arrays of small holes are desirable for other applications [17]. Thus, stability questions also include: Is it possible to produce a thin film containing a stable array of holes?

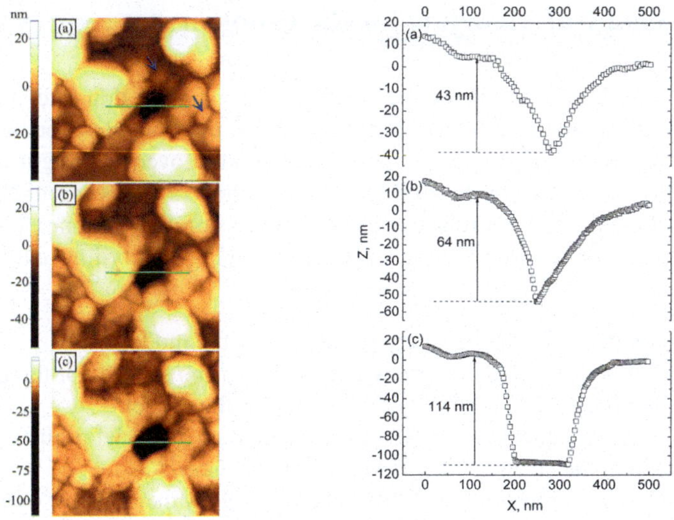

Fig. 2 **a–c**, left. AFM images of a Cu thin film on sapphire substrate after three consecutive anneals for 10 min at 300 °C; the lateral dimensions of the image are $1 \times 1 \mu m^2$. The two small grains, indicated by blue arrows in (**a**, left), later shrink and are no longer visible in (**c**, left), and a pit visible in (**a**, left), later sinks, leading to hole formation by (**c**, left). (**a–c**, right). The line topography profiles taken along the respective green lines in (**a–c**, left)

Stability phenomena are illustrated in Figs. 1[1] and 2.[2] In the upper panels of Fig. 1, SEM images are shown of thin Au films on KCl whiskers after annealing at 3, 10, and 60 min, respectively; it can be seen that holes appear in the film and then grow, with preferential growth along the whisker edge, until the thin film agglomerates into an array of islands. Preferential growth of holes along highly curved surfaces was also seen numerically [1], as shown in the lower panel in Fig. 1. In Fig. 2, AFM images of a thin Cu film on a sapphire substrate are shown following three consecutive anneals; during the annealing process, a hole can be seen to form and penetrate the film, and two small grains initially present in panel (a, left), later vanish and do not appear in panel (c, left).

[1] Reproduced from the graphical abstract of [1], Almog, Y., Derkach, V., Sharma, A., Novick-Cohen, A., Greer, J., Rabkin, E., Thermal stability of thin Au films deposited on salt whiskers, Acta Mat. **205**, 116537, (2021), doi:10.1016/j.actamat.2020.116537, with permission from Elsevier.

[2] Reproduced from [5], Derkach, V., Novick-Cohen, A., Vilenkin, A., Rabkin, E., Grain boundary migration and grooving in thin 3-D systems, Acta Mat. **65**, 194–206. (2014), doi:10.1016/j.actamat.2013.10.061, with permission from Elsevier.

2 Mathematical Modeling for the Problem

Suppose we wish to know, given the specifications of some thin (metallic) film on a substrate, if a hole should form in the film, will the hole close or will the hole grow possibly breaking the film up into islands? Moreover, we might also like to know whether is it possible to produce a stable array of holes in our film? As an applied mathematician, in order to develop useful stability criteria which address such questions, a reasonable approach could be to proceed according to the following steps:

- to decide on the basic evolutionary equations for the model,
- to choose some special geometry, which is neither too simple nor too complex,
- to formulate reasonable boundary conditions for the model,
- to prove well-posedness for the resultant mathematical model problem,
- to characterize the steady states of the model problem,
- to examine the stability of the steady states as well as the attractor dynamics.

2.1 Governing Equations for Thin Solid State Film Dynamics

Following our earlier discussion, we follow Mullins' modelling approach from 1956–1958, basing our model on coupling motion by mean curvature with motion by surface diffusion [11]. More specifically, we assume that the grain boundaries evolve by *Mean Curvature Motion* [9],

$$\mathcal{V} = A\kappa, \tag{1}$$

and that the exterior film surface evolves by *Motion by Surface Diffusion* [10]

$$\mathcal{V} = -B\Delta_X\kappa. \tag{2}$$

In (1) and (2), $X = X(t)$ corresponds to an evolving hypersurface in R^3, with normal velocity, $\mathcal{V}$, and mean curvature, κ. Moreover, Δ_X denotes the Laplace-Beltrami operator, and A and B are physical coefficients known, respectively, as the *reduced mobility* and the *Mullins' coefficient*. The Eqs. (1) and (2) are isotropic, and in considering models based on (1), (2), for the sake of simplicity, we have neglected possible anisotropy and elastic effects, which may at times be highly significant. Nevertheless, as we shall see, models based on (1), (2) may be highly nontrivial and can provide new insights.

To attain a fully developed working model, the governing equations need to be complemented by reasonable boundary and initial conditions. Following [8, 11], boundary conditions may be chosen based on considerations of force balance and flux balance, as well as holonomic constraints. However, in order to obtain a well-posed problem formulation, it is necessary to choose some specific geometry.

2.2 Some Special Geometries

1. Mullins [9–11] in 1956–1958 initiated the study of motion by mean curvature, surface diffusion as well as their coupled motions. In 1957, Mullins [10] focused on the dynamics of *thermal grooves* which form between two neighboring grains, then deepen self-similarly at a rate $\propto t^{1/4}$, by considering a special geometry.
2. Safran and Srolovitz [12, 13] studied the stability of holes in thin films with respect to wetting (hole shrinkage) and dewetting (hole growth), in 1986. In particular, they considered a polycrystalline hexagonal array of grains, possibly containing holes. They then focused on a single axisymmetric thin film disk with a hole, and developed the notion of a *critical radius* for hole growth or shrinkage based on energetic and kinetic criteria.
3. Wong et al. [16] in 1997 focused on the process of wetting/dewetting of a thin film on substrate by considering the dynamics of a semi-infinite solid state wedge whose exterior surface is governed by surface diffusion.

The early analyses, as in 1.–3. above, often relied heavily on the *small slope approximation*, which replaces the surface diffusion description given in (2) by its linear approximation, $y_t = -By_{xxxx}$, which is known as *fourth order diffusion* or the *Mullins equation* [10]. Nevertheless, many small slope approximation predictions are at least qualitatively correct. In certain cases, as in the context of thermal grooving, the small slope approximation is often reasonable for realistic values of the physical parameters, and hence the predictions of the small slope approximation are reliable.

In examining Eqs. (1), (2), it is readily seen that surfaces with vanishing mean curvature, $\kappa = 0$, constitute steady states for (1), and surfaces with constant mean curvature, $\kappa = C$, constitute steady states for (2). Accordingly, many analyses have focused on the stability of surfaces with constant (including vanishing) mean curvature, such as planes, spheres, cylinders, unduloids, catenoids, and nodoids. In coupling the two motions, (1) and (2), it follows from the considerations above that for appropriately formulated boundary conditions, hexagonal arrays of spherically capped grains as considered in [4, 12] constitute steady states. By considering cylinders with a small radius to length ratio or for hexagonal arrays of spherically capped grains with sufficiently large cap curvatures, we see that steady states can have large curvatures, and then the small slope approximation is unappropriate. Notably, spheres, nodoids, catenoids, and unduloids do not constitute steady states for the Mullins equation.

In a series of recent papers [7, 18, 19], we have considered geometries describing the evolution of one or more grains which are supported by a substrate and contained within an inert cylinder. If grains intersect, the surfaces of intersection (*grain boundaries*) were assumed to evolve by motion by mean curvature, (1), and the exterior surfaces of the grains were assumed to evolve by surface diffusion, (2). In the subsections which follow, we focus primarily on the geometry in the analysis in [18], returning with some remarks regarding the other geometries, as well as other possible directions, in the Summary in Sect. 3.

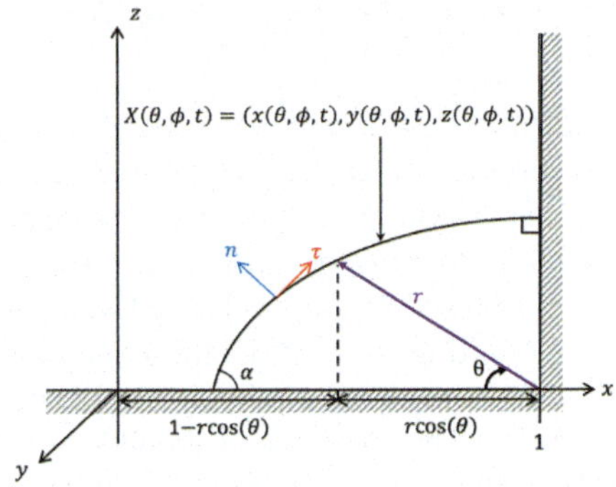

Fig. 3 The profile of an evolving grain exterior surface in the meridian cross-section. Here, the angle α denotes the contact angle of the exterior surface of the grain with the substrate

2.3 A Particular Axi-Symmetric Geometry with One Grain and One Hole

We now consider a particular geometry, namely an axi-symmetric system, which contains one annular grain and a central hole, which lie on a substrate and are bounded within an inert cylinder, [3, 4, 18]. The geometry that we have in mind is portrayed schematically in Fig. 3. As this geometry contains only one grain, we only need to consider the evolution of its exterior surface, which we take to evolve by surface diffusion, (2), in accordance with our earlier discussion. With regard to boundary conditions, the exterior surface should normally intersect the bounding cylinder since the surface of the bounding cylinder is assumed to be inert, and the contact angle made by the exterior surface with the substrate, denoted by α in Fig. 3, constitutes a physical parameter which is determined by the relative surface free energies. Additionally, no flux boundary conditions are imposed at the intersections of the exterior grain surface with both the bounding cylinder and the substrate. In [18], the governing equation, as well as boundary and initial conditions, were formulated in terms of a polar parametrization, via the radius, $r = r(\theta, t)$, indicated in Fig. 3.

In terms of $r(\theta, t)$ in the polar parametrization, following normalization our problem may be expressed as

$$\begin{cases} r_t = \mathcal{A}\kappa_{\theta\theta} + \mathcal{B}\kappa_\theta, & (\theta, t) \in (0, \pi/2) \times (0, T), \\ r_\theta\big|_{\theta=0} = -r\big|_{\theta=0} \cot(\alpha), & t \in (0, T), \\ r_\theta\big|_{\theta=\pi/2} = 0, & t \in (0, T), \\ \kappa_\theta\big|_{\theta=0,\pi/2} = 0, & t \in (0, T), \\ 0 < r(\theta, t)\cos(\theta) \le 1, & (\theta, t) \in [0, \pi/2] \times (0, T), \\ r(\theta, 0) = r_0(\theta), & \theta \in [0, \pi/2], \end{cases}$$

where

$$\mathcal{A} = -\frac{1}{r(r^2 + r_\theta^2)^{1/2}},$$

$$\mathcal{B} = -\frac{(r \sin(\theta) - r_\theta \cos(\theta))}{r(r^2 + r_\theta^2)^{1/2}(1 - r\cos(\theta))} + \frac{r_\theta(r + r_{\theta\theta})}{r(r^2 + r_\theta)^{3/2}},$$

$$\kappa = \frac{r r_{\theta\theta} - 2r_\theta^2 - r^2}{2(r^2 + r_\theta^2)^{3/2}} + \frac{r_\theta \sin(\theta) + r\cos(\theta)}{2(1 - r\cos(\theta))(r^2 + r_\theta^2)^{1/2}}.$$

In the problem formulation above, the cylinder radius, L, has been normalized to unity, and the Mullins' coefficient, B, in (2) together with the cylinder radius, L, have been used in scaling time. The first and second boundary conditions reflect the angle conditions. The third boundary condition, imposed at $\theta = 0$ and at $\theta = \pi/2$, reflects the no flux boundary condition. The bounds on $r(\theta, t)\cos(\theta)$ reflect the geometry. Note that in the scaled(dimensionless) problem formulation prescribed above, the contact angle α is the only physical parameter. For further details, see [18].

2.4 *Delaunay Surfaces, 1841*

We remarked above that surfaces with constant mean curvature constitute steady states for surface diffusion. In fact, for surfaces $X(t)$ evolving by (2), with no flux boundary conditions $n \cdot \nabla_X \kappa = 0$, imposed along the curve $\Gamma(t) = \partial X$, the steady states necessarily satisfy $\kappa = C, \quad C \in R$, which includes $\kappa = 0$ as a particular case. Upon assuming moreover axi-symmetry of the surface $X(t)$, the set of all such surfaces, known as *Delaunay surfaces* after Delaunay [2] in 1841, consists of planes, cylinders, spheres, catenoids, unduloids, and nodoids.

2.5 *Steady States for the Axi Symmetric Model with One Grain and a Hole*

For the axi-symmetric geometry described in detail above, the set of steady states can be described [3, 18] as *nodoids* with (constant) negative mean curvature $\lambda < 0$. More specifically, given any contact angle $\alpha \in (0, \pi)$, the possible nodoidal steady states may be prescribed explicitly via their meridian cross sectional profiles

$$\begin{cases} x(\varphi; \alpha, \lambda) = \dfrac{1}{2\lambda}\left[\sin(\varphi) - \sqrt{\sin^2(\varphi) + 4\lambda^2}\right], & \varphi \in [0, \alpha], \\ z(\varphi; \alpha, \lambda) = -\dfrac{1}{2\lambda}\displaystyle\int_\alpha^\varphi \left(\frac{\sin^2(\xi)}{\sqrt{\sin^2(\xi) + 4\lambda^2}} - \sin(\xi)\right) d\xi, & \varphi \in [0, \alpha], \end{cases}$$

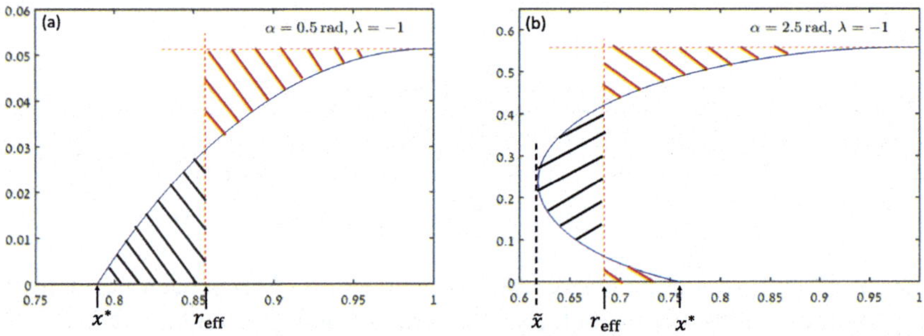

Fig. 4 View of the meridian profiles of the nodoidal steady states for $\alpha = 0.5$ rad, $\lambda = -1$ (left) and $\alpha = 2.5$ rad, $\lambda = -1$ (right). Here $x^* := x(0; \alpha, \lambda)$, $\tilde{x} := \min_{\theta \in [0,\pi]} x(\theta; \alpha, \lambda)$ and $r_{\text{eff}} = r_{\text{eff}}(\alpha, \lambda)$ are indicated. The volume corresponding to the area marked in black equals the volume corresponding to the area marked in maroon

where $\lambda \in (-\infty, 0)$ is arbitrary, and $\varphi \in [0, \alpha]$ denotes the angle between the tangents on the free surface and the substrate, which vanishes at the cylinder and then increases along the surface in the counter-clockwise direction until $\varphi = \alpha$, the contact angle, is attained where the free surface meets the substrate.

Since α is the only physical parameter in the (scaled) model equations and $\lambda \in (-\infty, 0)$, which denotes the (scaled) mean curvature, parameterizes the set of steady state solutions for a given value of $\alpha \in (0, \pi]$, the volume, V, of any given steady state can be computed from the expressions, with $V = V(\alpha, \lambda)$ depending continuously on α and λ. The question now arises: for a given contact angle $\alpha \in (0, \pi]$ and for a given grain volume, $V \in (0, \infty)$, how many steady states are there and what do they look like?

Claim 1 For any contact angle, $\alpha \in (0, \pi)$, $V_{\max}(\alpha) := \max_{\lambda \in (-\infty,0)}\{V(\lambda, \alpha)\} < \infty$, and for any volume V such that $0 < V \leq V_{\max}(\alpha)$, there exists a nodoid with contact angle α and volume V. Furthermore, if $0 < V < V_{\max}(\alpha)$, then there exist precisely two distinct nodoids with contact angle α and volume V.

The claim can be understood graphically by looking at the upper panel in Fig. 4,[3] upon noting that $V(\lambda, \alpha_1) < V(\lambda, \alpha_2)$ for $0 < \alpha_1 < \alpha_2 < \pi$ and $\lambda \in (-\infty, 0)$, and that $\lim_{\lambda \downarrow -\infty} V(\lambda, \alpha) = \lim_{\lambda \uparrow 0} V(\lambda, \alpha) = 0$ for $0 < \alpha < \pi$. See [6] for further details. Some typical steady state profiles are also portrayed in Fig. 4.

[3] Reproduced from [18], Zigelman, A. & Novick-Cohen, A., Critical effective radius for holes in thin films: Energetic and dynamic considerations. J. Appl. Phys. **130**, 175301 (2021), doi:10.1063/5.0053444, with the permission of AIP Publishing.

2.6 Introducing the Notion of an Effective Radius, r_{eff}

Given a nodoidal steady state which can be prescribed in terms of the parameters α and λ, we find it convenient to define the notion of an *effective radius*, $r_{\text{eff}} = r_{\text{eff}}(\alpha, \lambda)$, as the inner radius of a annular grain which is flat with height $z = z(0; \alpha, \lambda)$ and volume $V = V(\alpha, \lambda)$. Analytically, r_{eff} can be prescribed (without dimensions) by setting

$$r_{\text{eff}} := \sqrt{1 - V(\alpha, \lambda)/(\pi z(0; \alpha, \lambda))}. \tag{3}$$

See Fig. 4. Numerical computations in [18] indicate that $r_{\text{eff}} = r_{\text{eff}}(\alpha, \lambda)$ depends monotonically on λ for all $\alpha \in (0, \pi)$.

2.7 Energetic Considerations and the Critical Effect Radii: r^c_{eff}, $\tilde{r}^c_{\text{eff}}$

In the context of our axi-symmetric system with one grain and a hole, it is reasonable to define the total energy of the system as

$$E(t) = \gamma_{\text{ex}} A_{\text{ex}}(t) + \gamma_{\text{gas}} A_{\text{gas}}(t) + \gamma_{\text{grs}} A_{\text{grs}}(t),$$

where γ_{ex}, γ_{gas}, and γ_{grs}, which satisfy $\gamma_{\text{gas}} - \gamma_{\text{grs}} = \gamma_{\text{ex}} \cos(\alpha)$, are the surface free energies of the exterior surface, of the hole, and of the substrate covered by the grain, and $A_{\text{ex}}(t)$, $A_{\text{gas}}(t)$, $A_{\text{grs}}(t)$ are the corresponding surface areas. Since the cylinder surface is inert, $\gamma_{\text{cylinder-gas}} = \gamma_{\text{cylinder-grain}}$, and the exterior surface normally intersects the cylinder. In order to facilitate energetic comparison among the various steady states, it is convenient to define an *effective energy*, E_{eff}, as follows

$$E_{\text{eff}} := \frac{E(t) - \pi \gamma_{\text{grs}}}{\pi \gamma_{\text{ex}}}. \tag{4}$$

We remark that if the hole closes, then the resulting configuration is a flat grain without a hole, which strictly speaking does not constitute a steady state for our model problem, as the boundary conditions at the contact point are not satisfied. Nevertheless, it is easy to verify that $E_{\text{eff}} = 1$ for a flat grain without a hole, [18].

Based on the definitions (4) and (3) above, and noting that r_{eff} depends monotonically on λ (see [18]), we may calculate E_{eff} and V as functions of r_{eff} for various values of α. Such calculations were undertaken in and are portrayed in Fig. 5.[4] The following two claims follow upon examining Fig. 5.

---

[4] Reproduced from [18], Zigelman, A. & Novick-Cohen, A., Critical effective radius for holes in thin films: Energetic and dynamic considerations. J. Appl. Phys. **130**, 175301 (2021), doi:10.1063/5.0053444, with the permission of AIP Publishing.

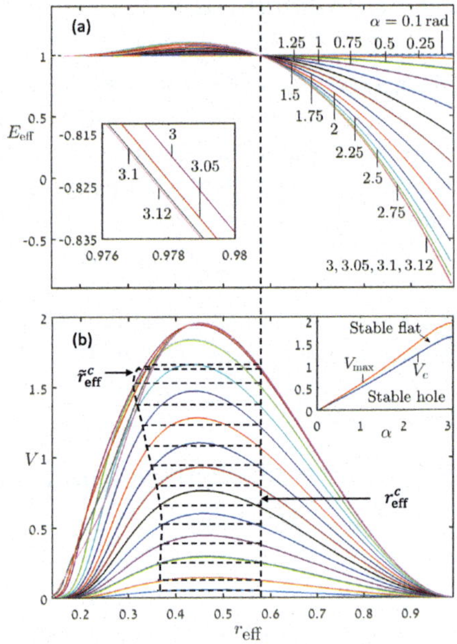

Fig. 5 Plots of E_{eff} and V versus r_{eff} are portrayed in panels (**a**) and (**b**), respectively, for various values of α. In (**b**), r^c_{eff} (α independent) and $\tilde{r}^c_{\text{eff}}(\alpha)$ are indicated, and it can be seen that $V = V_{\max}$ is attained at some unique value of $r_{\text{eff}} \in (\tilde{r}^c_{\text{eff}}, r^c_{\text{eff}})$

Claim 2 For all $\alpha \in (0, \pi)$, there exists a unique *critical effective radius*, $r^c_{\text{eff}} \in (0, 1)$, which satisfies $E_{\text{eff}}(\alpha, r^c_{\text{eff}}) = 1$. Moreover, r^c_{eff} is independent of α.
Claim 3 For all $\alpha \in (0, \pi)$, there exists a unique $\tilde{r}^c_{\text{eff}}(\alpha) \in (0, 1)$ such that $V(\alpha, \tilde{r}^c_{\text{eff}}(\alpha)) = V(\alpha, r^c_{\text{eff}})$ and $\tilde{r}^c_{\text{eff}}(\alpha) \neq r^c_{\text{eff}}$.

2.8 Energetic and Dynamic Stability

Based on Fig. 5 and the claims above, we can partition the rectangular α-r_{eff} domain, $(0, \pi) \times (0, 1)$, as a *phase diagram* containing four regions, L, C_L,C_R, R, in accordance with $\tilde{r}^c_{\text{eff}}$, r^c_{eff} and the value of r_{eff} where $V_{\max}$ is attained; see Fig. 6.[5] Note that only in region R are the steady states are energetically stable, since only there does $E_{\text{eff}} < 1$ hold.

Dynamic stability was also studied numerically for small perturbations [18]. The results of both stability analyses are summarized in Table 1. The steady states in region R, which are both energetically and dynamically stable, have large holes (large r_{eff} values) and a *hillock* like appearance, as has been reported experimentally [15]. The less stable steady states have smaller r_{eff} values, in accordance with [12, 13].

[5] Reproduced from [18], Zigelman, A. & Novick-Cohen, A., Critical effective radius for holes in thin films: Energetic and dynamic considerations. J. Appl. Phys. **130**, 175301 (2021), doi:10.1063/5.0053444, with the permission of AIP Publishing.

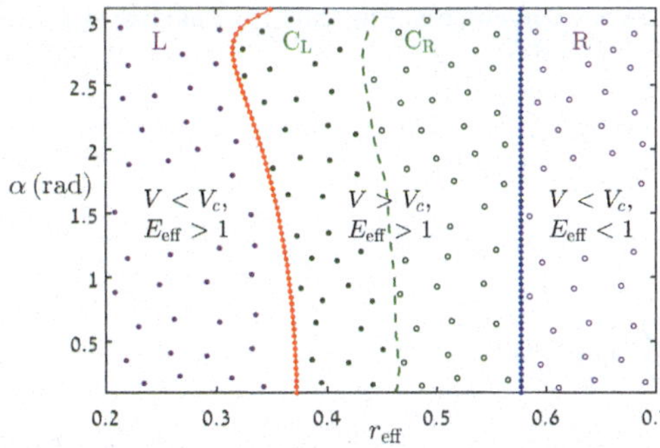

Fig. 6 The phase diagram can be partitioned into regions: L, C_L, C_R and R

Table 1 Energetic, linear, and asymptotic stability regimes for a volume preserving perturbation of a nodoid corresponding to the various regions, L C_{L}, C_{R}, and R, portrayed in Fig. 6

Region	Energy	Stability	Attractor
L	Unstable	Linearly unstable	Flat, flat+void or the second nodoid
C_{L}	Unstable	Linearly unstable	Flat, flat+void, self-intersecting[a] or the second nodoid[a]
C_{R}	Unstable	Linearly stable	
R	Stable	Stable	

[a] Self-intersecting or the second nodoid refer, respectively, to regimes in which self-intersections (loss of embeddedness) occur, and to approach to an equal volume lower energy steady state; the question mark indicates that while such behavior was not seen in our simulations, apparently it could occur

3 Summary

The stability of holes in solid thin films is crucial, as an absence of holes is necessary in some applications and holes are needed in others. Here we have primarily focused on an axisymmetric single grain model with a hole at the center [18], for which energetic and dynamic stability show that there exists a critical effective radius which is independent of the contact angle. More recently [7, 19] we studied an axi-symmetric two grain model with a central hole, with surface diffusion governing the exterior surfaces and mean curvature motion governing the grain boundary, which can exhibit grooving, wetting, dewetting, as well as void, hole, and hillock formation. Its steady states consist of coupled nodoidal and catenoidal surfaces, and it is again possible to define an *effective energy* and an *effective radius*. Modes of instability include: growth of one grain at the expense of the other, breakup induced by grooving, and hole closure. In the future we hope to include additional effects, to focus more on stability issues, in particular on the stability of arrays of holes, and to better understand the nonlinear effects not included in the small slope approximation [6, 7, 14].

Acknowledgements The authors would like to acknowledge support from the Israel Science Foundation #1200/16, #890/16. Rawan Tarabeh would like to acknowledge support from the Neubauer Family Foundation.
Competing Interests The authors have no conflicts of interest to declare that are relevant to the content of this chapter.

References

1. Almog, Y., Derkach, V., Sharma, A., Novick-Cohen, A., Greer, J., Rabkin, E.: Thermal stability of thin Au films deposited on salt whiskers. Acta Mat. **205**, 116537 (2021). https://doi.org/10.1016/j.actamat.2020.116537
2. Delaunay, C.: Sur la surface de révolution dont la courbure moyenee est constante. J. Math. Pures Appl. **6**, 309–320 (1841)
3. Derkach, V.: Surface and grain evolution in thin single- and poly-crystalline films. Ph.D. Thesis, Technion, Haifa (2017)
4. Derkach, V., McCuan, J., Novick-Cohen, A., Vilenkin, A.:Geometric interfacial motion: coupling surface diffusion and mean curvature motion. In: Mathematics for Nonlinear Phenomena-Analysis and Computation. Springer Proceedings in Mathematics & Statistics, vols. 28, 215 (2017)
5. Derkach, V., Novick-Cohen, A., Vilenkin, A., Rabkin, E.: Grain boundary migration and grooving in thin 3-D systems. Acta Mat. **65**, 194–206 (2014). https://doi.org/10.1016/j.actamat.2013.10.061
6. Goldberg, D.: Ph.D. Thesis, Technion, Haifa, in preparation
7. Golubkov, K., Novick-Cohen, A., Vaknin, Y.: Two grain axisymmetric system with a hole: steady states. Quart. Appl. Math. published electronically (2024). https://doi.org/10/1090/qam/1691
8. Kanel, J., Novick-Cohen, A., Vilenkin, A.: A traveling wave solution for coupled surface and grain boundary motion. Acta Mat. **51**, 1981–1989 (2003)
9. Mullins, W.W.: Two dimensional motion of idealized grain boundaries. J. Appl. Phys. **27**, 900–904 (1956)
10. Mullins, W.W.: Theory of thermal grooving. J. Appl. Phys. **28**, 333–339 (1957)
11. Mullins, W.W.: The effect of thermal grooving on grain boundary motion. Acta Metall. **6**, 414–427 (1958)
12. Srolovitz, D., Safran, S.: Capillary instabilities in thin films. I. Energetics. J. Appl. Phys. **60**, 247–254 (1986)
13. Srolovitz, D., Safran, S.: Capillary instabilities in thin films. II. Kinetics. J. Appl. Phys. **60**, 255–260 (1986)
14. Tarabeh, R.: Ph.D. Thesis, Technion, Haifa, in preparation
15. Thompson, V.: Solid-state dewetting of thin films. Acta Mater. Res. **42**, 399–434 (2012)
16. Wong, H., Miksis, M., Voorhees, P., Davis, S.: Capillary driven motion of solid film wedges. Acta Mater. **45**, 2477–2484 (1997)
17. Ye, J.: Fabrication of ordered arrays of micro and nanoscale features with control over their shape and size via templated solid-state dewetting. Sci. Rep. **5**, 09823 (2015)
18. Zigelman, A., Novick-Cohen, A.: Critical effective radius for holes in thin films: Energetic and dynamic considerations. J. Appl. Phys. **130**, 175301 (2021). https://doi.org/10.1063/5.0053444
19. Zigelman, A., Novick-Cohen, A.: Stability of an axisymmetric two-grain system with a hole. J. Appl. Phys. **134**, 135302 (2023). https://doi.org/10.1063/5.0153427

Choquet Capacity Networks for Random Point Process Classification and Regression

Mehran Mohammadi, Santiago Velasco-Forero, François Willot, Mateus Sangalli, and Jesus Angulo

Abstract In this study, we revisit the Choquet capacity in the framework of convolutional neural networks, in (max, +)-algebra. By incorporating a discrete and learnable Choquet capacity model, we enhance the ability to represent the spatial arrangement and density variations in random point processes of convolutional neural networks. To validate the effectiveness of our approach, numerical experiments are conducted on synthetic datasets simulating diverse spatial point patterns of the Neyman-Scott process. When compared to classical convolutional neural networks, the proposed approach exhibits comparable or improved performances in terms of classification. Superior results are also observed in regression problems involving the Neyman-Scott parameter that monitors the point patterns spatial dispersion.

Keywords Choquet capacity · Deep learning · Random point process

1 Introduction

There is a significant demand for analytical techniques that can handle vast amounts of image data in particular in engineering science [1]. For instance, images from satellites, geological maps, microscope images of metals and materials, cellular tissue, and data from computerized tomography [2] all require automatic and quantitative methods due to the sheer volume of data involved. In some scenarios, the information observed is set of coordinates in 2D or 3D, and the feature of interest depends on the spatial relationship between these points. Examples include individual brain activation maps [3, 4], RNA Localization patterns [5], positions on random sensor networks [6], realizations of random processes [7], among others. On the one hand, mathematical morphology [8–10] offers ideas that are applicable in this context, particularly due to its interpretation in terms of the size/shape relationship of the object of interest. Cord and co-authors combined textural descriptors and statistical learning

M. Mohammadi · S. Velasco-Forero (✉) · F. Willot · M. Sangalli · J. Angulo
Center for Mathematical Morphology PSL, 35 rue Saint Honoré, 77305 Fontainebleau, France
e-mail: santiago.velasco@minesparis.psl.eu

F. Willot et al. (eds.), *Continuum Models and Discrete Systems*,
Springer Proceedings in Mathematics & Statistics 457,
https://doi.org/10.1007/978-3-031-58665-1_18

to address segmentation problems [11] and extract defects merged in complex real textures [12]. On the other hand, data-driven learning methods have changed the way image problems are solved. Especially the deep models allow to solve recognition problems with performance sometimes superior to humans [13]. In this paper, we follow the results of [14–17], to present a new application of morphological models learned from data for the characterization of random point processes. The Neyman-Scott point process is presented in Sect. 2. The proposed Choquet Capacity model is described in Sect. 3. Section 4 demonstrate the benefits of the proposed model in comparison with classical convolutional neural networks (CNNs) for classification and regression problems on images generated by Neyman-Scott point process. We conclude in Sect. 5.

2 Neyman-Scott Point Processes

In the present work, we consider Neyman-Scott point processes [18–21] as a way to generate point patterns that are statistically different from the classical Poisson model of points [22–24]. The Neyman-Scott model, like Poisson's one, is a particular type of Boolean random set [8, 25–27]. Poisson point processes on $\mathbb{R}^2$ depend on a single parameter θ, called the *intensity*. The intensity θ represents the mean number number of points per unit area. In the Poisson point model, the number of points in any two disjoint domains are independent random variables which lead to the following Poisson-distributed law for the number of points $\eta(\Omega)$ in a domain Ω,

$$\mathbb{P}\{\eta(\Omega) = k\} = \frac{(\theta|\Omega|)^k}{k!} \exp^{-\theta|\Omega|}, \tag{1}$$

where $|\Omega|$ is the area of Ω. This leads to the formula for the *Choquet capacity* $T(K)$ of the Poisson point process $X = \{x_1, x_2, \ldots\}$:

$$T_X(K) := \mathbb{P}(X \cap K \neq \emptyset) = 1 - \mathbb{P}\{K \subseteq X^c\} = 1 - \exp^{-\theta|K|}, \tag{2}$$

where K is a compact set [28].

Neyman-Scott point processes are different from Poisson point processes in that they generate *clustered* sets of points. This is achieved in two steps. In the first step, a standard Poisson point process of intensity θ_p is generated, leading to a set of *parent points* $x^{(p)} \in \mathbb{R}^2$ ($p \in \mathbb{N}$). In the second step, we associate each parent point with a set of n_p *daughter points* $x_1^{(p)}, \ldots, x_{n_p}^{(p)}$. The Neyman-Scott point process is made of all daughter points; parent points are not included. Neyman-Scott models are accordingly parametrized by θ_p, the probability law for the n_p and that of the daughter points, given in term of the multivariate probability law for the $x_i^{(p)} - x^{(p)}$. For simplicity, we consider a constant distribution $n_p \equiv n$. We also assume that the vectors $x_i^{(p)} - x^{(p)}$ are i.i.d. variables drawn from a multivariate random distribution

Fig. 1 Realizations of a Poisson point process (**a**) and Neyman-Scott models (**b–c**) with the same density of points. **b** Parameters $r_m = 0.1, \alpha = 100$. **c** $r_m = 0.1, \alpha = 3$

in $\mathbb{R}^2$, with uniform angular distribution in $[0; 2\pi]$. Distances from parent to daughter points are presently drawn from a Pareto law with parameters α, r_m such as:

$$\mathbb{P}\{|x_i^{(p)} - x^{(p)}| > r\} = \left(\frac{r_m}{r}\right)^{\alpha} H(r - r_m), \qquad \alpha > 0, \quad r_m > 0, \tag{3}$$

where $H(\cdot)$ is the Heavyside function, equal to one on $\mathbb{R}^+$ and zero elsewhere. Here r_m controls the *typical size of the clusters* whereas the parameter α monitors the *spatial dispersion*. Thus, r has infinite variance when $\alpha \leq 2$ and infinite mean when $\alpha \leq 1$. Examples of realizations are shown in Fig. 1. The points spatial dispersion in the Neyman-Scott model is markedly more heterogeneous when α is large (see Fig. 1b). When α approaches one, the Neyman-Scott and Poisson models are difficult to distinguish (see Fig. 1).

Importantly, the Choquet-Matheron-Kendall theorem established independently by [29, 30] states that any stationary and ergodic random set, such as Poisson or Neyman-Scott point processes, is completely determined by the functional $T(K)$. In the next section, we propose a model that uses (2) to find optimal K in a discrete setting (denoted as K), in the sense of learning from labeled data for supervised problems. Note that this learning process makes use of convolutional layer where $\mathbf{K}$ are optimized in the real set. Accordingly, we propose regularization terms to encourage the network to learn binary kernels.

3 Choquet Capacity Models

3.1 Discrete Random Set

Usually random sets are defined in the sense of Euclidean geometry, as subsets of $\mathbb{R}^d$. However, in many applications, as in image processing, the domain is discrete, defined in pixels structures in 2D, $\Omega = \mathbb{Z}^2$. The discrete random set (DRS) X on $\mathbb{Z}^2$

is now defined by $\mathsf{X} := \{(i, j) \in \mathbb{Z}^2 : x_{i,j} = 1\}$. The *discrete-capacity functional* of a random set X is defined by [30]:

$$T_{\mathsf{X}}(\mathsf{K}) := \mathbb{P}(\mathsf{X} \cap \mathsf{K} \neq \emptyset) = 1 - \mathbb{P}\{\mathsf{K} \subseteq \mathsf{X}^c\} := 1 - Q_{\mathsf{X}}(\mathsf{K}) \tag{4}$$

for every $\mathsf{K} \in S$ the collection of all bounded subsets of $\mathbb{Z}^2$. From (4), it is easy to see that: a) $0 \leq T_{\mathsf{X}}(\mathsf{K}) \leq 1$ for every $\mathsf{K} \in S$, b) $T_{\mathsf{X}}(\emptyset) = 0$, and c) $T_{\mathsf{X}}(\mathsf{K}_1) \leq T_{\mathsf{X}}(\mathsf{K}_2)$, $\forall \mathsf{K}_1, \mathsf{K}_2 \in S$ such that $\mathsf{K}_1 \subseteq \mathsf{K}_2$. Other important properties can be shown for DRS [31, 32]. According to the Choquet-Kendall-Matheron theorem, any discrete random set X is uniquely characterized by its discrete-capacity functional $T(\mathsf{K})$, defined over all bounded subsets $\mathsf{K} \in S$ of $\mathbb{Z}^2$. Furthermore, we note that:

$$\mathbb{P}(\mathsf{X} \cap \mathsf{K} \neq \emptyset) = \mathbb{P}(x \in \delta_{\check{\mathsf{K}}}(\mathsf{X})) \tag{5}$$

where $\check{\mathsf{K}} = \{-x; x \in \mathsf{K}\}$ denotes the set K mirrored w.r.t. the origin. The capacity in (4) is accordingly obtained from the morphological dilation $\delta_{\check{\mathsf{K}}}(X)$ of X by K [27, 33].

Hereafter, the points $(i, j) \in \mathbb{Z}^2$ of the DRS X are obtained from the set X in $\mathbb{R}^2$ so that: $(i, j) \in \mathsf{X}$ whenever $([i; i+1[\times[j; j+1[) \cap X \neq \emptyset$. Accordingly, the DRS X is a partial representation of the Neyman-Scott point process defined by Eq. (3). The representation of X by X is however excellent when $r_m \gg 1$, i.e. when r_m is much larger than the pixel size.

3.2 Choquet Capacity Layer

This section describes a layer that can be used for the computation of (5), in the context of neural networks. To do so, we use the extension of convolutional models in the $(\bigvee, +)$ algebra, called max-plus convolutions [34], or *dilation layer* [15]. The dilation layer replaces discrete classical convolutions and can be compared to fuzzy measures in so-called fuzzy integral neural networks [35]. For an input image f, the dilation layer is defined as:

$$\delta_{\mathbf{K}}(f)(x) := \bigvee_{y \in \mathbb{Z}^2} [f(x-y) + \mathbf{K}(y)] = \bigvee_{y \in \mathbb{Z}^2} [f(y) + \mathbf{K}(x-y)] \tag{6}$$

where $\mathbf{K}$ is a matrix of real values called *kernel* or *structuring function*. In the cases where the image f represents a DRS (denoted by f), and the kernel $\mathbf{K}$ takes on values in $\{-1; 0\}$,[1] we have that $\mathsf{f}(y) + \mathbf{K}(x-y) = 0$ if $\mathbf{K}(x-y) \leq \mathsf{f}(y)$ ($y \in \mathbb{Z}^2$) and 1 elsewhere. By sliding the kernel through the domain and calculating its average, assuming ergodicity, we obtain an estimation for the value of $Q_{\mathsf{f}}(\mathbf{K})$ in (4):

[1] Hereafter, we consider input images lying in the interval [0, 1], so that in practice, -1 plays the same role as $-\infty$ for kernel.

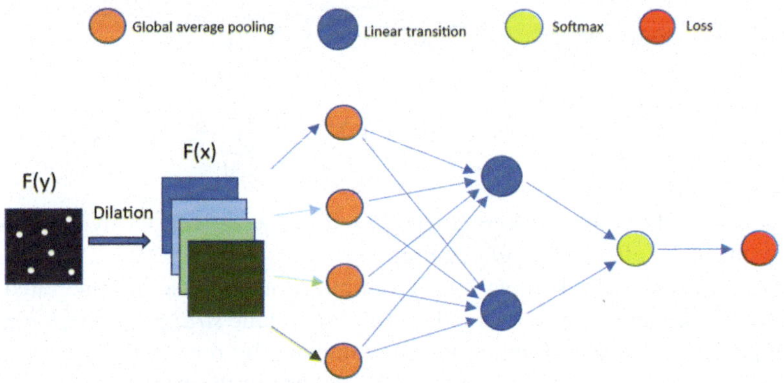

Fig. 2 Schematic of the proposed choquet capacity model

$$\hat{Q}_{\mathsf{f}}(\mathbf{K}) := \frac{1}{M}\sum_{j=1}^{M} \delta_{\mathbf{K}}(\mathsf{f})(x_j) \tag{7}$$

where M is the number of pixels in the image. In the following, Eq. (7) is called a *Choquet Capacity layer* with convolutional kernels $\mathbf{K}$.

Hereafter, we propose to study the performance of the Choquet Capacity layers to learn n_k kernels, $\mathbf{K}s := \{\mathbf{K}_i\}_{i=1}^{n_k}$, in the framework of neural networks. Accordingly, for each input image f, a vector containing the n_k values computed by (7) is used as features, transformed linearly by a dense layer (or many for multilayer perceptrons (MLP) models) with parameter $\mathbf{W}$, and finally passed trough an activation function adapted to the regression or classification problem. Figure 2 shows a schematic of the proposed Choquet Capacity model for a classification problem. The classical training mechanism on neural networks seeks to find the optimal model's parameters $(\mathbf{W}, \mathbf{K}s)$,

$$\underset{(\mathbf{K}s,\mathbf{W})}{\arg\min} \left\{ \sum_{\ell=1}^{N} \mathtt{loss}(Y_\ell, \widehat{Y_\ell}) + \mathtt{Reg}(\mathbf{W}, \mathbf{K}s) \right\} \tag{8}$$

where $\widehat{Y_\ell}$ is the prediction of the network for the ℓ-th sample, Y_ℓ is the ground truth, N is the number of samples and $\mathtt{Reg}(\mathbf{W}, \mathbf{K}s)$ is a regularization term on the parameters of the model. In its simplest form, the parameters $(\mathbf{W}, \mathbf{K}s)$ are initialized to random values and updated by means of a stochastic gradient descend, the gradient being computed by backpropagation [13]. Note that the shape K in (5) is a bounded subset, whereas each of the matrix $\mathbf{K}$ in (6) is a max-plus convolutional kernel with real values. Therefore, we propose two alternatives that address this difference during the training of the model:

(i) *Regularization*: the term $\mathtt{Reg}$ in (8) is set to penalize kernel values that are far from a binary equivalent, in our case of max-plus algebra on binary images, 0 or -1. Two different regularization functions are considered for R:

$$\mathtt{Reg}_1(\mathbf{K}s) = \lambda_1 \sum_{i=1}^{n_k} (\mathbf{K}_i + 1) + \lambda_2 \sum_{i=1}^{n_k} (\mathbf{K}_i + 1)^2, \tag{9a}$$

$$\mathtt{Reg}_2(\mathbf{K}s) = \lambda_1 \sum_{i=1}^{n_k} (\mathbf{K}_i + 1)\mathbf{K}_i + \lambda_2 \sum_{i=1}^{n_k} (\mathbf{K}_i + 1)^2 \mathbf{K}_i^2, \tag{9b}$$

where n_k in the number of kernels and $\lambda_1 \geq 0$ and $\lambda_2 \geq 0$ are the hyperparameters to control the effect of regularization.

(ii) *Quantization* of the kernels during optimization. We use an integrated value filter to enforce kernel values in $[-1, 0]$. We use the *straight-through estimator* proposed in [36, 37]. Predictions are computed by rounding the kernel to the closest integer (-1 or 0) whereas the gradient in the backpropagation is left unchanged.

The different approaches above are compared to one another and to classical convolutional networks in the next section. We would like to study the effect of including more layers into the model. This is called the *depth* of the network in deep learning [38]

4 Numerical Experiments

In the present section, we quantify the performances of the various convolutional networks. Two problems are addressed: first, that of classification of realizations of Neyman-Scott point patterns w.r.t. Poisson; second, the prediction of the dispersion parameter α in a regression problem.[2]

4.1 Classification of Point Pattern

In this experiment, we set the image size to 500×500 pixels and $\theta = 0.05$ in number of points per pixel unit. We choose $n = 50$ and let $\theta_p = \theta/n$ so that the density of points is the same irrespective of the model chosen. We also fix $r_m = 0.1 \times 500 = 50$ and let α varies between 1 and 100. All point patterns are statistically homogeneous. We perform the classification task on two classes of Poisson and Neyman-Scott point patterns with different values of the parameter α and conduct a comparative analysis with the CNNs. The data set contains 500 images for each class, with a train-to-validation ratio of 4:1. For simplicity, we use one layer in each neural network, with four filters of size 7×7 pixels. For supervised classification, a common choice of loss function is the *binary cross-entropy* also known as *log loss*, defined by

[2] The source code of this section is available at https://github.com/Jacobiano/ChoquetLayer.

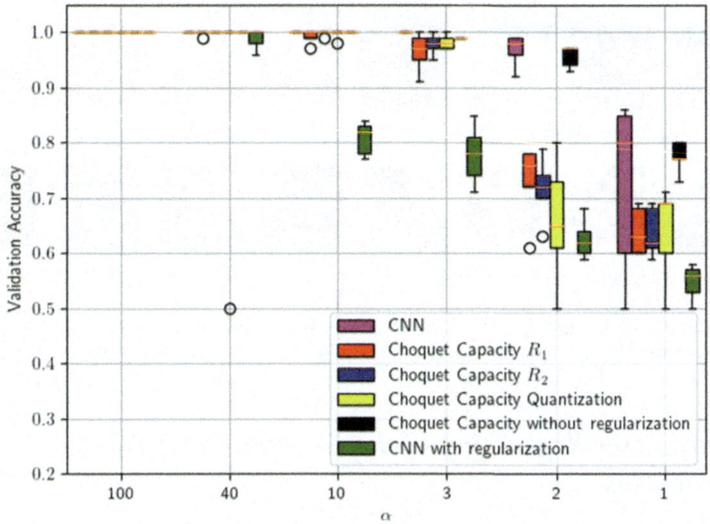

Fig. 3 Validation accuracy for the different neural network models in the two-classes classification problem (Poisson and Neyman-Scott point processes) versus the parameter α that monitors spatial dispersion

$$\texttt{logloss}(Y, \widehat{Y}) := -\frac{1}{N}\sum_{\ell=1}^{N}\left[Y_\ell \log q_0^\ell + (1 - Y_\ell)\log q_1^\ell\right] \tag{10}$$

where (q_0^ℓ, q_1^ℓ) denote the prediction of the network for the ℓ-th samples, Y_ℓ the ground truth of the ℓ-th samples and N is the number of samples. The models are trained to predict the class between images produced by Poisson processes and Neyman-Scott models with different values of α parameter. Figure 3 shows the boxplot of validation accuracy for models training from different initialization values. Interestingly, the Choquet Capacity layer demonstrates competitive performance, especially when the parameter α decreases, so that realizations of the Neyman-Scott point pattern are similar to that of a Poisson distribution. When regularization is applied, the model tends to have values $\{-1, 0\}$, but they are still values that prevent us from indicating that the solution is binary. For quantization, $\mathbf{K}s$ contain only values in $\{-1, 0\}$ in Fig. 4c, although no distinctive shape is observed. However, performance significantly degraded, when regularization or quantization is used, specially in more difficult cases.

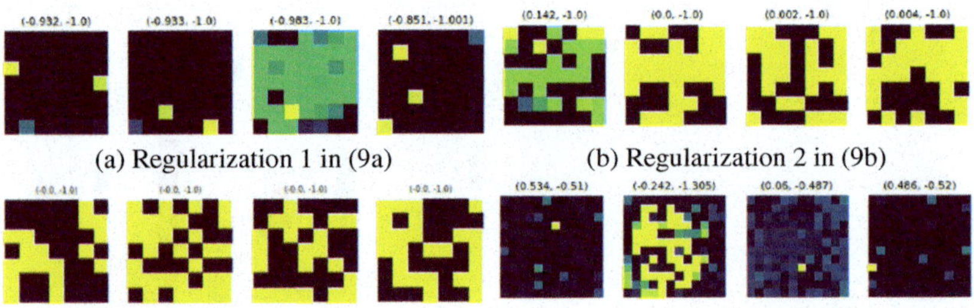

(a) Regularization 1 in (9a) (b) Regularization 2 in (9b)

(c) Quantization operator (d) No regularization

Fig. 4 Example of learned kernel $\mathbf{K}_S = \{\mathbf{K}_1, \ldots, \mathbf{K}_4\}$ using different proposed regularization alternatives of the choquet capacity layer. Max and min values indicated at the top of each image

4.2 Parameter Regression of the Neyman-Scott Point Processes

In the present subsection, we compare the performances of different neural network models for predicting the value of the parameter α, in a regression problem. We set the image size to 128×128 pixels, the density of points to $\theta = 100/(128 \times 128)$ and $r_m = 0.02$. We also let α vary between 0 and 9. This last one is the value that we would like to predict. For the regression problem, the *mean absolute error* is the mean absolute difference between true value and prediction:

$$\mathtt{MAE}(Y, \widehat{Y}) := -\frac{1}{N}\sum_{\ell=1}^{N} |Y_\ell - \widehat{Y}_\ell| \tag{11}$$

is used as loss function. During training the α values are scaled to the interval [0, 1] as it is standard practice in regression models. As a first example, we compare the result between Choquet Capacity and a classical convolutional layer followed by a multi-layer perceptron that consists of two layers of dimension ten and batch normalization [39], followed by a sigmoid activation, see Fig. 5. Note that the convolutional layer and our proposition have exactly the same number of parameters. In this case, our Choquet Capacity model obtains predictions that are clearly quantitatively better.

As a second illustration in the regression case we studied the effect of network depth, prior to the use of the Choquet Capacity model. Usually, increasing both depth and width (number of filters per layer) helps to improve the model performance until the number of parameters becomes too high and stronger regularization is needed [40]. For that, we consider convolutional layers with kernel size of 3×3 pixels and 48 filters are concatenated before the use of global pooling and the MLP. This architecture is represented in Fig. 6. We compare different convolutions in the last layer before the global pooling: proposed Choquet Capacity model, classical CNN, and depthwise convolutional (DWCNN). The latter is included because it contains exactly the same number of parameters compared to our Choquet Capacity

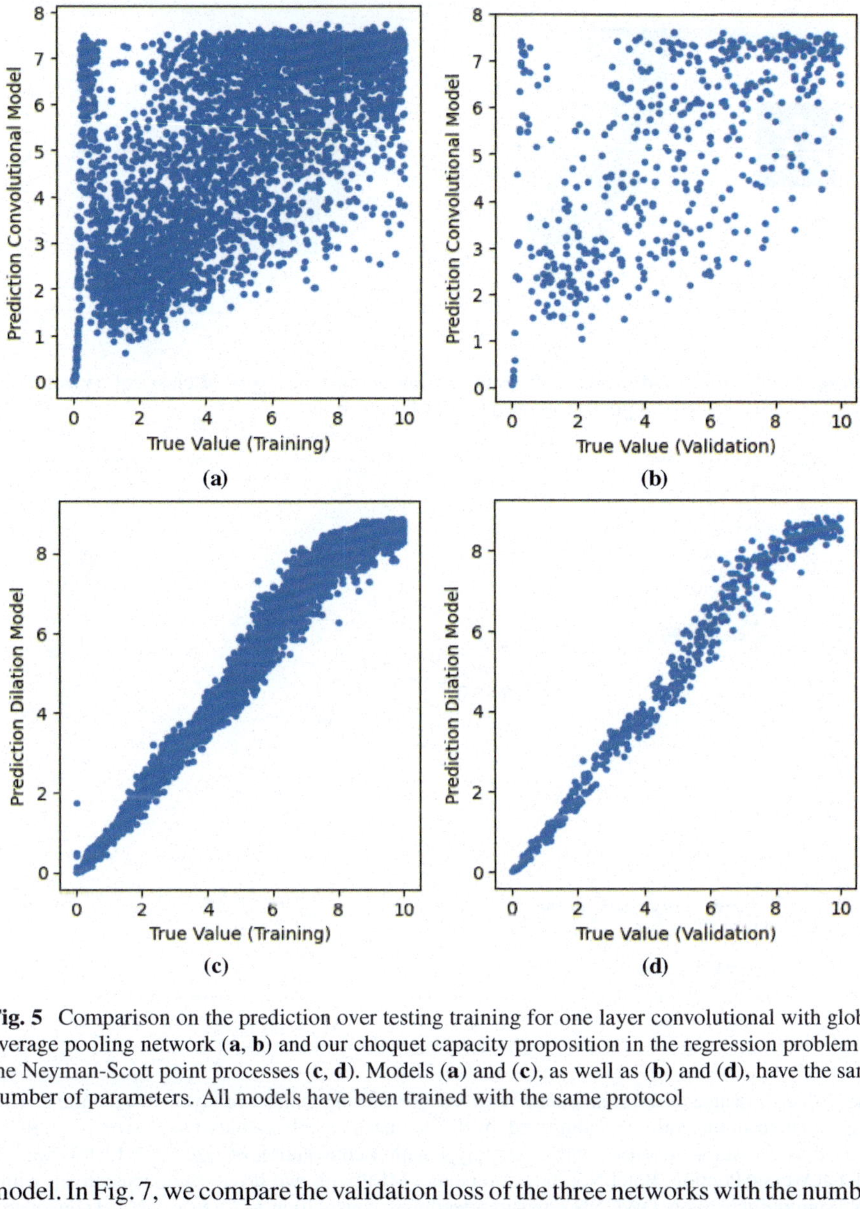

Fig. 5 Comparison on the prediction over testing training for one layer convolutional with global average pooling network (**a**, **b**) and our choquet capacity proposition in the regression problem of the Neyman-Scott point processes (**c**, **d**). Models (**a**) and (**c**), as well as (**b**) and (**d**), have the same number of parameters. All models have been trained with the same protocol

model. In Fig. 7, we compare the validation loss of the three networks with the number of parameters in each model shown in logarithmic scale. The number of degrees of freedom is increased by augmenting the depth from one to seven in the convolutional part in each architecture, leading to models with deeper feature extractions. The Choquet Capacity model improves the regression results in all explored cases, for a fixed number of parameters, compared to DWCNN. Finally, Fig. 7 also shows than comparable performances to the Choquet model are achieved by classical CNNs at the cost of a much higher number of parameters.

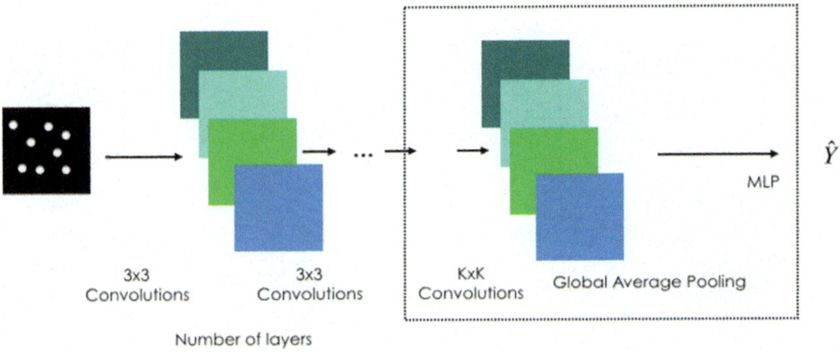

Fig. 6 Architecture used in Sect. 4.2, with varying number of layers of classical convolutional layers of kernel size 3 × 3 pixels followed by ReLU activation

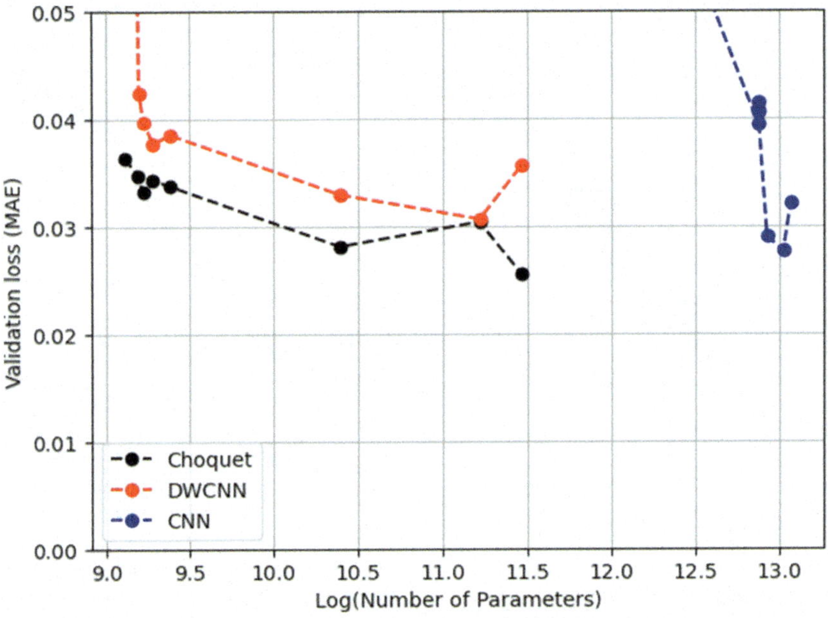

Fig. 7 Performance of different neural networks for varying number of layers in each model (See Fig. 6). Three architectures are compared. In all cases each layer has 48 kernels of size 3 × 3, ReLU is used as activation function, and a final layer with a convolution of size 13 × 13 followed by GlobalAveragePooling, BatchNormalization and a MLP with two layers of dimension ten with a final sigmoid activation. DWCNN refers to a depthwise convolution, CNN to a classical convolution and choquet capacity to the max-plus convolutions in Sect. 3. DWCNN and choquet capacity models have exactly the same number of parameters

5 Conclusion

The results obtained in this study are based on a novel approach that makes use of dilation layers with global pooling, inspired from the definition of the Choquet capacity of random sets. Numerical experiments performed on synthetic datasets which simulate diverse Neyman-Scott point patterns, demonstrate the efficiency of the method, as well as seemingly superior performances in particular in regression problems involving the dispersion parameter in the Neyman-Scott point process. Compared to classical convolutions, the Choquet Capacity networks are somewhat more difficult to train, due to the presence of a nonlinear convolution filters. Yet they are versatile in that they may be combined with classical convolutional networks. As a negative result, the alternatives studied to motivate the estimation of binary kernels result in very important losses in the performance of the models. This topic, as well as the adequacy of Choquet capacity networks to analyse real-world data, will be addressed in future research works.

Acknowledgements This work was granted access to the HPC resources of IDRIS under the allocation 2023-AD011012212R2 made by GENCI. The authors gratefully acknowledge the financial support of Institut Carnot (grant 220000496) and thank Thomas Walter and Christian Lantuéjoul for helpful discussions.
Competing Interests The authors have no conflicts of interest to declare that are relevant to the content of this chapter.

References

1. Chinesta, F., Cueto, E., Abisset-Chavanne, E., Duval, J.L., Khaldi, F.E.: Virtual, digital and hybrid twins: a new paradigm in data-based engineering and engineered data. Arch. Comput. Methods Eng. **27**, 105–134 (2020)
2. Etiemble, A., Douillard, T., Valentin, O., Badot, J.-C., Maire, E.: Numerical prediction of multiscale electronic conductivity of lithium-ion battery positive electrodes. J. Electrochem. Soc. **166**(8), A1692 (2019)
3. Worsley, K.J., Marrett, S., Neelin, P., Evans, A.: Searching scale space for activation in pet images. Hum. Brain Mapp. **4**(1), 74–90 (1996)
4. Poline, J.-B., Mazoyer, B.: Analysis of individual brain activation maps using hierarchical description and multiscale detection. IEEE Transac. Med. Imaging **13**(4), 702–710 (1994)
5. Imbert, A., Mueller, F., Walter, T.: Pointfish: learning point cloud representations for RNA localization patterns. In: ECCV, Springer, Berlin, pp. 487–502 (2022)
6. Ram, S.S., Manjunath, D., Iyer, S.K., Yogeshwaran, D.: On the path coverage properties of random sensor networks. IEEE Transac. Mob. Comput. **6**(5), 494–506 (2007)
7. Pawlasová, K., Dvořák, J.: Supervised nonparametric classification in the context of replicated point patterns. Image Anal. Ster. **41**(2), 57–109 (2022)
8. Serra, J.: Image Analysis and Mathematical Morphology. Academic Press Inc. (1983)
9. Soille, P.: Morphological Image Analysis: Principles and Applications, 2nd edn. Springer, Berlin, Heidelberg (2003)
10. Najman, L., Talbot, H.: Mathematical Morphology: From Theory to Applications. Wiley (2013)

11. Cord, A., Jeulin, D., Bach, F.R.: Segmentation of random textures by morphological and linear operators. In: Proceedings of the 8th International Symposium on Mathematical Morphology, Rio de Janeiro, Brazil, 10–13 Oct 2007, pp. 387–398 (2007)
12. Cord, A., Bach, F., Jeulin, D.: Texture classification by statistical learning from morphological image processing: application to metallic surfaces. J. Microsc. **239**(2), 159–166 (2010)
13. Le Cun, Y., Bengio, Y., Hinton, G.: Deep learning. Nature **521**(7553), 436–444 (2015)
14. Sangalli, M., Blusseau, S., Velasco-Forero, S., Angulo, J.: Scale equivariant neural networks with morphological scale-spaces. In: International Conference on Discrete Geometry and Mathematical Morphology, Springer, 2021, pp. 483–495 (2021)
15. Velasco-Forero, S., Pagès, R., Angulo, J.: Learnable empirical mode decomposition based on mathematical morphology. SIAM J. Imag. Sci. **15**(1), 23–44 (2022)
16. Velasco-Forero, S., Rhim, A., Angulo, J.: Fixed point layers for geodesic morphological operations. In: British Machine Vision Conference (2022)
17. Mondal, R., Santra, S., Mukherjee, S., Chanda, B.: Morphological network: how far can we go with morphological neurons?. In: 33rd British Machine Vision Conference 2022, BMVC 2022, London, UK, 21–24 Nov 2022. BMVA Press (2022)
18. Waagepetersen, R.P.: An estimating function approach to inference for inhomogeneous Neyman-Scott processes. Biometrics **63**(1), 252–258 (2007)
19. Vere-Jones, D.: Stochastic models for earthquake occurrence. J. Roy. Stat. Soc.: Ser. B (Methodol.) **32**(1), 1–45 (1970)
20. Westcott, M.: Results in the asymptotic and equilibrium theory of Poisson cluster processes. J. Appl. Probab. **10**(4), 807–823 (1973)
21. Ripley, B.D.: Modelling spatial patterns. J. Roy. Stat. Soc.: Ser. B (Methodol.) **39**(2), 172–192 (1977)
22. Daley, D., Vere-Jones, D.: An Introduction to the Theory of Point Processes, vol. I: Elementary Theory and Methods, 2nd Edition, Springer, Berlin (2003)
23. Daley, D., Vere-Jones, D.: An introduction to the theory of point processes: vol. II: General Theory and Structure. Springer Science & Business Media (2007)
24. Moller, J., Waagepetersen, R.P.: Statistical Inference and Simulation for Spatial Point Processes. CRC Press (2003)
25. Stoyan, D., Kendall, W.S., Mecke, J.: Stochastic Geometry and Its Applications, 2nd edn. Wiley, Chichester (1995)
26. Jeulin, D.: Morphological Models of Random Structures. Springer, Berlin (2021)
27. Serra, J.: Image Analysis and Mathematical Morphology, vol. 2, Theoretical Advances. Academic Press (1988)
28. Choquet, G.: Theory of capacities. In: Annales de l'institut Fourier, vol. 5, pp. 131–295 (1954)
29. Kendall, D.G.: Foundation of a theory of random sets, Stochastic geometry (1974)
30. Matheron, G.: Random Sets and Integral Geometry. Probability and Statistics Series. Wiley (1974)
31. Singh, B., Siddiqi, M.: Discrete random functions: modeling and analysis using mathematical morphology. In: Mathematical Morphology and its Applications to Image and Signal Processing, pp. 65–72 (1996)
32. Goutsias, J.: Morphological analysis of discrete random shapes. J. Math. Imaging Vis. **2**, 193–215 (1992)
33. Jeulin, D.: Random texture models for material structures. Stat. Comput. **10**, 121–132 (2000)
34. Akian, M., Bapat, R., Gaubert, S., et al.: Max-plus algebra. Handbook of Linear Algebra, vol. 39 (2006)
35. Islam, M.A., Anderson, D.T., Pinar, A.J., Havens, T.C., Scott, G., Keller, J.M.: Enabling explainable fusion in deep learning with fuzzy integral neural networks. IEEE Trans. Fuzzy Syst. **28**(7), 1291–1300 (2019)
36. Bengio, Y., Léonard, N., Courville, A.: Estimating or propagating gradients through stochastic neurons for conditional computation (2013). arXiv:1308.3432
37. Yin, P., Lyu, J., Zhang, S., Osher, S., Qi, Y., Xin, J.: Understanding straight-through estimator in training activation quantized neural nets. In: ICLR (2019)

38. Goodfellow, I., Bengio, Y., Courville, A.: Deep Learning. MIT Press (2016)
39. Santurkar, S., Tsipras, D., Ilyas, A., Madry, A.: How does batch normalization help optimization? In: Advances in Neural Information Processing Systems, vol. 31 (2018)
40. He, K., Zhang, X., Ren, S., Sun, J.: Deep residual learning for image recognition. In: Proceedings of the IEEE CVPR, pp. 770–778 (2016)

Simulation of Microstructures and Machine Learning

Katja Schladitz, Claudia Redenbach, Tin Barisin, Christian Jung, Natascha Jeziorski, Lovro Bosnar, Juraj Fulir, and Petra Gospodnetić

Abstract Machine learning offers attractive solutions to challenging image processing tasks. Tedious development and parametrization of algorithmic solutions can be replaced by training a convolutional neural network or a random forest with a high potential to generalize. However, machine learning methods rely on huge amounts of representative image data along with a ground truth, usually obtained by manual annotation. Thus, limited availability of training data is a critical bottleneck. We discuss two use cases: optical quality control in industrial production and segmenting crack structures in 3D images of concrete. For optical quality control, all defect types have to be trained but are typically not evenly represented in the training data. Addi-

K. Schladitz (✉) · J. Fulir · P. Gospodnetić
Fraunhofer ITWM, Image Processing, Kaiserslautern, Germany
e-mail: katja.schladitz@itwm.fraunhofer.de

J. Fulir
e-mail: juraj.fulir@itwm.fraunhofer.de

P. Gospodnetić
e-mail: petra.gospodnetic@itwm.fraunhofer.de

C. Redenbach · T. Barisin · C. Jung · N. Jeziorski
RPTU, Department of Mathematics, Kaiserslautern, Germany
e-mail: claudia.redenbach@rptu.de

T. Barisin
e-mail: barisin@rptu.de

C. Jung
e-mail: christian.jung@rptu.de

N. Jeziorski
e-mail: n.jeziorski@rptu.de

L. Bosnar
RPTU, Department of Computer Science, Kaiserslautern, Germany
e-mail: bosnar@rptu.de

F. Willot et al. (eds.), *Continuum Models and Discrete Systems*,
Springer Proceedings in Mathematics & Statistics 457,
https://doi.org/10.1007/978-3-031-58665-1_19

tionally, manual annotation is costly and often inconsistent. It is nearly impossible in the second case: segmentation of crack systems in 3D images of concrete. Synthetic images, generated based on realizations of stochastic geometry models, offer an elegant way out. A wide variety of structure types can be generated. The within structure variation is naturally captured by the stochastic nature of the models and the ground truth is for free. Many new questions arise. In particular, which characteristics of the real image data have to be met to which degree of fidelity.

Keywords Synthetic training data · Machine learning models · Consistent ground truth · Synthetic image data · Visual surface inspection · Surface texture modeling · Surface defect modeling · Inspection planning

1 Introduction

Machine learning offers attractive solutions to challenging image processing tasks. Tedious development and parametrization of algorithmic solutions can be replaced by training a convolutional neural network (CNN) or a random forest. Pre-trained CNN architectures are easily available. Moreover, machine learning (ML) methods have higher potential to generalize than classical, tailor-suited solutions. However, ML methods rely on huge amounts of representative image data along with a ground truth, usually obtained by manual annotation. Limited availability of training data is therefore a critical bottleneck in a variety of image processing problems.

Here, we discuss three use cases of synthetic images for ML solutions of image processing challenges: reconstruction of highly porous spatial structures from stacks of scanning electron microscopy (SEM) images of samples serially sectioned by a focused ion beam (FIB), segmenting crack structures in 3D images of concrete, and optical quality control in industrial production.

We start off with the reconstruction of highly porous structures from FIB-SEM stacks. This is a semantic segmentation task rendered exceptionally demanding by the high depth of view of the SEM leading to so-called shine-through-artifacts in the 3D image. This challenge is a rather old example of successful use of synthetic image data derived from stochastic geometry models [31, 32]. FIB-SEM is an established imaging method for materials' structures at the nano-scale [19]. However, deriving a truly spatial representation of the solid component of a highly porous material from a stack of SEM images of consecutive slices involves a tedious segmentation (or reconstruction) step. Several solutions combining morphological transformations, thresholding, and exploiting the specific anisotropic nature of the shine-through artifacts have been suggested [31, 38], but tailor-made algorithms prevail. For instance in [28], a tailor-suited decurtaining exploiting a homogeneous layer of the imaged thin tantalum films is followed by a difference of Gaussians, morphological fill-holes, and a white-top hat with elongated structuring element to finally enable segmentation by global gray value thresholding. Adaption of reconstruction algorithms like [31, 38] to new structures or materials is still challenging. Synthetic training data can be generated [32] and have been used successfully to train a 3d U-net [16]. We summarize the findings from [16] as our use case one below. Generalization to unseen data is subject of current research.

The second use case is motivated by the rise of Gulliver at RPTU in Kaiserslautern—a dedicated computed tomography (CT) device for observing concrete beams under load [34, 35]. Concrete beams of up to 6 m length will be imaged there spatially while undergoing bending tests. Gulliver will generate up to $10\,000 \times 10\,000 \times 2\,000$ voxels per scan and several scans per in-situ bending experiment. Consistent manual annotation of cracks in CT images of concrete is particularly challenging if not even impossible due to the cracks being very thin and dark. Moreover, for the Gulliver generated images, interactive crack detection and segmentation is completely out of reach due to the mere amount of data. Hence, algorithms for automatic crack segmentation have been collected from previous work [14, 29], modified, and newly developed [2]. For a fair quantitative comparison as well as for the training of ML models, CT images of cracked concrete samples with a ground truth were needed. Simulated crack structures [23] impressed on real CT images of concrete samples yield both—a ground truth and a degree of realism allowing for generalization to real crack structures. Additional effort is however needed to adapt to the strong variation in thickness of real crack structures [3, 21]. The scale invariant RieszNet [4] is the most innovative answer to this challenge.

In the previous two cases, the images are acquired by highly controlled, calibrated methods and are used for further analysis of material properties. The third case focuses only on the 2D imaging of the object's surface, using non-penetrating light. Here the images are acquired during an industrial production process which may vary over time and are further used for automated surface quality inspection. Visual surface inspection is a popular research field due to its relatively low cost of data acquisition, potentially leading to an abundance of data [11, 13, 40, 42, 43]. However, the obtained data is extremely diverse due to the large variety of production processes and object geometries, which impedes the development of one-fits-all solutions. Instead, dedicated data has to be collected and annotated, and new ML model must be trained for each specific use-case. In addition, visual appearance of the surface defects to be detected as well as their occurrence frequency within the collected dataset vary enormously. Surface defects are often classified based on their look, shape, tactile feedback or material [13]. However, the more safety-critical a defect is, the less frequently it is observed because it is avoided by all means in the manufacturing process. This introduces a strong imbalance in the dataset in the amount of non-defected and defected samples as well as between defect types. Additionally, manual annotation and inspection planning are costly processes limiting the available annotated data and inducing label inconsistencies due to subjective decisions or just plain fatigue of the inspector.

Generation of synthetic image datasets offers an elegant way out since a large variety of events can be simulated, together with a precise ground truth annotation. Despite the wide range of synthetic data applications [26], the tradeoffs have not yet been systematically evaluated. We observe two main research directions in synthetic data generation: **generative**, using AI models to produce synthetic data, and **rule-**

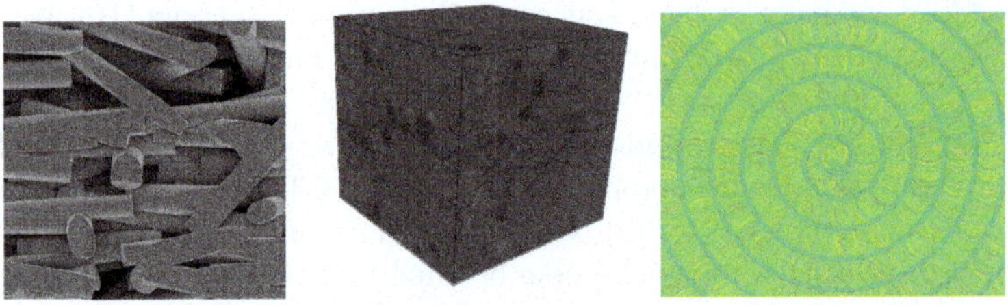

Fig. 1 Realizations of stochastic geometry models. From left to right: Simulated SEM image of a Boolean model of cylinders, Sect. 2. CT image of a concrete sample with impressed crack generated based on a Voronoi tessellation, Sect. 3. Color coded height map of a simulated milling pattern on a metal surface, Sect. 4

based, using computer graphics simulation based on physics described by a well-defined set of rules. With a generative approach, we control a single realization, whereas with a rule- based approach, we control the realization context, giving us far better control, reproducibility, and reliability.

Generation of images using stochastic geometry models, see Fig. 1, is a rule-based approach, where the stochastic geometry model parameters are used to control the realization context. This enables generation of a wide variety of structure types. The within-structure variation is naturally captured by the stochastic nature of the models. The ground truth is obtained for free since we know exactly where and how structure is placed in the scene, making it almost effortless to produce the labels. At the same time, many new questions arise. In particular, it is not clear at all, which characteristics of the real image data have to be met to which degree of fidelity. To address these questions at least partially, we describe our approaches to generate realistic optical images of metallic surfaces and computed tomography (CT) images of concrete with cracks. For the latter, ML methods are trained using dedicated crack structure models. Moreover, a new scale invariant neural network type based on Riesz transforms instead of convolutions drastically reduces the number of parameters to be trained and consequently the amount of training data needed.

2 Challenge 1: Reconstruct Highly Porous Structures from FIB-SEM Images

SEM images feature an exceptional depth of view. For porous structures, this causes rather bright gray values originating from solid structure much deeper than the currently imaged plane. When stacking SEM images of consecutive slices cut by a FIB, these shine-through-effects turn into artifacts, rendering the decision which pixels really represent solid structure a very tedious one.

Synthetic images simulated based on very simple geometries, namely realizations of Boolean models of spheres or circular cylinders, but correctly modeling

Fig. 2 Synthetic SEM images of realizations of Boolean models of spheres, cylinders, and cubes, a Cox-Boolean sphere model, and Altendorf-Jeulin's curved fiber model [1]. Images generated using Prill's method [32]

major physical effects of the imaging method [32] turned out to be very helpful. First, the insight gained from simulating the SEM imaging led to the morphological solution of the original reconstruction task thriving on detecting the true first appearance as sudden jump in the gray value profiles in the direction the electrons are flying in [31]. Second, the ground truth coming along with the synthetic structures enabled the quantitative comparison of Prill's segmentation [31] with local threshold backpropagation [37] and valley detection [38] reported in [36]. In terms of ML, reconstruction is a semantic segmentation task. Having the synthetic image data at hand, the question came naturally: Why not use them to train and test an ML solution?

Indeed, training the 3D version of the ready made, easily available convolutional neural network 3d U-net [12], yielded a usable solution [16]. The original U-net [33] features maps weighting each pixel in the loss function guiding the learning. These weight maps were adapted to the FIB-SEM reconstruction task: Voxels at the interface between pore space and solid component were particularly emphasized assigning the maximal weight 11 to voxels exactly at the interface and decreasing the weights exponentially to 1 with squared Euclidean distance to the interface. Moreover, edge effects due to tiling were reduced by a wide overlap of 40 voxels of the rather small 64^3 voxel patches used for training. For learning how to segment real FIB-SEM images, the synthetic training data nevertheless turned out to be too clean. This issue was solved by gray value histogram matching. However, it does not shed a good light on generalizability of the solution. A more systematic study on which features of the structures or the images have to be captured how well is currently under way. This envolves extension of the simulation [32] to new shapes like cuboids as in the central image of Fig. 2.

3 Challenge 2: Detect and Segment Cracks in 3D Images of Concrete

Semi-synthetic image data is currently the only practically feasible way to get 3D images of realistic sizes with cracks accompanied by a consistent ground truth. Generating 3D training data by interactive annotation is not really feasible as cracks

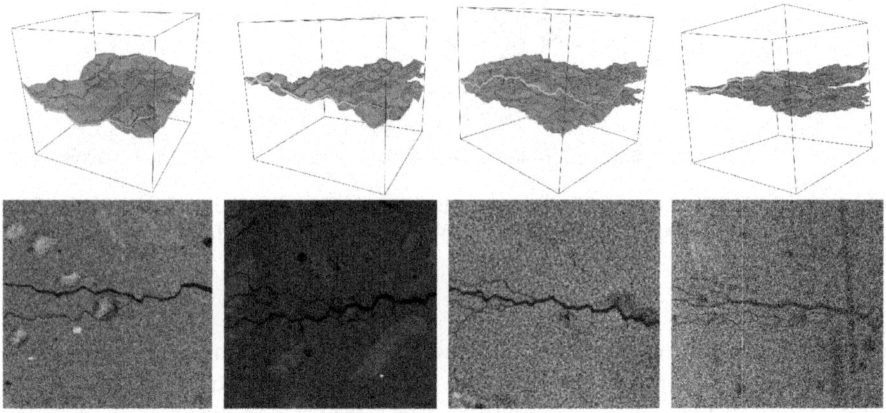

Fig. 3 Examples of simulated cracks in 400^3 voxel images. Minimal surfaces from spatial Voronoi tessellations generated by a Matérn cluster process (left), force-biased sphere packing (center left), a homogeneous Poisson point process (center right), and an inhomogeneous Poisson point process resulting from stretching a homogeneous one in x and z directions (right). The crack widths vary according to a Bernoulli random walk [23] with parameter $p = 0.01$. Top: volume renderings. Bottom: 2D slices of the 3D images of the cracks superimposed on CT images of normal, high performance, and aerated concrete and the polypropylene fiber reinforced concrete from [22]. Voxel sizes are 23, 20, 2.8, and 60 µm. Hence, cube and slice edge lengths vary between 1 and 24 mm

in CT images can be very thin and appear rather dark without a strong contrast to the surrounding material. The concrete matrix is heterogeneous, the gray value distribution there may therefore vary strongly. Completely synthetic images are far from the real ones. Simulation tools capturing this heterogeneity properly would have to incorporate scattering due to local variation in the chemical composition of the concrete on finer scales.

Two types of synthetic cracks have proven to model real cracks sufficiently realistically to train ML models based on them—realizations of fractional Brownian surfaces [9] and minimal surfaces formed by facets of a Voronoi tessellation [23]. The thickness can be nicely controlled by adaptive dilation. Synthetic cracks blend into the real background by distributing gray values as observed in air pores and smoothing the edges to mimic the partial volume effect, see Fig. 3.

So far, there is no provable advantage of ML over classical methods for crack segmentation. Hessian-based percolation [2, 14, 29] and template matching [30] still have their advantages [22]. However, from the application point of view, fine-tuning a trained ML model to deal with a yet unseen concrete type is probably easier than optimizing the parameters of the classical methods.

The RieszNet [4] is a completely new network type whose development was motivated by the crack segmentation task and the multi-scale nature of real cracks which can easily span a thickness range from 1 to 200 voxel edge lengths [22]. These multi-scale cracks can be taken care of by downscaling the image, applying the algorithm on each scale, and combining the results by upscaling again [3, 21],

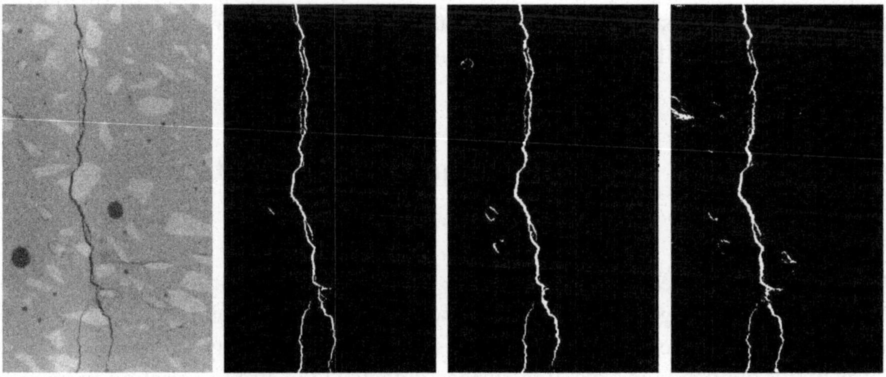

(a) original (b) Hessian-based percolation (c) 3d U-net (d) RieszNet, median filtered

Fig. 4 Segmentation results for the high performance concrete sample from [21]

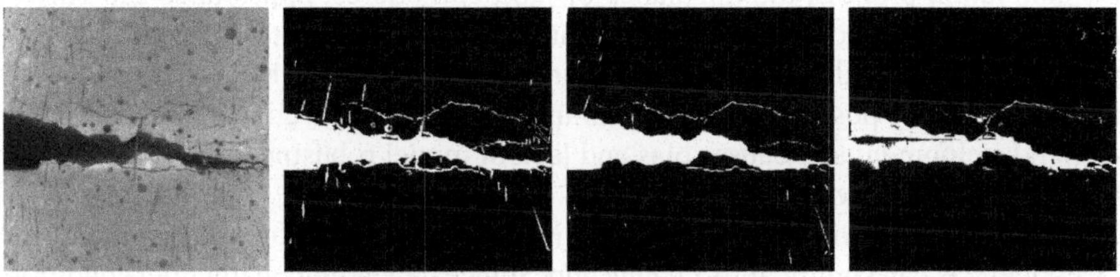

(a) original (b) Hessian-based per-colation (c) 3d U-net (d) RieszNet

Fig. 5 Segmentation results for the polypropylene fiber reinforced concrete sample from [22]. The RieszNet result is obtained on a downscaled version of the image and stitched. The missing part in the center of the crack is due to edge effects and the stitching

or by training additionally on synthetic multiscale cracks generated by adaptive dilation [27], see Figs. 4 and 5 for exemplary results. Both strategies are costly and have obvious limitations in the thickness range that can be covered. RieszNet instead replaces the convolutions in a typical convolutional neural network like 3d U-net by 1st and 2nd order Riesz transforms. The scale invariance of the latter two carries over to the network [4]. This results not only in a significant reduction in computation time, it also drastically reduces the number of trainable parameters by two magnitudes and thus allows for much smaller training data sets. The scale invariance of the RieszNet is not just a theoretical promise but proves to work in practice. The exemplary RieszNet results shown in Figs. 4 and 5 are achieved by training solely on synthetic cracks of fixed width 3 voxel edge lengths.

To summarize, well-known CNN, properly trained and fine-tuned do the job, are however costly if image scaling is needed and require fine-tuning for every new concrete type. The strong potential of Barisin's RieszNet has to be explored more deeply and better used by an implementation of the Riesz transform allowing to process full-scaled 3D images and dedicated post-processing.

4 Challenge 3: Virtually Plan and Simulate Optical Surface Inspection

Optical surface inspection tackles a large variety of defect types, depending greatly on the manufacturing process. Several research studies investigating surface defects from different perspectives give an impression of the variability [13, 15, 40].

A major challenge in detecting surface defects is the strong variation in appearance and occurrence frequencies of certain defect types in production. This can cause severe imbalances in datasets acquired in real industrial settings. While the imbalance of defect classes can be discovered through statistical analysis of classes upon the annotation process, the imbalance of intra-class defect appearance can only be detected by manual data analysis. Such analysis is however quite impractical since the image count is ideally in thousands and it would require the evaluator to imagine all potentially possible defect appearance scenarios. This complicates inspection model development and causes bias and lack of model robustness for defects which have not been observed sufficiently often.

One particularly demanding use case is visual inspection of specular metal surfaces. Defects can significantly change appearance depending on imaging conditions like light and camera placement and parameters. This increases the number of detection cases that need to be handled even further. Thus, deep learning models are needed that offer higher recognition robustness in such complex cases. The robustness can only be achieved by representing a wide variety of defect appearances for each defect class. As explained earlier, collecting such a dataset is highly impractical, introducing the need for synthetic generation of industrial data.

Generative models have been used for creating surface inspection data [41, 42]. However, hallucination is an inherent behavior which can not be reliably controlled. Schemedemann [39] combined rule-based generation with the generative approach to increase the realism of the images and reported both hallucination and possible dataset degradation. It can happen that the image looses the defect after the generative domain adaption, although the defect label is still present in the ground truth. Or the generative model creates a defect where the ground truth does not account for the defect. There is no automatic way to catch these artifacts. Hence, a human operator would have to be employed once again to manually check the images.

Procedural computer graphic models on the other hand enable explicit control over the geometry and microstructure of defects and surface finishes [8, 24] and thus rule-based synthetic data creation. Procedural methods enable the design of functions which can realistically model defect geometry [7] and micro-structures of surface texture [8]. The models themselves can be created using any mathematical modeling approach, and image rendering techniques can use these methods to generate realistic images, with variability controlled by fitting parameters to physical measurements (e.g. elevation and density of the surface pattern). This has been introduced and well documented in the pipeline introduced by Bosnar et al. [6]. This pipeline combines for the first time a 3D model of the inspected product with procedural texture generation, procedural defect geometry modeling, and detailed description of the inspection acquisition setup in order to recreate the scene virtually and generate a photorealistic synthetic image through tristimulus (RGB) rendering [25], together with defect segmentation masks. Commonly used texture and defect models are based on empirical findings from observing the surface texture by naked eye and examining its behavior in the acquired inspection image. Bosnar et al.'s work [8] focuses on results enabled by introducing procedural texture stochastic geometry models, which have been developed based on physical surface topography measurements.

So far, the use of stochastic geometry models has shown to be very promising for achieving a high degree of realism. The work of SynosIs [18] aims to create synthetic images of machined aluminum surfaces. There the focus lies on milling which is a common method to form objects. Material is removed by blades installed at the milling head while the head rotates. Meanwhile, the tool moves along a predetermined path over the surface. On the final surface, the milling creates a pattern of micrometer-scale surface scratches. This pattern as well as how it interacts with optical light depend on size, path, and speed of the cutting tool.

Additionally, due to material inhomogeneities, cutting tool wear, and environmental influence, no two surfaces are completely the same, even though they might be produced using exactly the same parameters. To recreate this behavior, stochastic geometry models are integrated into Bosnar's pipeline for modeling both the defects and the textures procedurally. The defects are modeled as described in [7].

The texture model is designed to mimic the physical change of the surface due to milling [20]. The scratches defining a face-milled pattern are ring-shaped caused by the cycloidal paths of the blades. To simplify that movement in the model, rings are used instead. A multi-step model is developed describing the positioning of rings, the appearance of an individual ring, and the interaction of overlapping rings. Moreover, the model is parameterized including the knowledge of real-world parameters from the manufacturing process. The lateral cutting depth, the feed rate, and the spindle speed determine the tool-path and are therefore decisive for the placement of rings. The diameter and tilting of the milling head and the width of its blades directly describe the fundamental shape of the rings. All parameters not known from the manufacturing process are estimated based on visually comparing the synthetic textures and real height measurements of the milled surfaces. The observed variation within textures generated using the same parameter configuration is captured selecting their values from suitable random distributions.

Fitting the parameter values to the machining parameters and to meet microscopic images of the surfaces, it is possible to synthesize an image in which the object's surface texture is causing a very similar light response and pattern when compared with the real image (see Fig. 6). Note that the illumination intensity for these images has been matched approximately. By increasing the exposure manually, the similarity

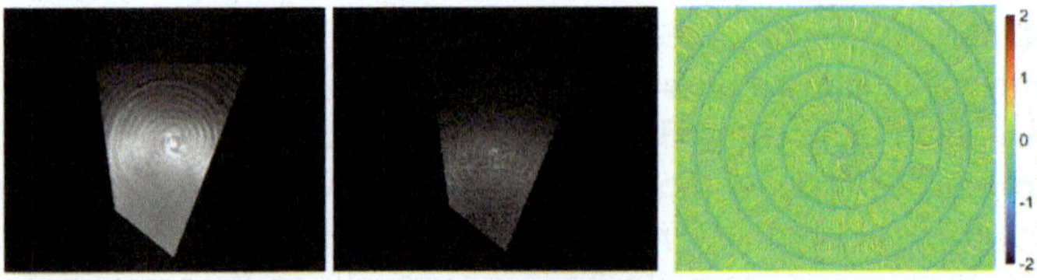

(a) Spiral milling

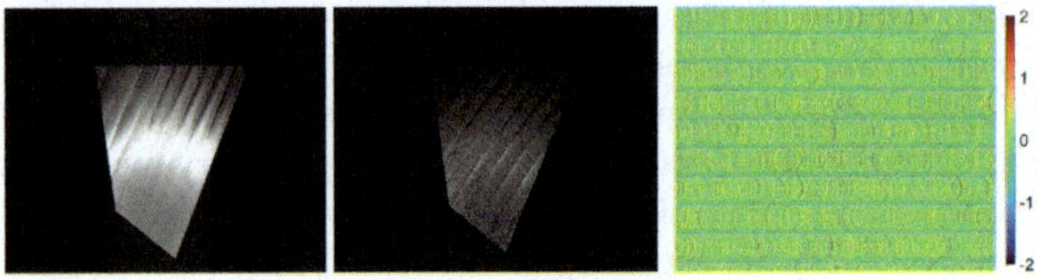

(b) Parallel milling

Fig. 6 Comparison of real image, synthetic image, and synthetic image with additional exposure adjustment for two milling paths. The rightmost images feature the procedural textures used to generate the synthetic data, cropped to make the texture structures more visible. The textures are applied onto the surface of the 3D object at rendering time through texture mapping [10]

of overexposed regions becomes more prominent, rendering the visual appearance even more similar.

5 Conclusion and Discussion

Synthetic data has proven to benefit the training of machine learning models for segmentation and classification problems [16, 17, 24, 27]. Here, we emphasized in particular the advantages of building the synthetic data on stochastic geometry models. There are nevertheless new challenges arising. In general, we still do not know what degree of realism has to be reached, neither in terms of structure, texture, or pattern nor in terms of the overall synthetic image appearance and acquisition-specific image features such as noise pattern or the shine-through effects in SEM. It is however also a question, how much effort it takes to reach a certain degree of realism, whether it is always necessary, and whether we can afford it.

It is possible to visually asses the synthetic image data and to state that it is similar to the real data. However, there are so far no general similarity metrics quantifying the deviations of synthetic and real images for our challenges. Commonly used image comparison metrics typically focus on super-pixel level features. Even when they do consider high frequency texture details as well (e.g. SSIM), they do not have the capacity to correlate them to the global image pattern. This means for example that the metric would show very little difference between the parallel and spiral milling patterns in Challenge 3 because those two textures are locally very similar.

By developing dedicated metrics, it will be possible to evaluate synthetic dataset quality beyond the empirical measure "the network has successfully managed to generalize from it". This will open the opportunity to quantify the so called domain gap – something that has not been achieved so far. Domain gap refers to differences in data characteristics and distribution between source and target domains [5]. In the general context of machine learning, it is typically the difference between the training dataset and the real data the model should be used on. In a more specific context of synthetic data it describes how realistic the synthetic data is.

In surface inspection, a particular challenge beyond realism measurement is over-labeling. This term describes the situation where the ground truth reports defects that are unrecognizable for human inspectors. For instance, defects might be located in very dark areas or hidden by heavy blur, such that they should be detected with low confidence or not at all. However, defining a suitable defect visibility metric is still an open problem. Hence, the over-labeling problem has to be handled specifically for each use-case.

For the cracks in concrete, it is too early to discuss structure similarity. For now, the tessellation based models are visually convincing and sufficiently versatile for cracks induced by mechanical stress. A systematic geometric analysis of 3D crack structures, including cracks induced by other effects like chemical ones [22] is subject of further research and might lead to more sophisticated models tailored more closely to the mechanism causing the cracks. The concrete example teaches us also that the degree of realism needed heavily depends on the used ML model: The RieszNet model is capable of segmenting multiscale cracks although having been trained on cracks of one fixed width, only.

The answer for how much of the imaging physics has to be simulated correctly seems to depend crucially on both imaging method and application: For the SEM, more than the correct electron matter interaction seems to be needed. For the CT images of concrete samples with cracks on the other hand, no explicit simulation of the X-ray imaging is involved at all. Finally, for the optical images of structured metal surfaces, surely more parametric models of surface finishing are needed, while the chosen micro-scale optical properties seem suitable. However for all other materials beside metal, this is yet to be explored, with particular interest on subsurface scattering approximation for transparent materials and the use of correct refraction data when simulating glass. Finally, to maximize the image appearance similarity, spectral rendering [25] should be used instead of classical RGB rendering.

Acknowledgements This work was supported by the German Federal Ministry of Education and Research (BMBF) [grant numbers 05M2020 (DAnoBi), 01IS21058 (SynosIs), 01IS21054 (poST)]. We thank Matthias Pahn and Szymon Grezsiak from CE of RPTU for concrete samples and experimental design, Franz Schreiber from Fraunhofer ITWM and Michael Salamon from Fraunhofer EZRT for CT imaging, Markus Kronenberger from Fraunhofer ITWM for the cubes in Fig. 2, Anna Nowacka from Fraunhofer ITWM and RPTU-CE for the fine-tuned 3d U-net results featured in Figs. 4 and 5, and Fraunhofer IOF for machined surface production and topography measurements.

References

1. Altendorf, H., Jeulin, D.: Modeling fiber systems using random walks. In: 3rd International Workshop: 3D Imaging, Analysis, Modeling and Simulation of Macroscopic Properties. Fontainebleau (2010)
2. Barisin, T., Jung, C., Müsebeck, F., Redenbach, C., Schladitz, K.: Methods for segmenting cracks in 3D images of concrete: a comparison based on semi-synthetic images. Pattern Recognit. **129**, 108747 (2022)
3. Barisin, T., Jung, C., Nowacka, A., Schladitz, K., Redenbach, C.: Crack segmentation in 3D concrete images: perspectives and challenges. In: International Symposium on Non-Destructive Testing in Civil Engineering (NDT-CE 2022) (2022)
4. Barisin, T., Redenbach, C., Schladitz, K.: Riesz networks: scale invariant neural networks in single forward pass. J. Math. Imaging Vis. **66**, 246–270 (2024). https://doi.org/10.1007/s10851-024-01171-4
5. Ben-David, S., Blitzer, J., Crammer, K., Pereira, F.: Analysis of representations for domain adaptation. In: B. Schölkopf, J. Platt, T. Hoffman (eds.) Advances in Neural Information Processing Systems, vol. 19. MIT Press (2006)
6. Bosnar, L., Dutta, S., Saric, D., Weibel, T., Rauhut, M., Hagen, H., Gospodnetić, P.: Image synthesis pipeline for surface inspection. In: LEVIA'20: Leipzig Symposium on Visualization in Applications (2020). https://doi.org/10.31219/osf.io/kqt8w
7. Bosnar, L., Hagen, H., Gospodnetić, P.: Procedural defect modeling for virtual surface inspection environments. IEEE Comput. Graphics Appl. **43**(2), 13–22 (2023). https://doi.org/10.1109/MCG.2023.3243276
8. Bosnar, L., Rauhut, M., Hagen, H., Gospodnetić, P.: Texture synthesis for surface inspection. In: LEVIA 22: Leipzig Symposium on Visualization in Applications. (2022). https://doi.org/10.36730/2022.1.levia.4
9. Botev, Z.: Fractional Brownian field or surface generator. https://de.mathworks.com/matlabcentral/fileexchange/38945-fractional-brownian-field-or-surface-generator. Accessed 02 Feb 2022
10. Catmull, E.: Computer Display of Curved Surfaces, pp. 35–41. Association for Computing Machinery, New York, NY, USA (1998). https://doi.org/10.1145/280811.280920
11. Chen, Y., Ding, Y., Zhao, F., Zhang, E., Wu, Z., Shao, L.: Surface defect detection methods for industrial products: a review. Appl. Sci. **11**(16) (2021). https://doi.org/10.3390/app11167657, https://www.mdpi.com/2076-3417/11/16/7657
12. Çiçek, Ö., Abdulkadir, A., Lienkamp, S.S., Brox, T., Ronneberger, O.: 3D U-Net: learning dense volumetric segmentation from sparse annotation. In: Lecture Notes in Computer Science, vol. 9901, pp. 424–432 (2016). https://doi.org/10.1007/978-3-319-46723-8_49
13. Czimmermann, T., Ciuti, G., Milazzo, M., Chiurazzi, M., Roccella, S., Oddo, C.M., Dario, P.: Visual-based defect detection and classification approaches for industrial applications—a survey. Sensors **20**(5), 2505–2523 (2020). https://doi.org/10.1007/s10845-018-1412-0
14. Ehrig, K., Goebbels, J., Meinel, D., Paetsch, O., Prohaska, S., Zobel, V.: Comparison of crack detection methods for analyzing damage processes in concrete with computed tomography. In: International Symposium on Digital Industrial Radiology and Computed Tomography (2011)
15. Erdogan, V.: Ein systematischer Zugang für die Beschreibung von Inspektionsaufgaben am Beispiel der Oberflächeninspektion in der Automobilindustrie. Ph.D. thesis, Universität Stuttgart (2015)
16. Fend, C., Moghiseh, A., Redenbach, C., Schladitz, K.: Reconstruction of highly porous structures from FIB-SEM using a deep neural network trained on synthetic images. J. Microsc. **281**(1), 16–27 (2021)
17. Fulir, J., Bosnar, L., Hagen, H., Gospodnetić, P.: Synthetic data for defect segmentation on complex metal surfaces. In: 2023 IEEE/CVF Conference on Computer Vision and Pattern Recognition Workshops (CVPRW), pp. 4424–4434 (2023). https://doi.org/10.1109/CVPRW59228.2023.00465

18. Fulir, J., Jeziorski, N., Bosnar, L., Herffurth, T., Gischkat, T., Redenbach, C., Hagen, H., Gosodnetic, P.: SynosIs: Image synthesis pipeline for machine vision in metal surface inspection. In preparation
19. Holzer, L., Stenzel, O., Pecho, O., Ott, T., Boiger, G., Gorbar, M., de Hazan, Y., Penner, D., Schneider, I., Cervera, R., Gasser, P.: Fundamental relationships between 3D pore topology, electrolyte conduction and flow properties: towards knowledge-based design of ceramic diaphragms for sensor applications. Mater. Des. **99**, 314–327 (2016)
20. Jeziorski, N., Redenbach, C.: Stochastic geometry models for texture synthesis of machined metallic surfaces: sandblasting and milling (2024). arxiv:2403.13439
21. Jung, C., Müsebeck, F., Barisin, T., Schladitz, K., Redenbach, C., Kiesche, M., Pahn, M.: Towards automatic crack segmentation in 3D concrete images. In: 11th Conference on Industrial Computed Tomography, Wels, Austria (iCT 2022) (2022)
22. Jung, C., Nowacka, A., Barisin, T., Meinel, D., Paetsch, O., Grzesiak, S., Salamon, M., Schladitz, K., Redenbach, C., Pahn, M.: 3D imaging and analysis of cracks in loaded concrete samples. In: 12th Conference on Industrial Computed Tomography, Fürth, Germany (iCT 2023) (2023)
23. Jung, C., Redenbach, C.: Crack modeling via minimum-weight surfaces in 3D Voronoi diagrams. J. Math. Ind. **13**(10) (2023)
24. Moonen, S., Vanherle, B., de Hoog, J., Bourgana, T., Bey-Temsamani, A., Michiels, N.: CAD2Render: a modular toolkit for GPU-accelerated photorealistic synthetic data generation for the manufacturing industry. In: Proceedings of the IEEE/CVF Winter Conference on Applications of Computer Vision (WACV) Workshops, pp. 583–592 (2023)
25. Murray, D., Fichet, A., Pacanowski, R.: Efficient spectral rendering on the gpu for predictive rendering. In: A. Marrs, P. Shirley, I. Wald (eds.) Ray Tracing Gems II: Next Generation Real-Time Rendering with DXR, Vulkan, and OptiX, pp. 673–698. Apress, Berkeley, CA (2021). https://doi.org/10.1007/978-1-4842-7185-8_42
26. Nikolenko, S.: Synthetic Data for Deep Learning. Springer, Cham (2021)
27. Nowacka, A., Schladitz, K., Grzesiak, S., Pahn, M.: Segmentation of cracks in 3D images of fiber reinforced concrete using deep learning. To appear in J. Soft. Comput. Civil. Eng. (2023)
28. Ott, T., Roldán, D., Redenbach, C., Schladitz, K., Godehardt, M., Höhn, S.: Three-dimensional structural comparison of tantalum glancing angle deposition thin films by FIB-SEM. J. Sens. Sens. Syst. **8**(2), 305–315 (2019)
29. Paetsch, O., Baum, D., Ehrig, K., Meinel, D., Prohaska, S.: Automated 3D crack detection for analyzing damage processes in concrete with computed tomography. In: Proceedings of the Conference on Industrial Computed Tomography, pp. 321–330 (2012)
30. Paetsch, O.: Possibilities and limitations of automated feature extraction shown by the example of crack detection in 3D-CT images of concrete specimen. In: 9th Conference on Industrial Computed Tomography (iCT) (2019)
31. Prill, T., Schladitz, K., Jeulin, D., Faessel, M., Wieser, C.: Morphological segmentation of FIB-SEM data of highly porous media. J. Microsc. **250**(2), 77–87 (2013)
32. Prill, T., Schladitz, K.: Simulation of FIB-SEM images for analysis of porous microstructures. Scanning **35**, 189–195 (2013)
33. Ronneberger, O., Fischer, P., Brox, T.: U-net: convolutional networks for biomedical image segmentation. In: Lecture Notes in Computer Science, pp. 234–241 (2015). https://doi.org/10.1007/978-3-319-24574-4_28
34. RPTU Kaiserslautern-Landau, Department of Civil Engineering: Tomography portal Gulliver. https://bauing.rptu.de/en/massivbau/research/dfg-large-equipment-initiative
35. Salamon, M., Reims, N., Hanke, R.: High energy X-ray tomography for large concrete structures. In: M. Pahn, C. Thiele, C. Glock, J. Schnell (eds.) Vielfalt im Massivbau: Festschrift zum 65. Geburtstag von Prof. Dr.-Ing. Jürgen Schnell, pp. 595–609. Ernst & Sohn (2018)
36. Salzer, M., Prill, T., Spettl, A., Jeulin, D., Schladitz, K., Schmidt, V.: Quantitative comparison of segmentation algorithms for FIB-SEM images of porous media. J. Microsc. **257**(1), 23–30 (2015)

37. Salzer, M., Spettl, A., Stenzel, O., Smått, J.H., Lindén, M., Manke, I., Schmidt, V.: A two-stage approach to the segmentation of FIB-SEM images of highly porous materials. Mater. Charact. **69**, 115–126 (2012)
38. Salzer, M., Thiele, S., Zengerle, R., Schmidt, V.: On the importance of FIB-SEM specific segmentation algorithms for porous media. Mater. Charact. **95**, 36–43 (2014)
39. Schmedemann, O.: Adapting synthetic training data in deep learning-based visual surface inspection to improve transferability of simulations to real-world environments. In: Proceedings of Automated Visual Inspection and Machine Vision V (2023)
40. Tang, B., Chen, L., Sun, W., Lin, Z.k.: Review of surface defect detection of steel products based on machine vision. IET Image Process. **17**(2), 303–322 (2023). https://doi.org/10.1049/ipr2.12647. https://ietresearch.onlinelibrary.wiley.com/doi/abs/10.1049/ipr2.12647
41. Wang, R., Hoppe, S., Monari, E., Huber, M.: Defect transfer GAN: diverse defect synthesis for data augmentation. In: 33rd British Machine Vision Conference 2022, BMVC 2022, London, UK, 21–24 Nov 2022. BMVA Press (2022). https://bmvc2022.mpi-inf.mpg.de/0445.pdf
42. Wen, X., Shan, J., He, Y., Song, K.: Steel surface defect recognition: a survey. Coatings **13**(1) (2023). https://doi.org/10.3390/coatings13010017. https://www.mdpi.com/2079-6412/13/1/17
43. Yang, J., Li, S., Wang, Z., Dong, H., Wang, J., Tang, S.: Using deep learning to detect defects in manufacturing: a comprehensive survey and current challenges. Materials **13**(24) (2020). https://doi.org/10.3390/ma13245755. https://www.mdpi.com/1996-1944/13/24/5755

Fuzzy Inference System for Fatigue Parameters Prediction in Metals: from Strength to Fatigue

Inna M. Gitman, Ruixuan Tu, and Luca Susmel

Abstract In order to enable engineers to make informed decisions about material selection, design, and maintenance, contributing to the safety, reliability, and longevity of components and structures, subjected to cyclic (fatigue) loading, it is essential to have knowledge of the threshold value of the stress intensity factor range and the range of plain fatigue limit of a material. Traditional experimental approaches, although offering an accurate determination of these parameters, are, however, expensive and time-consuming. It is thus evident, that there is a need for an alternative methodology, offering accurate and reliable on one hand, but fast and cheap on the other, predictions of aforementioned parameters. The main focus of this chapter is to analyse the ability of the data driven fuzzy inference system (FIS) approach to serve this goal and predict fatigue parameters of a material, knowing material's strength (static) characteristics. Results, reported in this chapter, for aluminium alloys and different steels data sets demonstrate capabilities of FIS to estimate, with the high degree of accuracy, the range of the plain fatigue limit and the range of the threshold value of the stress intensity factor, based on provided ultimate tensile strength and yield strength of a metal.

Keywords Data-driven approach · Fuzzy inference system · Yield strength · Ultimate tensile strength · Fatigue limit · Stress intensity factor

I. M. Gitman (✉)
University of Twente, Enschede, The Netherlands
e-mail: i.m.gitman@utwente.nl

R. Tu · L. Susmel
University of Sheffield, Sheffield, United Kingdom
e-mail: rtu1@sheffield.ac.uk

L. Susmel
e-mail: l.susmel@sheffield.ac.uk

F. Willot et al. (eds.), *Continuum Models and Discrete Systems*,
Springer Proceedings in Mathematics & Statistics 457,
https://doi.org/10.1007/978-3-031-58665-1_20

1 Introduction

The threshold value of the stress intensity factor range and the fatigue limit are critical factors in assessing the structural integrity and durability of structural components subjected to in-service cyclic (i.e., fatigue) loading.

The threshold value of the stress intensity factor range, ΔK_{th}, is a fundamental parameter used in fracture mechanics-driven design. It represents the minimum range of the stress intensity factor required for a crack to propagate in a material.

The concept of the fatigue limit, $\Delta\sigma_0$, refers to the maximum stress level that a material or a component subjected to fatigue loading can withstand indefinitely without developing any fatigue cracks. Since, very often, a fatigue limit cannot be identified unambiguously or it disappears [1], it is common engineering practice to perform the high-cycle fatigue assessment in terms of endurance limit. The endurance limit is a reference stress range (or amplitude) extrapolated in the high-cycle fatigue regime (i.e., typically, in the range 10^6–10^7 cycles to failure).

$\Delta\sigma_0$ and ΔK_{th} enable engineers to make informed decisions about material selection, design, and maintenance, ultimately contributing to the safety, reliability, and longevity of components and structures.

Since, the above two material properties play a role of primary importance in the structural engineering field, a number of specific standard codes and recommendations have been made available in order to determine $\Delta\sigma_0$ and ΔK_{th} experimentally. While these standard approaches allow these two material properties to be determined in a very accurate way, the experimental approach is by its nature very expensive and time-consuming. This is why structural engineers always look for empirical rules allowing $\Delta\sigma_0$ and ΔK_{th} to be estimated by always reaching an adequate level of safety and accuracy.

1.1 From Strength to Fatigue Properties

The main goal of the chapter is to check the hypothesis that fatigue (time-variable) parameters of a material: ΔK_{th}–the threshold value of the Linear Elastic Fracture Mechanics (LEFM) stress intensity factor range, and $\Delta\sigma_0$–the range of the plain fatigue limit (experimentally determined under a particular load ratio $R = \sigma_{min}/\sigma_{max}$) are related to the static mechanical properties, i.e. yield strength σ_Y and ultimate tensile strength σ_{UTS}. To this end, the link between fatigue and static material properties is established using a fuzzy interference system-based approach. The systematic validation exercise based on the large database reported in Ref. [2] makes it evident that data-driven methodologies are successful in estimating fatigue strength from the corresponding static material properties.

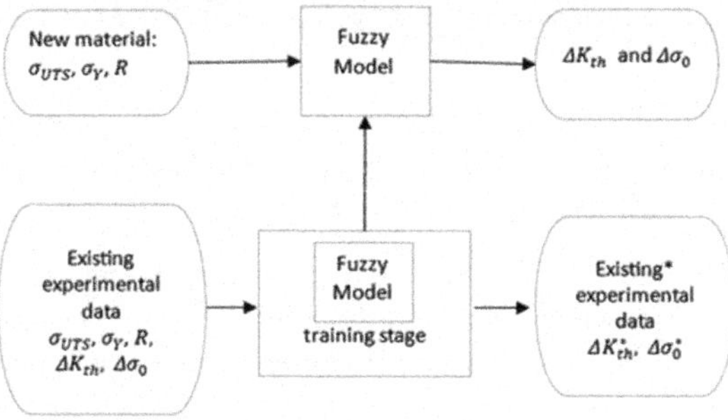

Fig. 1 A generic flow of the data-driven methodology-illustration of the process

1.2 Data-Driven Methodology

Modern data-driven methodologies have shown their capabilities to overcome the data imprecision and require fewer experimental tests in the engineering applications (see among others [3–5]). The main idea behind such approaches is to

- **collect**: by gaining knowledge from existing data, for example from experiments,
- **train and validate**: create patterns (*training stage*) linking input and output parameters, and subsequently,
- **test and use**: adopt these patterns for estimations and prediction (*testing stage*) of a new input data.

Previous studies have shown that the data-driven techniques are promising alternatives for fatigue analysis and fracture estimations, see [6] for an overview of these methodologies.

The main focus of the present work is to apply a particular data-driven methodology (based on the fuzzy sets theory) in order to characterize complex relationships between fatigue parameters and strength characteristics (see Fig. 1 for illustration).

2 Fuzzy Inference System

One of the existing in the literature and promising data-driven methodologies, the fuzzy inference system (FIS) is based on the fuzzy logic and the fuzzy sets theory, introduced in Ref. [7]. Fuzzy logic offers solutions to problems with imprecision and indeterministic relationships between parameters. Complex systems can be modelled using simple fuzzy logic rules, similar to human reasoning process. Conventionally, such imprecision is introduced via linguistic variables (words) in a natural or artificial language [8]. Each linguistic word summarizes the description of a fuzzy subset at a universe of discourse representing the meaning of this word [9]. In the present study, however, a data (obtained from experiments)—based variable determination strategy is adopted, replacing the conventional linguistic words.

2.1 Building Fuzzy Sets Based Data-Driven Algorithm

The algorithm of fuzzy sets based data-driven methodology has already been discussed above (see also Fig.1), mathematically the methodology can be summarized as a sequence of the following steps:

1. For a given number n of experiments (where both strength and fatigue parameters have been testes and recorded, so-called *known data*), build fuzzy relations (*rules*) S^i between parameters of each ith experiment:

$$S^i = X^i \times Z^i \tag{1}$$

here X is a set of strength (static) parameters (σ_Y and σ_{UTS}), and particular load ratios R for fatigue analysis; Z is a set of fatigue characteristics (ΔK_{th} and $\Delta\sigma_0$).

2. As the next step, define the operator F collecting together fuzzy *rules* S^i of all n of experiments:

$$F = U_{i=1}^{n} S^i \tag{2}$$

3. Finally, for the set of *unknown data* (i.e. for materials where there are experimentally available only strength (static) parameters $\hat{X}$), using the defined above operator F, estimate $\hat{Z}$ (set of *unknown* fatigue characteristics):

$$\hat{Z} = \hat{X} \cdot F \tag{3}$$

here "·" is the maximin product (an ordinary product of matrices, where min is substituted for the multiplication operation, and max is used instead of the addition operation.

2.2 Membership Functions

The idea behind fuzzy logic approach is the allowance of *partial* belongings, which can be described mathematically using a membership functions (MF), ranging from 0 to 1. For example, Fig.2 illustrates an example of membership functions for *small*, *medium* and *large* ultimate tensile strength.

The membership functions' definition adopted here is the triangular MF (see Eq. 4), which is commonly accepted choice due to its simplicity:

$$\mu(x) = \begin{cases} 0, \ x \leq a \\ \frac{x-a}{b-a}, \ a < x \leq b \\ \frac{c-x}{c-a}, \ b < x < c \\ 0, \ x \geq c \end{cases} \tag{4}$$

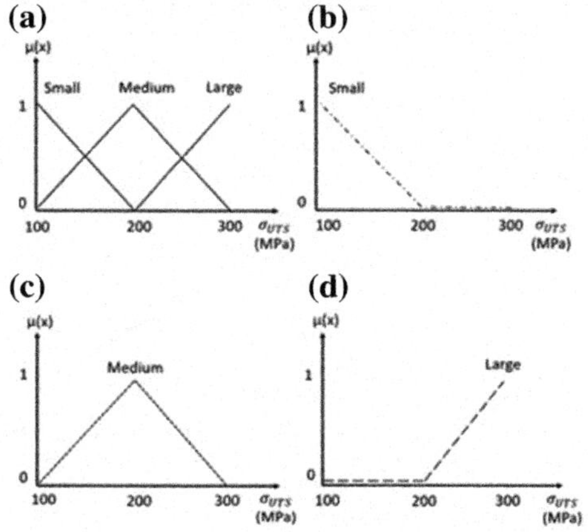

Fig. 2 **a** Triangular membership functions for UTS; **b**, **c** and **d** de-composition of membership functions for UTS

here $\mu(x)$ is the MF value of the analysed input x; a, b, and c are parameters of the function defined by the user, corresponding to the lower and upper limits and tips of the MF triangle. For example, assuming that σ_{UTS} is equal to 100 MPa, 200 MPa and 300 MPa, Fig. 2 shows that $\sigma_{UTS} = 100$ MPa is defined as *small* value, $\sigma_{UTS} = 200$ MPa is defined as *medium* value and $\sigma_{UTS} = 300$ MPa is defined as *large* value. However, an intermediate value (i.e. a potential new experimental data such as 150 MPa) *partially* belongs to *small* and *medium*. Hence, $\sigma_{UTS} = 150$ MPa has the membership value equal to 0.5 for both *small* and *medium*, but 0 for *large*. Such linguistic words and parameters of membership functions are normally determined based on experts' intuition, intelligence and understanding.

In this study, the linguistic words are replaced by the observed values from existing experimental data.

The membership values, calculated using MFs, are then brought to the fuzzy inference engine (FIS) and processed with relevant fuzzy rules (see [3, 4] for more details). Note here, that fuzzy rules are formulated for each data entry (experiment). Figure 3 illustrates the algorithm of FIS for an illustrative (simplified) case.

In this illustrative example the following two rules were assumed:

1. "**if** the ultimate tensile strength is 300 MPa **and** yield strength is 250 MPa, **then** ΔK_{th} is 7 MPa $m^{1/2}$".
2. "**if** the ultimate tensile strength is 100 MPa **and** yield strength is 50 MPa, **then** ΔK_{th} is 8 MPa $m^{1/2}$".

These two rules were used for building and training FIS. The output of the fuzzy inference engine is a weighted average value, which can be calculated as:

$$W_a = \frac{\Sigma_1^k \mu_i \times n_i}{\Sigma_1^k \mu_i} \quad (5)$$

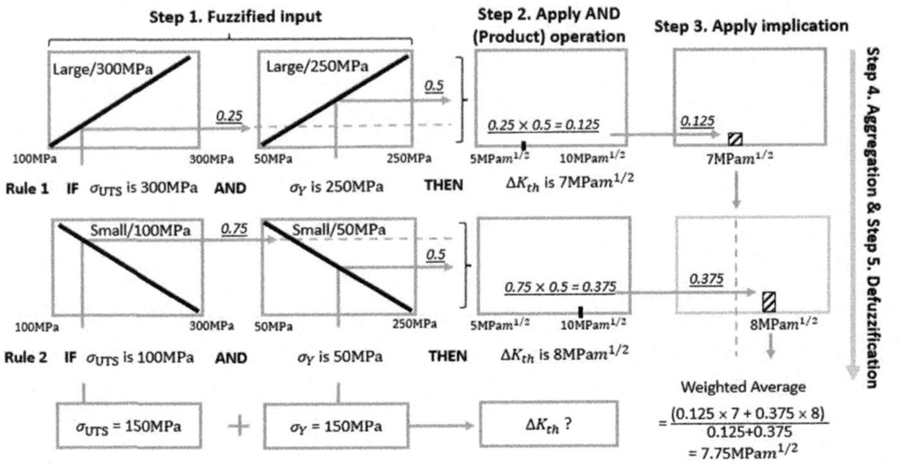

Fig. 3 Decomposition of fuzzy inference system

where k refers to the total amount of fuzzy rules (data entries, two in our example), W_a refers to the estimated output value for the new input, μ_i is the processed membership value of the new input for ith fuzzy rule and n_i is the *known output* value (from training data) for the same rule. Exercising the two aforementioned rules, the weighted average of all outcomes is calculated, using Eq. 5 and as shown in Fig. 3, resulting in the estimated ΔK_{th}. Hence, knowing data from rules 1 and 2, the estimated ΔK_{th} for a new material (with new ultimate tensile strength 150 MPa and yield strength 150 MPa) will be 7.75 MPa m$^{1/2}$.

Note: it is worth mentioning that if in the new, to be estimated, material one of the input parameters is equal to a boundary value, for example, a new material has $\sigma_{UTS} = 100\,\text{MPa}$ and $\sigma_Y = 150\,\text{MPa}$, MFs for σ_{UTS} will be 0 and 1 for rules 1 and 2, respectively. In this case, the product of membership values for rule 1 is 0 and, hence, the product of membership value for rule 2 is the only effective factor in the weighted average calculation. Eventually, changing yield strength will not cause any change in ΔK_{th}, causing the failure of estimation.

This scenario occurs in the case of a low number of factors included in the weighted average calculation (e.g. in the simplified example above: two input parameters with only two possible values each are considered). However, in later sections, where a realistic example with a higher number of more varied input parameters is analysed, a fuzzy system will have many more input rules. In such a system, there will always be at least two rules in effect; hence, the above failure is not envisaged.

An alternative strategy could include replacing the triangular MF with the Gaussian MFs [5, 10].

3 Results and Discussion

As discussed in the Introduction, the main aim of this chapter is to investigate the ability of a data-driven approach to predict fatigue parameters of materials knowing their strength characteristics.

To this end, by demonstrating the results of the discussed above data-driven fuzzy sets based model, literature experimental data, reported in and adapted from [2] have been be analysed.

We must note here, that special attention should be given to the fact that all experiments, collected from various literature sources and reported in [2], were performed in different laboratories, different countries, by different research groups and in different years. Consequently, the authors of the present chapter, cannot confirm and guarantee consistencies between all reported experiments. Undoubtedly, this *uncertainties* have influence the results of our analysis as will be discussed later.

The data collected in [2] were split in categories of different metals, with particular attention given to aluminium and steel; and *cleaned*–experiments with missing data (e.g. missing information about at least one of analysed parameters: either strength *or* fatigue characteristics) were taken out of consideration. In principle, fuzzy sets based methodology is capable of dealing with missing data, but at this initial analysis the attention was concentrated on a complete data set. The case of missing data could be a potential future study.

Aluminium alloys

The first set of data that has been analysed was collected for various aluminium alloys. The experimental data adopted from [2] are presented in Table 1. For reasons of simplicity of presentation, particular names of aluminium alloys, together with actual literature sources (references) are omitted from the current chapter. All relevant and important information and references about these data can be found in [2].

The data presented in Table 1 has then been split into two categories: the first set (approximately 70%) was used for training the model, and the rest (every 3rd row, approximately 30% of all data) was used for validation e.g. testing stage. This 70–30% split is generally accepted way to train and validate FIS models.

The results of FIS model–predicted values of fatigue parameters (ΔK_{th} and $\Delta\sigma_0$) are presented in Table 2, together with the actual experimentally obtained values, taken directly from Table 1.

In order to evaluate the accuracy of our predictions, the estimated output E_{est} is compared with the actual experimental output E_{exp}, here E represents either the threshold value of the LEFM stress intensity factor range ΔK_{th}, or the range of the plain fatigue limit $\Delta\sigma_0$. Percentage errors for both parameters (Pct Error in Table 2) are calculated as follows:

$$\text{Pct error} = \frac{|E_{est} - E_{exp}|}{E_{exp}} \times 100\% \tag{6}$$

Table 1 Experimental results for aluminium alloys, adapted from [2]

σ_{UTS} [MPa]	σ_Y [MPa]	R	ΔK_{th} [MPam$^{1/2}$]	$\Delta\sigma_0$ [MPa]
466	357	–1	3.5	248
483	322	–1	6.4	300
486	425	–1	4.2	300
460	352	–1	4.2	323.6
450	345	–1	2.2	256
523	467	–1	1.6	296
342	256	–1	1.6	125
260	192	–1	4.0	231
466	357	0	4.0	172
452	334	0.1	3.2	130
475	442	0.1	2.2	195
533	417	0.1	2.6	168
454	330	0.1	3.3	138
541	479	0.1	2.8	125
510	456	0.1	2.4	125
498	432	0.1	2.17	165
340	175	0.1	3.2	150
431	282	0.1	4.2	250
442	340	0.1	2.5	135

Table 2 Experimental output with the estimation results for aluminium samples

σ_{UTS} [MPa]	σ_Y [MPa]	R	ΔK_{th} [MPam$^{1/2}$]	$\Delta\sigma_0$ [MPa]	Pred ΔK_{th} [MPam$^{1/2}$]	Pred $\Delta\sigma_0$ [MPa]	Pct error ΔK_{th}	Pct error $\Delta\sigma_0$
486	425	–1	4.2	300	3.50	248.00	16.67	17.33
523	467	–1	1.6	296	4.02	224.30	148.15	24.22
466	357	0	4.0	172	3.75	276.61	6.34	60.82
533	417	0.1	2.6	168	2.10	165.01	19.22	1.78
510	456	0.1	2.4	125	2.43	168.32	1.19	34.66
431	282	0.1	4.2	250	4.02	212.50	4.29	15.00
						Average	32.64	25.64

The average estimation error for aluminium samples is reported to be 32.64% for ΔK_{th} and 25.64% for $\Delta\sigma_0$, respectively.

It is interesting to point out that although the proposed FIS approach can estimate both output parameters accurately in most cases, some samples show large estimation errors. More specifically, as shown in Table 2, the second validation aluminium alloy sample (row 2) has unacceptably large estimation error of 148.15%.

Scatter plots in Fig. 4 demonstrate the predicted values of ΔK_{th} and $\Delta\sigma_0$ versus the experimental values. The dotted lines represent the trend lines, accompanied with

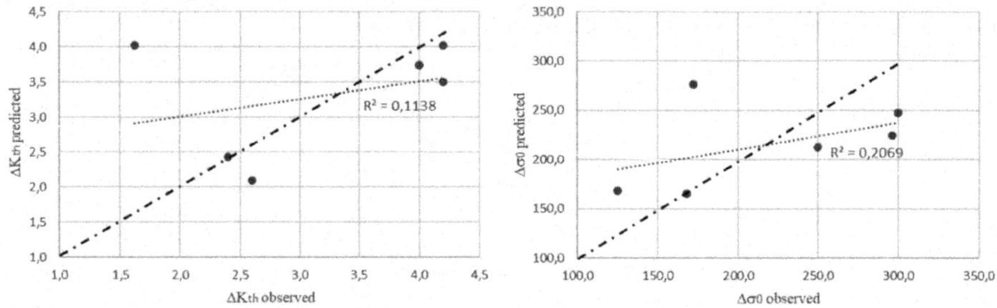

Fig. 4 Predicted versus observed ΔK_{th} and $\Delta\sigma_0$

the calculated R^2 values; and the dash-dotted lines show the ideal case scenarios when the predicted values are equal to the experimental ones.

It can be seen from Fig. 4 that the aforementioned sample with an estimation error of 148% is clearly an outlier, however as there were only six samples in the validation set, this outlier strongly influences the overall *average* accuracy of predictions, see also R^2 value. It is evident, however, that removing the outlier would result in trend lines being much closer to the ideal case with higher R^2 values.

While trying to explain this error, we noticed, that *experimentally observed* 1.6 MPa m$^{1/2}$ is the minimum ΔK_{th} in the whole aluminium data set in Table 1. It is known that when, mathematically, estimating data on a boundary of the observed data range, it is highly possible to get large errors as in this case the problem of extrapolation is being solved. In this case we would recommend to be extra careful with such data points.

For the completeness of statistical analysis, FIS methodology has been re-trained and re-validated for several (five in total), randomly selected realisations, maintaining 70%–30% split in samples used for training and validations respectively. This statistical analysis yielded a mean (between realisations) of Pct errors 27.14% for ΔK_{th} and 27.38% for $\Delta\sigma_0$ respectively; the standard deviation was reported as 8.46 for ΔK_{th} and 2.76 for $\Delta\sigma_0$, respectively. Overall, the FIS approach has shown a stable performance.

Steel

Next, different steels were analysed as presented in Table 3. Note, that available experimental data (again adapted from [2]) ranges from mild steels to high strength steels. That makes it even more interesting to see if our data driven approach is capable of predicting properties of such diverse range of steel types.

Similar to aluminium alloys test, the complete data set of 35 steel experimental results, was split into two subsets: the first one, used for training stage and creating the fuzzy rule base (24 data entries, approximately 70%); and the second subset (again, every 3rd row, i.e. approximately 30% of data, from Table 3 – 11 data entries in total) was classified as the validation group whose outputs are assumed to be *unknown*. Following the logic of FIS methodology, the inputs of validation group are then fed into in the FIS for the predicted (estimated) results, which are then compared with the

Table 3 Experimental results for steels, adapted from [2]

σ_{UTS} [MPa]	σ_Y [MPa]	R	ΔK_{th} [MPam$^{1/2}$]	$\Delta\sigma_0$ [MPa]
530	380	–1	12.0	440
769	721	–1	10.5	750
600	350	–1	7.0	270
632	364	–1	8.1	582
623	377	–1	10.3	498
957	820	–1	12.8	1000
823	794	–1	13.0	800
433	286	–1	8.1	440
402	233	–1	8.2	380
440	334	–1	12.8	420
353	206	–1	11.0	362
543	284	–1	10.5	480
364	227	–1	11.6	234
720	390	–1	9.0	604
745	472	–1	6.9	606
798	645	–1	7.7	760
1984	1617	–1	7.5	1200
1182	1054	–1	7.0	1134
965	847	–1	6.9	1000
1011	900	–1	7.5	920
423	194	–1	12.4	326
445	280	–1	10.2	396
666	588	–1	12.7	630
516	339	–1	11.6	440
803	683	–1	7.4	860
476	360	–1	8.0	490
520	446	–1	9.8	570
483	251	0	6.5	318
720	390	0	13.8	448
423	194	0	8.4	274
1855	1540	0.1	4.5	450
1770	1620	0.1	4.2	1050
598	510	0.1	6.6	425
423	194	0.4	6.4	244
270	190	0.1	6.7	190

Table 4 Experimental output with the estimation results for steel samples

σ_{UTS} [MPa]	σ_Y [MPa]	R	ΔK_{th} [MPam$^{1/2}$]	$\Delta\sigma_0$ [MPa]	Pred ΔK_{th} [MPam$^{1/2}$]	Pred $\Delta\sigma_0$ [MPa]	Pct error ΔK_{th}	Pct error $\Delta\sigma_0$
600	350	–1	7.0	270	9.86	462.85	40.88	71.43
957	820	–1	12.8	1000	10.01	898.12	21.82	10.19
402	233	–1	8.2	380	11.60	234.00	41.46	38.42
543	284	–1	10.5	480	8.83	424.69	15.90	11.52
745	472	–1	6.9	606	10.53	614.77	52.65	1.45
1182	1054	–1	7.0	1134	7.50	920.00	7.45	18.87
423	194	–1	12.4	326	11.00	362.00	11.29	11.04
516	339	–1	11.6	440	11.84	434.00	2.07	1.36
520	446	–1	9.8	570	10.05	611.37	2.54	7.26
423	194	0	8.4	274	7.09	205.76	15.55	24.90
598	510	0.1	6.6	425	13.80	448.00	109.09	5.41
						Average	29.16	18.35

actual experimental results to analyse the accuracy of our estimations. The predicted results, together with the actual experimental data and percentage errors are shown in Table 4.

For steel samples, the average estimation error is reported as 29.16% for ΔK_{th} and 18.35% for $\Delta\sigma_0$, respectfully. Generally, it can be seen that the proposed FIS framework is capable of estimating both output parameters based on the given input parameters.

Again, a statistical analysis of five randomly selected realisations of training and validation sets, offered an average (between different realisations) Pct errors of 28.05% for ΔK_{th} and 19.65% for $\Delta\sigma_0$ respectively; the standard deviation was reported as 8.18 for ΔK_{th} and 3.35 for $\Delta\sigma_0$, respectively.

In the following study we have "exaggerated" the aforementioned statement and taken "all, but one" experimental data entries, reported in Table 3, as *known*, in order to estimate just one, assumed *unknown*, test data entry. The process was repeated for every single data point (i.e. for all experiments, one at a time), replicating the scenario of potentially existing large data-set, that is used for training of the model, and a new particular unknown material being analysed. The results of the analysis is presented in log-log graphs in Fig. 5.

Two sets of data are presented in each figure with blue dots representing experimental values (used for validation) and orange dots representing predicted by model parameters. As it can be seen, the model is successful in predicting trends of parameters relationships, and general accuracy (apart from some particular outliers) is fairly high. Every outlier needs to be analysed separately and the discussion needs to be held, whether the particular outlier appears as a result of extrapolation problem (when testing data appears *outside* of the previously observed range), or any other reasons needs to be investigated. We must repeat the comment mentioned earlier: experimental data, analysed here, was taken from very different sources, hence we do accept some degree of possible inaccuracy.

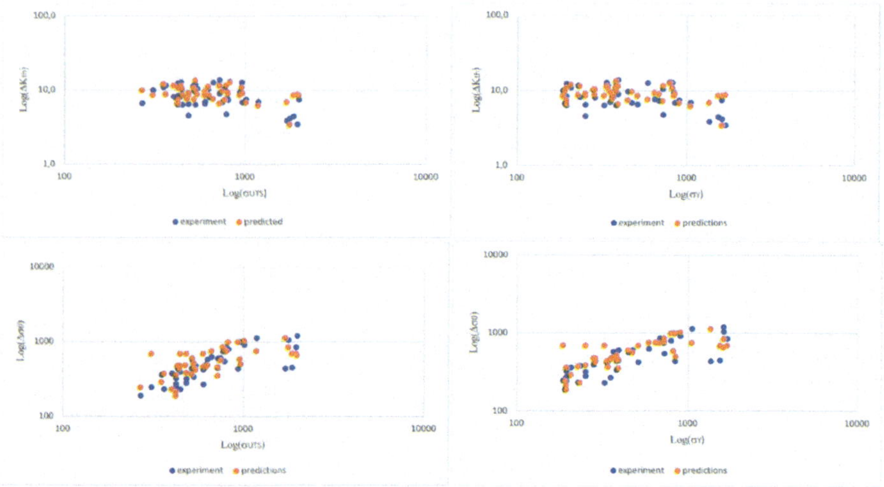

Fig. 5 Predicted (orange) versus observed (blue) ΔK_{th} and $\Delta\sigma_0$ – "all, but one" approach

4 Conclusions

The main aim of the present contribution was to analyse the ability of the data driven approach to predicting fatigue (time-variable) parameters of a material, knowing material's strength (static) characteristics. Based on discussed above tests, analysing aluminium alloys and different steels, we can conclude that FIS approach is capable to estimate the range of the plain fatigue limit and the range of the threshold value of the stress intensity factor, based on given ultimate tensile strength and yield strength. It has been shown that trends and actual estimated values are predicted with relatively high degree of accuracy. Considering the intrinsic versatility of the presented data-driven approach, FIS can deal with a number of different parameters. As a comprehensive solution, it has a great potential as a tool for characterising the relationship between strength parameters and fatigue parameters.

Competing Interests The authors have no conflicts of interest to declare that are relevant to the content of this chapter.

References

1. Odonnell, M.: The fatigue limit and its elimination. Fatigue Fract. Eng. Mater. Struct. **22**(7), 545–557 (2001)
2. Susmel, L.: Multiaxial Notch Fatigue. Elsevier Science (2022)
3. Tu, R., Gitman, I.M., Susmel, L.: Fuzzy inference system for failure strength estimation of plain and notched 3D-printed polylactide components. Fatigue Fract. Eng. Mater. Struct. **45**, 1663–1677 (2022)

4. Tu, R., Gitman, I.M., Susmel, L.: Fuzzy set-based methodology for manufacturing parameter determination of 3D-printed polylactide components with user-specified strength, geometrical design, and cost requirements. Fatigue Fract. Eng. Mater. Struct. **46**, 2754–2765 (2023)
5. Slodczyk, I., Fletcher, D., Gitman I.M., Whitney B.: Fuzzy inference model for railway track buckling prediction. Transp. Res. Rec. J. Transp. Res. Board (2023). https://doi.org/10.1177/03611981231184245.
6. Nasiri, S., Khosravani, M.R.: Applications of data-driven approaches in prediction of fatigue and fracture. Mater. Today Commun. **33**, 104437 (2022)
7. Zadeh, L.A.: Fuzzy Sets. Inf. Control. **8**, 338–353 (1965)
8. Çen, Z.: Fuzzy algorithm for estimation of solar irradiation from sunshine duration. Sol. Energy **63**, 39–49 (1998)
9. Demir, F.: A new way of prediction elastic modulus of normal and high strength concrete-fuzzy logic. Cem. Concr. Res. **35**, 1531–1538 (2005)
10. Tay, K.M., Lim, C.P.: Optimization of gaussian fuzzy membership functions and evaluation of the monotonicity property of fuzzy inference systems. In: IEEE International Conference on Fuzzy Systems, pp. 1219–1224 (2011). https://doi.org/10.1109/FUZZY.2011.6007387.

Using Neural Networks to Explore Structure-Property Relations in Bio-Inspired Impact-Resistant Structures

Shashank Kushwaha, Junyan He, Diab Abueidda, and Iwona Jasiuk

Abstract Inspired by biological structural designs observed in nature, this work explores the structure-property relationships in structures under dynamic transverse and longitudinal compression. The two primary structures analyzed are low porosity structures inspired by sheep horns and 2D extruded thin-walled structures inspired by design elements found in bamboo, beetles, and crabs. The low-porosity structures exhibit superior mechanical properties in nature, with porosity ranging from 1–5%, while the thin-walled structures provide insights into the effect of geometric feature interactions leading to high energy absorption during impact. The current work utilizes the gated recurrent unit (GRU) model to predict the mechanical response of the structures during the impact. The inputs used in GRU models were varied to present different techniques to predict stress-strain response for the structures at a given loading condition. The first method utilized the parametric representation of the geometric features, while the other used a combinatorial approach and autoencoders to prepare inputs for the GRU model. The ground-truth data was obtained using finite element simulations with the rate-dependent elastoplastic and Johnson-Cook material models. The trained models allow rapid evaluations of stress-strain response and allow the elimination of poor designs.

Keywords Neural networks · Structure-property relations · Thin-walled structures · Impact-resistant structures · Energy absorption · Stress-strain prediction

---

S. Kushwaha · J. He · D. Abueidda · I. Jasiuk (✉)
University of Illinois Urbana-Champaign, Urbana, IL, USA
e-mail: ijasiuk@illinois.edu

S. Kushwaha
e-mail: sk89@illinois.edu

J. He
e-mail: junyanh2@illinois.edu

D. Abueidda
e-mail: abueidd2@illinois.edu

F. Willot et al. (eds.), *Continuum Models and Discrete Systems*,
Springer Proceedings in Mathematics & Statistics 457,
https://doi.org/10.1007/978-3-031-58665-1_21

1 Introduction

Over millions of years, biological structures have evolved to be multifunctional, impact-resistant, and anatomically optimized. The biological designs derived from nature offer insights into fabricating engineering structures with exceptional mechanical properties. During collisions and combat, the biological structures withstand axial, lateral, or multi-directional loading at low-to-medium velocity impacts. One of the examples includes intricate architectures like trabecular-honeycomb structure in beetle elytra, capable of handling impacts up to 50 m/s [1]. Other examples include sheep horns, which can bear a huge impact force of 3400 N [2]. Similarly, the equine hoof sustains high-impact loading forces close to 9000 N while galloping [3]. A common feature in equine hoofs and sheep horns is the presence of long tubular arrays, resulting in low-porosity structures. Such structures have a porosity range of 1% to 5% and exhibit distinctive attributes that render them highly favorable in applications necessitating robust resistance to dynamic forces. Unlike equine hoofs, the sheep horns absorb more energy when tubules are compressed in a transverse direction [14]. Hence, this work explores the performance of 2D low-porosity structures under transverse compression loading.

At the same time, bio-inspired thin-walled lattices combine the benefit of robust mechanical performance under impact and being lightweight. Thin-walled structures with different cross-sections have been investigated under various loading conditions [4, 5]. The crushing strength in such structures is affected by the edge length and the wall thickness [6]. The studies by Zhao et al. [7] found that hierarchical honeycomb cells improved elastic properties and damage tolerance, compared to regular honeycombs. Further, the bio-inspired honeycombs inspired by beetle elytra have also been shown to have better compressive mechanical properties [1]. Such studies indicate that the properties of the thin-walled lattices are affected by their 2D cross-section and loading conditions. Hence, this work also explores structure-property relations in the 2D extruded thin-walled lattices.

Overall, this paper develops a framework to generate distinct cross-section designs for transverse and longitudinal compression. It uses a combinatorial framework to generate thin-walled lattices for longitudinal compression and parametric inputs for the low-porosity structures. Neural network (NN) models have been used to predict stress-strain curves and toughness of composites [8, 9]. However, the use of NN for bio-inspired structures remains scarce. Therefore, this work develops two NN models to approximate the structure-property relations, linking the input design parameters and loading conditions with the mechanical performance of the structure. Once trained, the NN can efficiently predict the mechanical performance of new designs much faster than classical numerical simulations, thus allowing a rapid preliminary survey of the design space.

2 Methods

2.1 Geometry Generation and Finite Element Analysis

The designs considered in this work can be uniquely characterized by their 2D, in-plane cross-sections. The two types of geometries considered in the study include 2D structures with polygon-shaped porosities and thin-walled lattices defined using geometric descriptors represented using a key-based system or simply using geometric parameters. In both cases, a Python script was developed to generate cross-sectional sketches in the finite element (FE) analysis package Abaqus [10]. For low-porosity structures, the cross-section was defined using volume fraction, tubule shape, tubule orientation, and the arrangement of the tubules within the structure. The ranges of the geometric parameters used to define low-porosity structures are defined in Table 1. However, all the tubules in a given configuration have the same shape, and the designs with non-intersecting tubules were considered valid. Other designs were excluded from the analysis. Some selected structures in the design space are shown in Fig. 1.

The thin-walled lattices were generated using geometric descriptors, including vertex, edge segment, and interior support. The geometric descriptors were inspired by truss descriptors used by Bastek et al. [11] and Zok et al. [12]. The geometric descriptors are defined as follows:

- Vertex. The vertices of the unit cell bounding box can take on different styles: it can be a vertex or a straight edge, or an arc can replace the vertex; see Fig. 2.
- Edge segment. Each edge of the unit cell bounding box can be divided into different edge segments. Each can take on a different style: it can be a straight edge or two

Table 1 Ranges of geometric parameters in low-porosity structures

Tubule volume fraction	Sides in tubule cross-section	Rotation angle (degrees)	Rows (ny)	Columns (nx)
[1%, 10%]	[3, 6] (Triangles to Hexagons)	[0, 360]	[1, 8]	[1, 8]

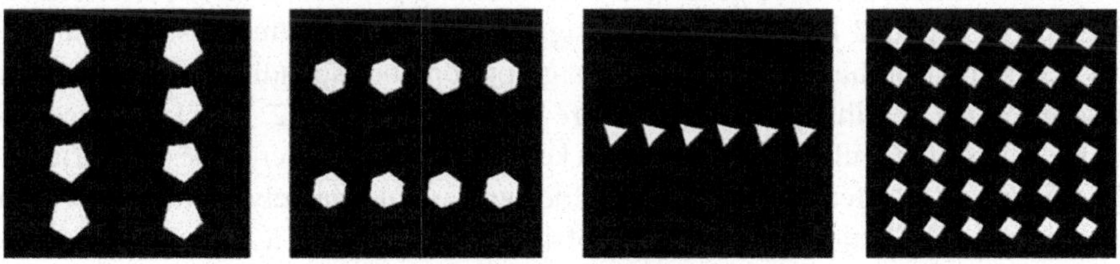

Fig. 1 Sample structures in the design space for low porosity structures

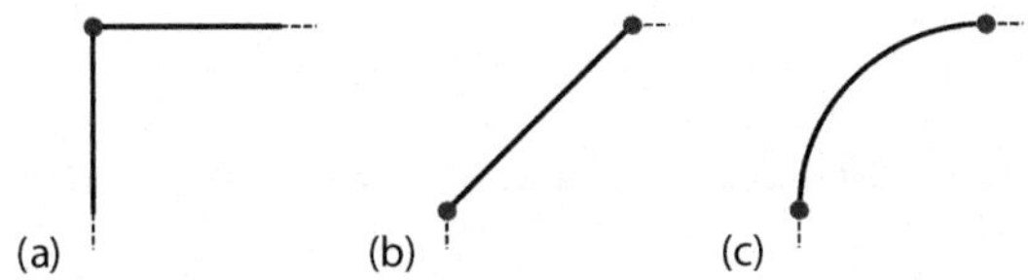

Fig. 2 **a** The vertex of the unit cell remains as-is. **b** A straight edge replaces the vertex. **c** An arc replaces the vertex

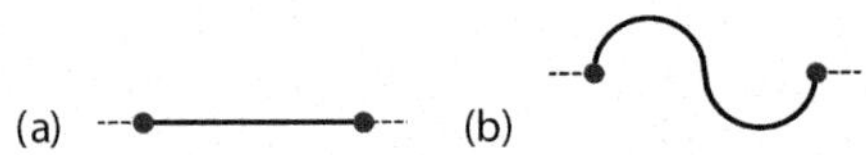

Fig. 3 **a** The edge segment of the unit cell remains as-is. **b** Two arcs replace the edge segment

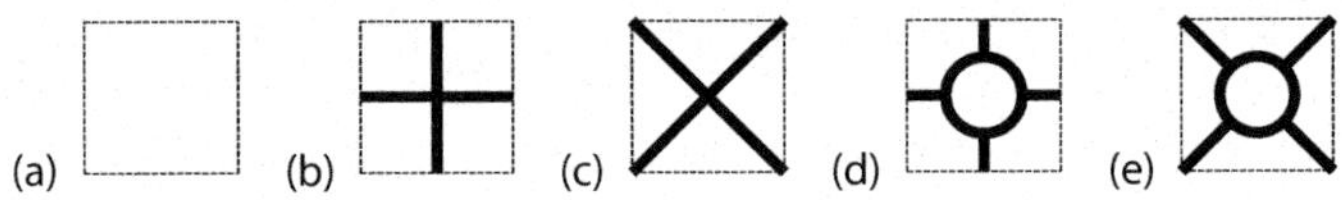

Fig. 4 **a** No interior support. **b** A +-shape support. **c** A X-shape support. **d** A +-shape support along with a circle. **e** A X-shape support is added along with a circle

Table 2 Explanation of the 8-digit design key system

Digit	1	2	3	4	5	6	7	8
Option	Vertex	Arc sub-opt	# segments horizontal	# segments vertical	Edge sub-opt horizontal	Edge sub-opt vertical	Interior	Interior sub-opt
Possible values	0 (Fig. 1a) 1 (Fig. 1b) 2 (Fig. 1c)	0 (in) 1 (out)	2 3 4	2 3 4	0 (Fig. 2a) 1 (Fig. 2b)	0 (Fig. 2a) 1 (Fig. 2b)	0 (Fig. 3a) 1 (Fig. 3b) 2 (Fig. 3c)	0 (Fig. 3d) 1 (Fig. 3e)

arcs (inspired by the bio-inspired honeycomb studied in the work of Ha et al. [13]), see Fig. 3.

- Interior support. Supports can be added to the unit cell's interior: no support, a +-shape, or an X-shape support. A circle can be added to the center of the unit cell; see Fig. 4.

Unique lattices could be generated by combining geometric descriptors. The lattice designs created by an array of design options were defined using an 8-digit key. Each digit in the key refers to a design option corresponding to the geometric descriptor. All possible design options are presented in Table 2. This key-based system consists of 660 unique lattice design keys. Besides the discrete design variables, the framework also features a continuous design variable, namely the thickness of the lattice walls. The lattice designs consist of 4 unit cells, forming a 2-by-2 arrangement. Sample designs defined using a key-based system are presented in Fig. 5.

McKittrick et al. [14] discussed that when rams butt heads, the horns are loaded in the transverse direction, which provides more energy absorption than in the longi-

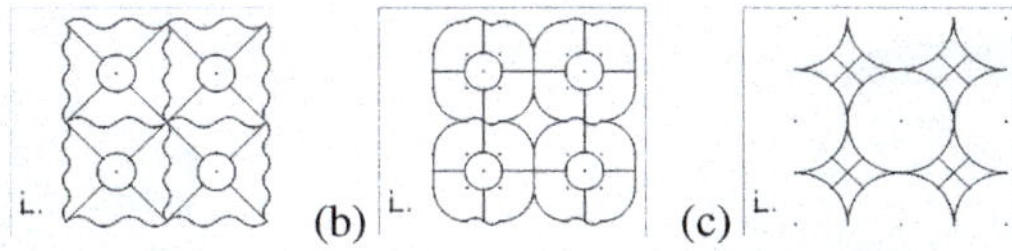

Fig. 5 A survey of some lattice designs generated by the key-based system. **a** Design 00231121 **b** Design 10331011 **c** Design 11220020

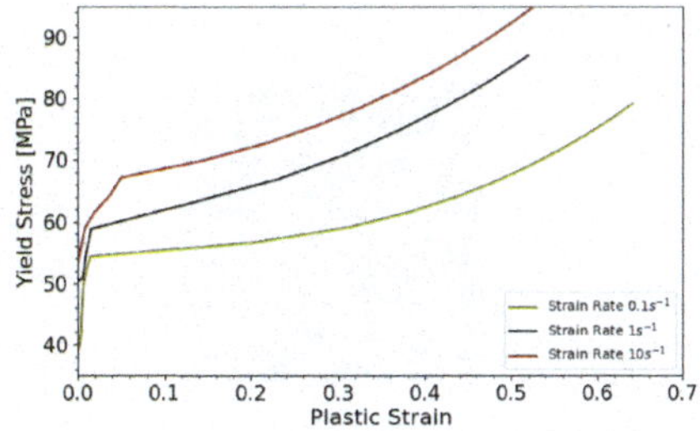

Fig. 6 Yield stress versus plastic strain at different strain rates

tudinal direction. Hence, low-porosity structures are loaded in a transverse direction in contrast to the thin-walled structures commonly used in crash-worthiness applications when compressed in the longitudinal direction. The Abaqus/Explicit dynamic simulations for low-porosity designs utilized a rate-dependent elastic-plastic material model to capture the structures' response at varying strain rates. The material properties of the base material chosen for the study are similar to polycarbonate-acrylonitrile butadiene styrene (PC-ABS). The Young's modulus and Poisson's ratio are 2.5 GPa and 0.35, respectively. The strain rate dependent yield stress versus plastic strain curves used to define the plastic region are included in Fig. 6. All the structures were discretized with 4-node bilinear plane strain quadrilateral elements with reduced integration. A nominal element edge length of 0.24 mm was chosen for meshing. The thin-walled lattices were modeled using shell elements and the material was Ti-6Al-4V. The Johnson-Cook plasticity and damage model [15] were used with material properties taken from literature [16, 17]. For simplicity, the thickness-dependent material properties [18] of thin-walled parts were not considered.

In this study, the boundary conditions for impact loading were approximated by sandwiching the structure between two rigid plates. The bottom plate was fixed, and the top plate traveled downward with a constant velocity. The nominal strain rate was uniformly sampled from the range [0.45, 90.9] s^{-1} corresponding to indenter velocity from the range [5, 1000] m/s for low-porosity structures. For the thin-walled lattice structures, the nominal strain rates were uniformly sampled on the log scale from the range [10^2, 10^5] s^{-1}, representing strain rates during blast loading [19]. The wall thicknesses for thin-walled lattice designs were uniformly sampled from the range [0.25, 0.75] mm. All sidewalls were traction-free and were free to deform. All simulations had a constant final nominal compressive strain of 25% for low-porosity structures and 20% for thin-walled lattices. The reaction force, displacement plastic dissipation, and elastic strain energy of the structures were outputs of the

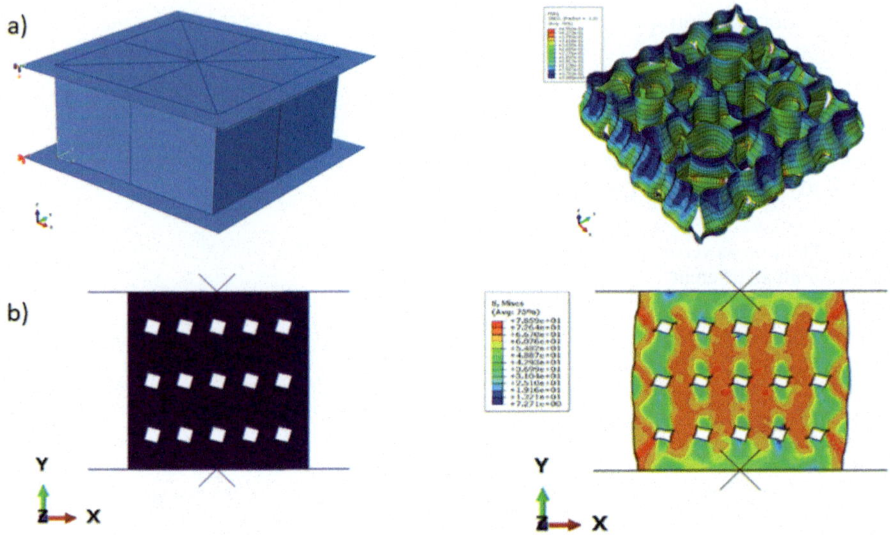

Fig. 7 FE model setup and deformed structures showing von Mises stress: **a** Typical thin-walled lattice under dynamic longitudinal compression. **b** Typical low-porosity structure under dynamic transverse compression

FE simulations. Figure 7 depicts the FE models and typical deformed structures at the end of dynamic compression. A total of 7196 simulations were conducted for low-porosity structures, whereas 15000 simulations were conducted for thin-walled lattices on an AMD Ryzen 7 5800H processor with 8 cores.

2.2 *Encoding Lattice Cross-Sectional Images*

The 2D cross-section of the thin-walled lattice was converted into a 128 × 128 binary image. The image did not include the shell thickness. Hence, the thickness of the material at the walls was a single pixel wide for all the structures. This means that image matrices are highly sparse, and it would be wasteful to include entire binary image information as input to the GRU model. Hence, an autoencoder extracted information from the binary image of the lattice. The autoencoder consists of two dense layers, each with 100 neurons and a rectified linear unit activation function. The trained autoencoder converted a 128 × 128 binary image into a 100 × 1 latent feature vector. The decoder consists of a single dense layer with 128^2 neurons with a sigmoid activation function. The autoencoder was developed in Keras [20] with Tensorflow [21] backend. The autoencoder was trained and tested using 600 randomly chosen design images with an 80–20 split. We utilized mean squared error (MSE) as the loss function and the Adam optimizer [22] to minimize the loss function. The autoencoder was trained for 80 epochs with a batch size of 50. The performance of the autoencoder was checked using 60 additional lattice designs not included in the training data set.

2.3 *Neural Network for Sequence Prediction*

2.3.1 Input Data, Data Augmentation, and Loss Function

The output arrays obtained from FE simulations were downsampled to 50-time steps for efficient NN training. The inputs are given as parameters discussed in Sect. 2.1 for the 2D plane strain problem, whereas for thin-walled structures, a 100×1 latent feature vector was used as input (defined in Sect. 2.2). Further, the inputs include physics-informed temporal information arrays. The 2D plane strain and longitudinal compression problem used the array of current time values, nominal compression strain, and nominal compression strain rate. The longitudinal compression also defined a few additional arrays as inputs. These include lattice shell thickness, $\log_{10}\bar{\epsilon}$ (the logarithm of the strain rate is used instead of the strain rate since the strain rate dependence in the Johnson-Cook model is logarithmic), and a binary elastic wave indicator is defined as:

$$I_e = \begin{cases} 0, & t \leq t_e \\ 1, & t > t_e \end{cases}, \tag{1}$$

where $t_e = \frac{H}{\sqrt{E/\rho}}$ is an estimated time for the elastic stress wave to travel through the height of the lattice H. This indicates that the reaction force at the rigid bottom plate should remain 0 before the impact stress wave arrives.

A standard scalar in Scikit-Learn normalized all the inputs. The available data was increased using data augmentation. For the 2D plane strain problem, multiple final nominal strains were randomly sampled in the range [10%, 25%], whereas, for longitudinal compression of thin-walled structures, the final nominal strains were sampled in the range [5%, 20%].

The mean absolute error (MAE) has been employed as the loss function in this study [23]. The loss function is defined as:

$$\text{MAE} = \frac{\sum_{i=1}^{N} |\mathbf{Y}_i - \hat{\mathbf{Y}}_i|}{N}, \tag{2}$$

where N, $\mathbf{Y}_i$, $\hat{\mathbf{Y}}_i$ denote the number of training data points, ground-truth outputs, and the NN predictions, respectively.

2.3.2 Neural Network Model

The current work utilizes a recurrent neural network (RNN) model, specifically, the gated recurrent unit (GRU) model, to train the forward model for output prediction. The model has been used for sequence predictions for various applications [24, 25]. Further, Abueidda et al. [26] compared the performance of various RNN models to predict the response of elastoplastic material under variable strain rates. Although GRU model is computationally expensive compared to other RNN models, it predicts sequential outputs with low error. Based on the capabilities of the GRU model to predict sequential data, this study utilizes the GRU model to predict stress-strain curves for transverse compression in a 2D plane-strain problem and longitudinal compression of thin-walled structures.

The developed GRU model was implemented and tested in Keras [20] with a TensorFlow [21] backend. The GRU models for the two problems had three stacked layers, each with a hyperbolic tangent (tanh) activation function. However, the first model for the 2D plane strain problem utilized 475 units in each layer, while the second model for longitudinal compression for the thin-walled lattices used 300 units in each layer. This led to 3.77 million trainable parameters for the first model compared to 1.45 million trainable parameters for the second model. While the second model utilized less trainable parameters, it also required the training of an additional autoencoder as discussed in Sect. 2.2. The NN architecture schematic [27] for the first model is shown in Fig. 8. The second GRU model uses 100×1 latent feature vector from autoencoder and temporal arrays as input [28]. The loss function was minimized using an Adam optimizer [22] with an initial learning rate of $1 \text{ x } 10^{-3}$. The model was trained with a batch size of 600 for 150 epochs. The training was repeated 10 times to obtain average training time and model accuracy. The dataset selection was randomized for each training as described in Sect. 2.3.1.

2.3.3 Global Optimization

A Python script was developed to traverse the input design space and evaluate specific energy absorption (SEA) using the trained neural network predictions. The input space was divided into grid points based on the input parameters discussed

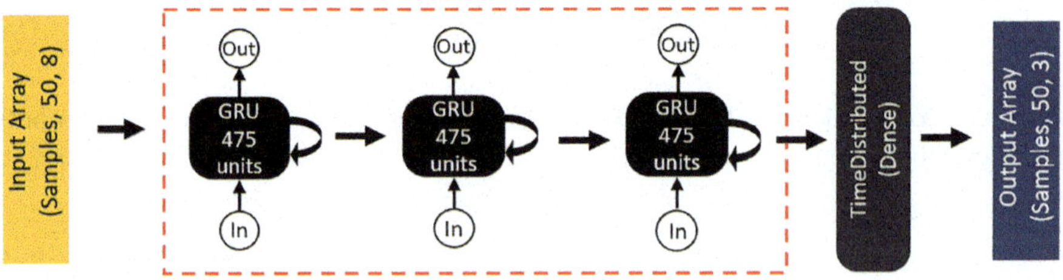

Fig. 8 Neural network architecture

in Sect. 2.3.1. Each grid point represents a unique structure within the input design space. The SEA was evaluated for each grid point by using the area under the load-displacement curve predicted by the trained GRU model. The continuous variables were divided into equally spaced intervals, while other variables could take discrete values within their respective input range.

3 Results and Discussion

3.1 Lattice Design Encoding and Reconstruction

The 2D cross-sectional lattice binary images were compared to the reconstructed ones generated by the trained autoencoder for three randomly chosen designs, see Fig. 9. The dice similarity coefficient (DSC) is used to quantify the reconstruction quality:

$$\mathrm{DSC} = \frac{2|\boldsymbol{I} \cap \boldsymbol{I}_r|}{|\boldsymbol{I}| + |\boldsymbol{I}_r|}, \tag{3}$$

where $\boldsymbol{I}$ and $\boldsymbol{I}_r$ are the ground-truth and reconstructed binary skeleton images (1 denotes the presence of material and 0 indicates void), respectively.

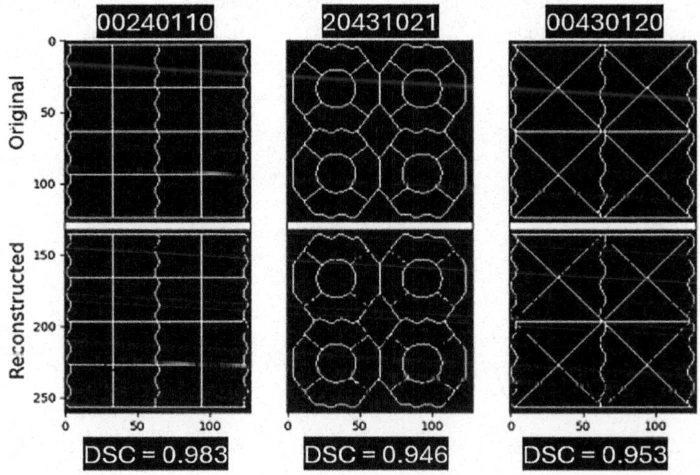

Fig. 9 Comparison of original and reconstructed lattice design skeletons for three randomly chosen designs

3.2 Predicting Stress-Strain Curves

The average response of the GRU model was measured by training the model 10 times, shuffling data for each iteration. The average times for FE simulations and training and inference times for the GRU model are presented in Table 3.

The NN predictions were compared to the ground truth obtained from FE simulations, ranked by the percentile of MAE (2) for each output array. The model with median MAE among the ten trained NNs was used to generate the plots in Fig. 10 for both problems. Up to the 75% percentile shown in the first three columns in Fig. 10, the GRU model could closely predict the stress-strain curves. Even in the worst case, the GRU model predicts the stress-strain curve's general shape. For the low-porosity structures, a cross-section image of the structure was parameterized by five design variables. In contrast, a 2D cross-section of a thin-walled lattice was converted to a 100×1 latent feature vector. The latter approach used an additional autoencoder apart from the GRU model, while for the low-porosity structures, a larger GRU model without using any other NN models. Although other output arrays like plastic dissipation, elastic strain energy, and damage dissipation were also predicted by the two trained neural networks, they are not presented here. Further details can be found in the works by He et al. [28] and Kushwaha et al. [27].

Table 3 Computational cost for GRU training, inference, and FE simulations

	GRU training	GRU inference	FE simulation
Low-porosity structure	5192.9s	1.63×10^{-4}s	300–1800s
Thin-walled lattice	3597.4s	1.87×10^{-4}s	60–120s

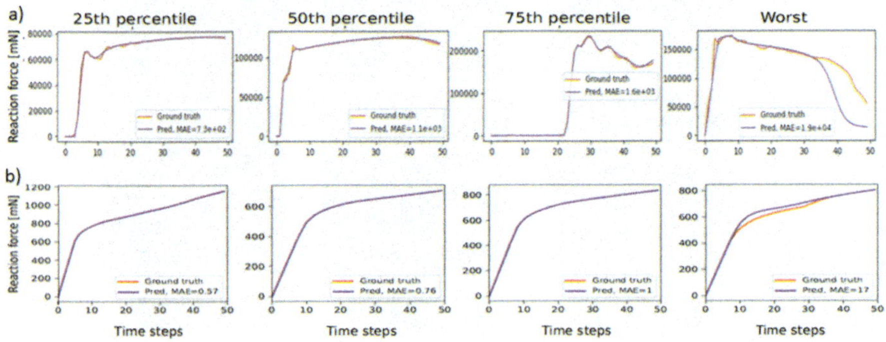

Fig. 10 Comparison of ground truths and GRU predictions for the data set, ranked by percentile of MAE (2) to provide a representative sampling **a** Thin-walled lattice under longitudinal compression. **b** Low-porosity structures under transverse compression

3.3 Structure-SEA Map

The Python script described in Sect. 2.3.3 was used to calculate SEA using the predicted load-displacement curve for each structure. Each design index for a low-porosity structure corresponds to one grid point defined by the five geometric parameters, while each thin-walled lattice is defined using an eight-letter key as discussed in Sect. 2.1. This work presents the structure-property relations for thin-walled lattices in Fig. 11. The figure shows the design index plotted against SEA for various wall thicknesses represented in the colormap. The figure shows that some 2D cross-sections in thin-walled lattices lead to higher SEA, even with low wall thickness.

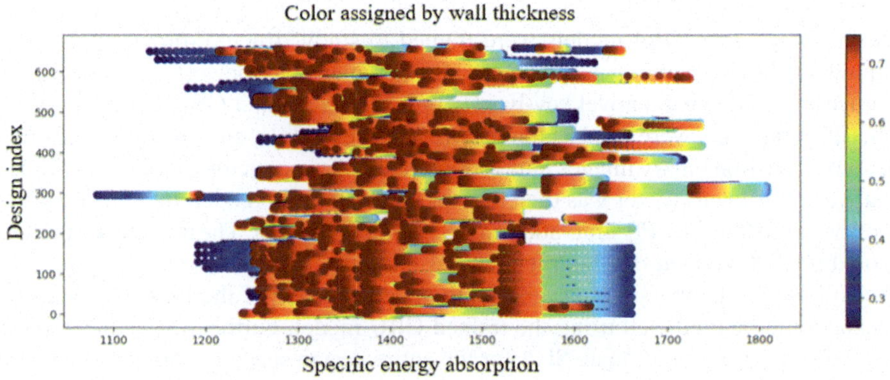

Fig. 11 Distribution of SEA for different designs and thicknesses

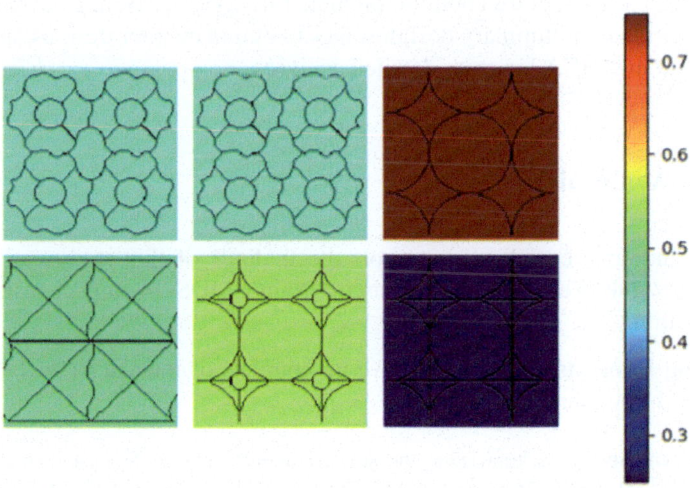

Fig. 12 Optimum designs for varying strain rates. The strain rate increases from left to right and from the first row to the second row. The background is colored by the wall thickness of the design

To gain further insight into design features that maximize SEA, we conducted a brute-force optimization using the trained GRU model. It fixes the final strain and maximum volume but varies the strain rate. After optimization, the optimum designs for various strain rates are presented in Fig. 12. We observe that the optimum designs are strain-rate sensitive since different optimum designs are obtained for different strain rates. Hence, this method is useful for rapid preliminary screening to identify designs with high SEA while eliminating low-performance designs. Further, the SEA maps for low-porosity structures can be found in work by Kushwaha et al. [27].

4 Conclusions

In this work, two GRU models were trained to study structure-property relations in low-porosity structures generated using a parametric framework and thin-walled lattices defined using a novel key-based approach. In the first GRU model, the geometric parameters were used as inputs directly. Meanwhile, an autocoder converted 2D cross-section binary images into 100×1 latent feature vectors for the thin-walled lattices. This latent vector was then used as input for GRU model. In both cases, the data was generated by FE simulations using Abaqus/ Explicit. The trained NN models could predict the load-displacement curves up to the 75% percentile for both problems. Even in the worst-case scenario, the general shape of the load-displacement curve could be predicted using the trained GRU model. Further, the trained neural network was used to evaluate SEA for the entire design space in both problems and identify the best-performing designs. Overall, the trained model NN could efficiently generate all the performance predictions, even on low-end laptops. The stress-strain response for each structure could be predicted in 0.18 ms. Hence, it renders itself a suitable guide in preliminary design stages to eliminate poor designs and identify designs with high SEA for more detailed analyses.

5 Data Availability

The data and source code that support the findings of this study can be found at: https://github.com/Jasiuk-Research-Group/CMDS-14-impact-resistantance-NN-prediction-

Competing Interests The authors have no conflicts of interest to declare that are relevant to the content of this chapter.

Acknowledgements We acknowledge the support of the National Science Foundation grant (MOMS-1926353) and the Army Research Office contract (No. W 911NF-18-2-0067).

References

1. Yu, X., Pan, L., Chen, J., Zhang, X., Wei, P.: Experimental and numerical study on the energy absorption abilities of trabecular-honeycomb biomimetic structures inspired by beetle elytra. J. Mater. Sci. **54**(3), 2193–2204 (2019)
2. Kitchener, A.: An analysis of the forces of fighting of the blackbuck (antilope cervicapra) and the bighorn sheep (ovis canadensis) and the mechanical design of the horn of bovids. J. Zool. **214**(1), 1–20 (1988)
3. Bertram, J.E.A., Gosline, J.M.: Fracture toughness design in horse hoof keratin. J. Exp. Biol. **125**(1), 29–47 (1986)
4. Wu, S., Zheng, G., Sun, G., Liu, Q., Li, G., Li, Q.: On design of multi-cell thin-wall structures for crashworthiness. Int. J. Impact Eng. **88**, 102–117 (2016)
5. Fengxiang, X.: Enhancing material efficiency of energy absorbers through graded thickness structures. Thin Walled Struct. **97**, 250–265 (2015)
6. Li, K., Gao, X.-L., Wang, J.: Dynamic crushing behavior of honeycomb structures with irregular cell shapes and non-uniform cell wall thickness. Int. J. Solids Struct. *44*(14–15), 5003–5026 (2007)
7. Zhao, L., Zheng, Q., Fan, H., Jin, F.: Hierarchical composite honeycombs. Mater. Des. **40**, 124–129 (2012)
8. Yang, C., Kim, Y., Ryu, S., Gu, G.X.: Prediction of composite microstructure stress-strain curves using convolutional neural networks. Mater. Des. **189**, 108509 (2020)
9. Abueidda, D.W., Almasri, M., Ammourah, R., Ravaioli, U., Jasiuk, I.M., Sobh, N.A.: Prediction and optimization of mechanical properties of composites using convolutional neural networks. Compos. Struct. **227**, 111264 (2019)
10. SIMULIA. Abaqus (2021)
11. Bastek, J.-H., Kumar, S., Telgen, B., Glaesener, R.N., Kochmann, D.M.: Inverting the structure-property map of truss metamaterials by deep learning. Proc. Natl. Acad. Sci. **119**(1), e2111505119 (2022)
12. Zok, F.W., Latture, R.M., Begley, M.R.: Periodic truss structures. J. Mech. Phys. Solids **96**, 184–203 (2016)
13. Ha, N.S., Lu, G., Xiang, X.: Energy absorption of a bio-inspired honeycomb sandwich panel. J. Mater. Sci. **54**(8), 6286–6300 (2019)
14. McKittrick, J., Chen, P.-Y., Tombolato, L., Novitskaya, E.E., Trim, M.W., Hirata, G.A., Olevsky, E.A., Horstemeyer, M.F., Meyers, M.A.: Energy absorbent natural materials and bioinspired design strategies: a review. Mater. Sci. Eng. C **30**(3), 331–342 (2010)
15. Johnson, G.R., Cook, W.H.: Fracture characteristics of three metals subjected to various strains, strain rates, temperatures and pressures. Eng. Fract. Mech. **21**(1), 31–48 (1985)
16. Kay, G.: Failure modeling of titanium 6AI-4V and aluminum 2024–T3 with the Johnson-Cook material model. Office of Aviation Research, Federal Aviation Administration (2003)
17. Leseur, D.: Experimental investigations of material models for Ti-6A1-4V and 2024-T3. https://doi.org/10.2172/11977.
18. He, J., Kushwaha, S., Mahrous, M.A., Abueidda, D., Faierson, E., Jasiuk, I.: Size-dependence of AM Ti-6Al-4V: Experimental characterization and applications in thin-walled structures simulations. Thin Walled Struct. **187**, (2023)
19. Othman, H., Marzouk, H.: Strain rate sensitivity of fiber-reinforced cementitious composites. Am. Concr. Inst. Mater. J. **113**(2), 143–150 (2016)
20. Francois, C., et al.: Keras (2015). https://github.com/fchollet/keras
21. Abadi, M., Agarwal, A., Barham, P., Brevdo, E., Chen, Z., Citro, C., Corrado, G.S., et al.: Tensorflow: Large-scale machine learning on heterogeneous distributed systems (2016). arXiv:1603.04467
22. Kingma, D.P., Ba, J.: Adam: A method for stochastic optimization (2014). arXiv:1412.6980
23. Willmott, C.J., Matsuura, K.: Advantages of the mean absolute error (MAE) over the root mean square error (RMSE) in assessing average model performance. Clim. Res. **30**(1), 79–82 (2005)

24. Gorji, M.B., Mozaffar, M., Heidenreich, J.N., Cao, J., Mohr, D.: On the potential of recurrent neural networks for modeling path dependent plasticity. J. Mech. Phys. Solids **143**, 103972 (2020)
25. Qu, T., Di, S., Feng, Y.T., Wang, M., Zhao, T.: Towards data-driven constitutive modelling for granular materials via micromechanics-informed deep learning. Int. J. Plast. **144**, 103046 (2021)
26. Abueidda, D.W., Koric, S., Sobh, N.A., Sehitoglu, H.: Deep learning for plasticity and thermo-viscoplasticity. Int. J. Plast. **136**, 102852 (2021)
27. Kushwaha, S., He, J., Abueidda, D., Jasiuk, I.: Designing impact-resistant bio-inspired low-porosity structures using neural networks. J. Mater. Res. Technol. **27**, 767–779 (2023)
28. He, J., Kushwaha, S., Abueidda, D., Jasiuk, I.: Exploring the structure-property relations of thin-walled, 2D extruded lattices using neural networks. Comput. Struct. **277**, 106940 (2023)

The manufacturer's authorised representative in the EU is Springer Nature Customer Service Centre GmbH, Europaplatz 3, 69115 Heidelberg, Germany. If you have any concerns regarding our products, please contact ProductSafety@springernature.com

Printed and bound by CPI Group (UK) Ltd, Croydon, CR0 4YY

30/09/2026

02225586-0008